WILDFLOWERS IN THE FIELD AND FOREST

Field Guide Series
edited by Jeffrey Glassberg

WILDFLOWERS IN THE
FIELD AND FOREST

A FIELD GUIDE TO THE
NORTHEASTERN UNITED STATES

Steven Clemants
and Carol Gracie

OXFORD
UNIVERSITY PRESS
2006

OXFORD
UNIVERSITY PRESS

Oxford University Press, Inc., publishes works that
further Oxford University's objective of excellence
in research, scholarship, and education.

Oxford   New York
Auckland   Cape Town   Dar es Salaam   Hong Kong   Karachi
Kuala Lumpur   Madrid   Melbourne   Mexico City   Nairobi
New Delhi   Shanghai   Taipei   Toronto

With offices in
Argentina   Austria   Brazil   Chile   Czech Republic   France   Greece
Guatemala   Hungary   Italy   Japan   Poland   Portugal   Singapore
South Korea   Switzerland   Thailand   Turkey   Ukraine   Vietnam

Copyright © 2006 by Glassberg Publications

Published by Oxford University Press, Inc.
198 Madison Avenue, New York, NY 10016

www.oup.com

Oxford is a registered trademark of Oxford University Press

Library of Congress Cataloging-in-Publication Data
Clemants, Steven.
Wildflowers in the field and forest:
a field guide to the northeastern United States/
by Steven Clemants and Carol Gracie.
p. cm.—(Glassberg field guide series)
Includes bibliographical references and indexes.
ISBN 978-0-19-530488-6 (cl)
ISBN 978-0-19-515005-6 (pbk)
QK117.C54 2005
582.13'0974—dc22
2005040866

9 8 7 6 5 4 3
Printed in China
on acid-free paper

For my mother, Doris, and my wife, Grace.
S.E.C.

For my grandsons, Cole, Blake, and Brandon
C.A.G.

Contents

Foreword

TO SOME OBSERVERS, forests, meadows, and roadsides are a monotonous green, punctuated occasionally by buildings, domestic and wild animals, and people. To other more fortunate people, nature presents a rich mosaic of individual plants and animals, differing in size, shape, appearance, and all manner of characteristics, but growing together to make up a tapestry that returns much to those who have developed an appreciation for it. Those who learn to know these plants and animals open for themselves a lifetime of enjoyment and education, and they therefore appreciate all the more superb efforts such as the book you now hold in your hands.

In this concise and original guide, Steve Clemants, of Brooklyn Botanic Garden, and Carol Gracie, who is associated with The New York Botanical Garden, present in a highly accessible format some of the fruits of their lifelong study of the plants of the northeastern United States. These talented authors introduce some 1,450 species of native and naturalized wildflowers found from Maine to central Wisconsin and from Virginia to central Illinois, including all of the most abundant plants and about two-thirds of the total number of species found in the area. With excellent, clear photographs and simple keys to finding the correct section of the book, the authors make the identities of these plants quite simple to determine, opening them up to further investigation in other books or on the World Wide Web. The very accurate maps are a special feature of this guide, presenting the ranges of the individual plant species exactly and, with the use of different colors, highlighting their season of flowering, which often differs in different parts of their range. A book of this kind provides a most welcome and inviting bridge between the scientifically technical literature and the interested non-scientist. Because of its accessibility, it enhances a reader's simple delight in learning about wildflowers and thus enjoying them more fully.

In addition, the availability of this information provides one of the most important approaches to gaining an appreciation of nature and learning how to conserve it. No one can rationally doubt that our world is changing rapidly, and few who think back on the surroundings in which they grew up would find them unchanged. Around our major cities, as for example the detailed investigations of Steve Clemants and his network of colleagues have demonstrated for metropolitan New York, there have been drastic changes over the past century—in fact continuous change has occurred in the native flora from the time the City was established three centuries ago. Most obvious has been the elimination of habitat; the mental image of Peter Stuyvesant retiring after the British had taken over New York and moving out to tend his fruit orchards beyond the city limits, in what is known today as Greenwich Village, tells us a lot about how extensive these losses have been.

There have been a number of additional trends that have also hastened the loss or decline of native plants and animals over wide areas. Among these have been the effects of gathering wild plants and animals for food or medicine, activities that have led to the elimination of ginseng, for example, over wide areas where it has been harvested. Invasive aliens such as Oriental bittersweet, garlic mustard, and giant fescue have spread, often explosively, at the expense of the native species. For those who breathe it on a daily basis, the polluted air around cities is a familiar and unpleasant—not to say unhealthy—complication of urban life; but this same pollution has led to the loss of many plants and animals in and around cities, in places where they were once abundant. For example, most members of the heather family, which require acid soil for healthy growth, have either been eliminated completely or greatly restricted in the New York area as the soils have been depleted by atmospheric pollution. The most serious threat of all may eventually be global warming, which simply eliminates the habitats or shifts the ranges of most species of plants and animals as time passes, and it will continue to do so as long as it remains unchecked.

Because of all of these effects, and more, many of our native plants and animals are disappearing or becoming scarce at a rate of which few people are aware. Learning to know and appreciate them is the key to preserving them, and this book will provide a solid basis for thousands of people to find the key to do just that—simply, directly, and in a very solid way. Botanical gardens, conservation organizations, and other institutions work hard to preserve the wealth of plant diversity that enriches our lives, and those who use this book are likely to want to join them in their activities and support them, both as members and volunteers, and by direct actions. Interested citizens can encourage the preservation and maintenance of open space and nature preserves; they can work to eliminate invasive alien plants and to grow native plants in their gardens as they encourage parks and public gardens to do the same. Citizen-stakeholders can also promote air and water quality as a means of overcoming the pollution that is so harmful, not only to human health but also to the plants and animals that so enrich our surroundings.

The cultivation of native plants is becoming more widespread as people come to understand them better, and this book will help to build the familiarity that will allow those who wish to grow them to select the ones best suited for the conditions found in their own gardens. No plants should be taken from nature, because such collecting threatens their future viability in the localities where they are native. There are, however, many nurseries that propagate native plants and make them available. Information about such sources can be obtained from local botanical gardens and, often, natural history museums as well. Planting native plants around our homes, where they often do better than exotics, may lower the need for chemical fertilizers and pesticides, another means of reducing harm to the environments in which they—and we—live.

This volume portrays the wildflower diversity of the northeastern and midwestern United States and southeastern Canada with a degree of clarity and precision that has been matched by no earlier book. It also contains useful information on how important plants are to the ecological communities they inhabit, and it includes interesting notes on pollinators and other relationships as well. It will have lasting value for those who simply want to appreciate wildflowers more fully and thus enjoy them more; for those who wish to cultivate them and observe the marvelous cycles of their growth throughout the year; and for those who wish to conserve them for future generations. This book is a treasure that can serve as a guide to all those who care about nature and wish to know more about it.

Users of this book owe a debt of gratitude to Steve Clemants and Carol Gracie for providing such a substantial and useful guide. Readers can express their appreciation each time they enthusiastically head out to the nearest woods, fields, hedges, or garden borders with this fine guide in their hands!

Missouri Botanical Garden   Peter H. Raven
St. Louis, Mo.   Director

Preface and Acknowledgments

THIS WORK IS THE CULMINATION of thirty years of study of the plants of the northeastern United States that culminated in five years of intensive research, writing, and assembly for the present work. My greatest hope for this book is that it will spur increased interest in the magnificent plant diversity found in our region. It is only after putting this book together that I am able to truly understand and appreciate what we have.

In writing this book I have tried to view the plants as a novice would. Accordingly there are many instances where I refer to something as a "petal," for instance, when in fact it is a *sepal*; or to a "leaf" that is actually a *phyllode*. I have done this because most users of this book will not be familiar with the intricate body of botanical morphological research that underlies our knowledge of the flower and leaf. In a few instances I have put quotation marks around the most glaring of these cases so that the botanist-reader will not be taken aback. Similarly, I have tried to use fewer botanical terms and, wherever possible, I employ everyday English, as in using "oval" instead of *ovate*, "triangular" instead of *deltate*, and "narrowly oval" instead of *lanceolate*.

In listing the people who have helped with this book I want to acknowledge first those who helped me learn to appreciate the flora. First and foremost is my mother, Doris Seward, whose love of flowers and gardening showed me a path to my career as a botanist.

Professors Gerald Ownbey and Thomas Morley of the University of Minnesota were most instrumental. Dr. Ownbey was the advisor for my Master's degree and Dr. Morley taught my first plant taxonomy class, "Spring Flora of Minnesota." Equally influential were Drs. James Luteyn and Arthur Cronquist of The New York Botanical Garden. Dr. Luteyn was the advisor for my Ph.D., and Dr. Cronquist taught a plant geography class and generally encouraged me in further study of the local flora. I also owe a debt of gratitude to Drs. Richard Mitchell and Charles Sheviak at the New York State Museum, who welcomed me as a colleague when I worked for the New York Natural Heritage Program. I learned a lot from all of these men.

I wish to thank Judith Zuk, President of Brooklyn Botanic Garden, for encouraging me to do this book and for putting up with this distraction from my duties. The curators and librarians of Brooklyn Botanic Garden and The New York Botanical Garden were helpful in giving me access to their collections and allowing me to photograph leaves from herbarium specimens.

The following people looked through the manuscript at various stages and gave their feedback. Dr. Gerry Moore, Dr. Kerry Barringer, Dr. Skip Blanchard, Dr. Charles Sheviak. Thanks to all of them.

Dr. Peter Raven kindly offered to write the Foreword. Thank you.

Jeff Glassberg asked Carol Gracie and me to write this book as part of the Field Guide series. From his long experience, he has guided us through this process, advising us on what is best for the end user, in terms of format and presentation. This book would not exist without his help. Thanks. Further, I thank Orland (Skip) Blanchard for suggesting to Jeff that he contact me to do this book.

Carol Gracie agreed to be co-author and field photographer. Without her detailed work, this book would be much inferior. In addition to her photographic contribution, Carol helped write the text, edited much of the material, and sought additional expert comment. However, I accept full responsibility for any errors in this book.

Lastly this work would not have been possible without the unremitting encouragement, help, and assistance of my wife, Grace Markman.

Brooklyn, N.Y.   S. E. C.

☙ ☙ ☙

A PROJECT SUCH AS THIS could not have been accomplished without the substantial assistance and input of others. As the principal photographer, I am indebted to numerous people who have provided help in so many ways.

I would first like to thank my co-author, Steve Clemants, for inviting me to join him in this endeavor back in 1998. When he told me that the book was to cover the wildflowers of the Northeast, I was eager to participate. I had long had an interest in identifying and photographing the wildflowers of this geographic area. At that point my concept of the Northeast included New England, New York, New Jersey, and Pennsylvania. It was only later that I learned that Steve was using Northeast in its broadest sense, i.e., the entire northeastern quarter of the United States and adjacent Canada. I realized then that this was to be a long-term project, but one filled with the excitement of finding new plants and visiting new localities. What I did not anticipate was the pleasure of meeting the many wonderful people who would, in one way or another, contribute to this book.

A debt of gratitude is due to Jeff Glassberg, who first proposed the idea for this book and who freely gave advice based on his success in writing field guides for the series of which this volume is a part. His guidance led us to choose the format that we have used, which provides succinct but thorough descriptions accompanied by detailed photographs.

Two people deserve special praise for their contributions. My long-time friend, Ginnie Weinland, has served as an inspiration and role-model since I first met her thirty years ago. Our shared interests in wildflowers and photography have given us much pleasure on local ramblings. Ginnie was an invaluable resource, remembering when and where she saw every species she ever photographed. In 2001, at the age of 88, she joined me in a climb up a rather steep slope on the Appalachian Trail and led me right to a site for Devil's bit that she had photographed years before. Ginnie also provided some of her photographs for species that I did not find.

Another person to whom I am enormously grateful is Linda Kelly. We joined forces frequently for field trips in Linda's home territory, the Pine Barrens and adjacent southern New Jersey. Linda's keen eye can spot something different along the roadside while traveling at 55 mph (or more). She is knowledgeable about plant habitats and always eager to explore new terrain in search of hard-to-find plants. Linda's e-mail messages about some new discovery would prompt me to change my plans for the following day and quickly prepare for an early morning departure to the Pine Barrens. My coverage of the plants of that area would not be as complete without Linda's help and hospitality. In addition, her four-wheel drive Chevy Blazer got us anywhere we needed to go.

Many other people shared information about plant localities and/or provided companionship on field trips: Karl Anderson, Bobbi Angell, John Askildsen, Bob Bartholomew, Ana and George Berry, Lauren Brown, Bill Buck, Mike Catania, Joan Coffey, Patrick Cooney, Guy Denny, Frank Dye, Ed Emerson, Paul Goldstein, Ted Gordon, Michael Gross, Chris Harmon, Dick Harris, Ed and Ellie Hecklau, Ursula Joachim, Jackie Kallunki, Kathy Kelly, Zell Kerr, Eric Lamont, Gary Lawton, Katie Lee, Deb Lievens, Carol Levine, Ruth Levitan, Rick Lewandowski and staff at the Mt. Cuba Center, Marion Lobstein, Ted Lockwood, Roz Lowen, Les Mehrhoff, Jane McKean, John Mitchell, Rob Naczi, Albert Paolini, Sondra Peterson, Gwynn Ramsey, Eleanor

(Sam) Saulys, Ted and Jane Settle, John and Lisa Skoric, Victor Soukup, Micky Spano, Bill Standaert, Nancy Staunton, the late Sara Stein, Dorothy Swan, Christina and George Tiedermann, Janet Townsend, Susan Tripp, Mike Van Clef, Nan and Sue Williams (who alerted me to the flowering of the rare and short-flowering three-birds orchid—and took me to the site), and Susan Yost.

The hospitality provided by the following people made my visits to other areas more comfortable and enjoyable. Thanks to George and Ana Berry, Ed and Ellie Hecklau, Linda Kelly, Eric and Mary Laura Lamont, Marion and George Lobstein, John and Beth Mitchell, Rob and Mary Naczi, Gwynn and Betty Ramsey, and Ted and Jane Settle.

I am most grateful to Walter Brust, Ken Cameron, Jackie Kallunki, Ed Kanze, Rob Naczi, Perry Peskin, and Virginia Weinland for allowing us to use photographs of species that I wasn't able to locate. Proper credit for their contributions is given in the Photographic Credits at the end of this book.

Of course, one's spouse always plays an important role in achieving a goal such as this. My husband, Scott Mori, provided encouragement and support throughout the four-year photographic portion of this project. His companionship on some of the longer forays added much to my enjoyment of the fieldwork, and his professional expertise in keying out species as I photographed others saved valuable time. I thank him, too, for often having a meal on the table when I returned, usually tired, wet, and muddy, from an all day field trip, and for his patience as I toiled late into the night over the microscope or labeling slides.

I am grateful to the following botanical societies and their knowledgeable members and leaders for providing an ongoing series of field trips to localities rich in interesting species: The Torrey Botanical Society, The Connecticut Botanical Society, the Philadelphia Botanical Club, and The Josselyn Botanical Society. Their field programs are highly recommended for anyone wishing to learn more about the flora of the tri-state area, Pennsylvania, and Maine. Garden in the Woods, the New England Wild Flower Society's beautiful wildflower preserve in Framingham, Mass., provided the opportunity to photograph plants that I was otherwise unable to find in the wild. This was also true of The New York Botanical Garden, where I have worked for the past twenty-five years.

Many people have reviewed portions of the manuscript or provided input on the format at various points during the preparation of the drafts. A sincere debt of gratitude is due to them for their thoughtful insights and comments: Bobbi Angell, Dr. Anne Bradburn, Dr. Garrett Crow, Dr. Barre Hellquist, Dr. John K. Morton, Dr. Rob Naczi, Charlotte Seidenberg, Nate Smith, and especially

Dr. Bill Standaert, whose careful critique of much of the text and photos helped to greatly improve both. Many of the people mentioned above as field companions or locality informants also shared their opinions on an informal basis. However, I join Steve Clemants in accepting full responsibility for any errors.

Finally, to all of those who "made shade" for my photographic subjects, I am deeply grateful.

South Salem, N.Y.   C. A. G.

Introduction

IDENTIFYING AN UNKNOWN PLANT is like solving a mystery. There are various clues, and each sleuth can develop a particular approach to piecing them together. In each case, however, the ultimate result will be the same—the satisfaction that comes from identifying a new species where it grows. As you use this book, we hope you will enjoy these small puzzles as much as we do.

This photographic field guide is intended primarily for people who want to identify wildflowers by looking at photos. The descriptions are intentionally brief and not fully formed to the species level. If you wish to use a detailed botanical key and see fully formed species descriptions, we recommend Gleason and Cronquist (2004).

The Species Accounts at the core of this book are arranged by flower color, the most readily apparent feature of the blooming plant. The purpose of this presentation is to provide the untrained user a handy way of finding and identifying plants without expert knowledge.

Also in the interest of informing without overwhelming, we have minimized botanical terminology. The Glossary briefly defines the key terms in simple language. For the professional botanist, this terminology may sometime be grating—for example, when *cymes* are not distinguished from *panicles*. But we hope that the use of everyday English will be of benefit for most readers.

Using This Book to Identify a Plant

There are several pathways to identifying a plant with this field guide. Browsing through a color section may lead to discovery. Likewise, using the "Quick Guide" to plant characters at the top left of each page can guide the user to a few pages devoted to plants with those characters. The indexes at the back of the book provide another path to finding relevant species accounts and photographs. Consulting the Index can be particularly useful for users considering a plant they believe they know.

Begin with Color

Classification by color alone can be problematic, first because many species have white forms, and also because people perceive or know colors differently. If you have a white flower in your sights and you can't find the plant in the white section, check the color sections. Also, colors that are close on the spectrum can be difficult to classify, so check adjoining color sections if you do not find a species where you expect it to be.

Outline Guide

The outline guide that follows will help locate a plant in the book. It works from a series of nested levels. To use this guide first identify the flower color then look at the leaf arrangements available indented below, choose the appropriate leaf arrangement, and look at the leaf lobing options indented below, and so on. The three levels below color are repeated on each page under the "Quick Guide" in the upper left-hand corner.

Other Significant Characteristics

It is inevitable that you will find a flower or plant that comes close to what is portrayed but that differs in some respects. It is important to keep in mind that patterns in nature vary, and individual specimens may not agree in every detail. This is why several characteristics are given in each description.

Missing Species

As mentioned earlier, this book provides descriptions and photographs of more wildflowers than any comparable field guide for the region; nevertheless, it is incomplete. We estimate that there are over 2,000 wildflowers in the region, but only about 1,450 are treated here. You may find a flower that is not in the book. The Bibliography will guide you to other informative sources.

Learning about Wildflowers

Flowers are around us every day, but we often don't notice them. To see the diversity of species presented in this book, it may be necessary to travel widely, but much enjoyment can be found close to home. When you are looking for a certain species, the most important thing to consider is location. Even in a relatively small geographic area, there is a wealth of different habitats. Each set of growing conditions, from cracks in the pavement to swamps and bogs, fosters its own collection of wildflowers ready to be identified and observed. Unlike birds and butterflies, plants stay still and can often be gently handled without being damaged. They are ideal subjects for teaching children about nature.

Beginning in early spring and extending into the late days of autumn, a stream of species come into bloom. Some have a very small window of flowering activity (a week or two), and others produce blooms nearly all year long.

Every interested seeker of wildflowers can make a valuable contribution to documenting species. A log recording the discovery of the flowering and fruiting of various species can show changes in plant behavior from year to year. Long-term dated records can be used by botanists and ecologists to analyze the impact of climate change on plant life. All that is needed is a notebook or a PDA and an interest in keeping an eye on the plants.

Parts of a Flower

Because the Species Accounts in this book are arranged by flower color, we begin with a brief description of the biology of flowers. The flower is the reproductive part of the plant. Because plants are stationary, they need help getting the sperm to the egg—a process known as pollination. Some flowers attract this help from insects and birds; others are not as showy and rely on pollination by wind or water.

The outermost series of floral parts are the *sepals*. Sepals are usually green and function mainly to protect the developing bud. In some flowers (such as the lilies), they are indistinguishable from the petals. The *petals* are the reason human observers have so much interest in flowers. Apparently, what attracts bees, butterflies, and hummingbirds also attracts us. The next inner group is the *stamens,* filaments that are tipped with anthers. Anthers are where the *pollen* is produced, and inside the pollen is the sperm cell. At the center of the flower is the *pistil*, or several pistils, in which the egg is found and from which the fruit is formed. A pistil is generally composed of three parts: the *ovary*, the *style*, and the *stigma*. The ovary, at the base, is often swollen, and it is where the eggs are found; the style is a stalk projecting from the top of the ovary, and at its tip is the stigma.

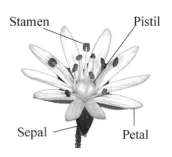

The four main parts of a flower.

A typical Composite head showing the rays, the disk, and, on the underside, the involucral bracts.

Sexual reproduction in plants occurs when the pollen is transported in some way from the anther to the stigma by an insect, an animal, or by wind or water. (See the following section, Understanding Pollination Biology.) At the stigma, the pollen starts to grow a tube down through the style to the ovary; it is through this tube that the sperm cell travels until it comes into contact with the egg.

The "flowers" of the Composite family, such as Asters, Goldenrods, Thistles, and Sunflowers, are not true flowers but clusters of flowers called *heads*. These clusters act in the same way as flowers do in attracting pollinators, but their "petals" are actually modified flowers called *rays*. The center of the "flower," if distinct from the rays, is called the *disk*, comprised of many small flowers, and the structures surrounding the base of the head, the "sepals" are, in fact, *involucral bracts*.

Understanding Pollination Biology

There are many different ways in which flowers attract pollinators and ensure that the pollen is distributed to the stigma. Some flowers allow for their own pollen to fall on the stigma; others have mechanisms to ensure that this doesn't happen.

There are some traits of flowers that make it easy for specific types of insects to pollinate them. As shown in the accompanying table, these traits make it obligatory that a specific species of insect or bird does the pollinating.

Flower Traits Associated with Different Pollinators

Pollinator	Plant Color	Character-istic Scent	Flower Structure	Flowering Time	Reward
Bee or long-tongued fly	Bright blue, yellow	Light, minty or flowery	Landing platform, often tubular	Day	Nectar
Beetle	White, yellowish	Fruity, very spicy	Large, often solitary, open	Day	Pollen

Flower Traits Associated with Different Pollinators
(continued)

Pollinator	Plant Color	Character-istic Scent	Flower Structure	Flowering Time	Reward
Bird	Orange, red	Usually none	Large, tubular, often without a platform	Day	Nectar
Butterfly	Purple, orange, red or pastel	Spicy or none	Landing platform, nectar at base of tubes	Day	Nectar
Short-tongued fly	Dull red, purple or brown, often veined	Dead meat or dung	Sturdy, often trapping the insect for a time	Day and Night	NA
Moth	White, dull	Heavy or flowery	No platform, nectar in tubes or spurs	Evening, night	Nectar
Water	Green, white	None	Floating, often without petals, little pollen	Anytime	NA
Wind	Greenish	None	Usually no petals, exposed pistils and stamens, light pollen	Anytime	NA

Wildflower Gardening

Gardening with wildflowers can be enjoyable, but it can also be beneficial to the wildflower populations nearby. As some native wildflower populations dwindle, their pollinators may have difficulty finding the nectar or pollen they seek. Furthermore, some butterflies and birds migrate and need sources of nourishment along the way. There are several good books on planting gardens to attract butterflies and hummingbirds. Two that we recommend are Brooklyn Botanic Garden's: *The Butterfly Gardener's Guide* (Dole, 2003) and *Hummingbird Gardens: Turning Your Yard into Hummingbird Heaven* (Kress, 2000).

Although wildflower gardening can be beneficial, it can also be detrimental to grow wildflowers in an inappropriate way. The two problems to be aware of are these: (1) collecting whole plants from the wild may be harmful to the native population; it may also be illegal; (2) some species (non-native species) may be invasive in your region, so it is inadvisable to propagate them. All wildflower lovers should live by the maxim: "First, do no harm."

Generally it is best to grow wildflowers from seed. Some favorites such as Trilliums, present a challenge, taking seven years to mature to flowering time. Therefore, removing a whole plant from the wild could have a negative impact on that population. Interested gardeners might benefit from the following references on the subject: *Growing and Propagating Wildflowers of the United States and Canada* (Cullina, 2000) and *Planting Noah's Garden: Further Adventures in Backyard Ecology* (Stein, 1997).

Wildflower Photography

The illustrations in this book were taken on 35-mm slide film with a single-lens reflex camera and a variety of lenses. The objective was to provide clear images that show the detail necessary for identification, so the type of lens most often used was the macro lens with a ring flash, which allowed even illumination of the subject plant while the background remained in shadow. Although it may appear that many of the images were taken at night, this was true in only a few cases. The black background serves to isolate the plant visually by eliminating distracting elements, and it allows the viewer to see the form of the flower or leaves more clearly.

Several factors can influence the process and the product of photography in the field. First among these is the weather. Field photographers should always carry protection from sun and rain, both for themselves and for their equipment. A vest with several pockets will aid in keeping things organized and at hand. Ensuring safe return to the parking area can also be a challenge. It's wise to carry a compass or Global Positioning System (and know how to use it), as well as a whistle and a flashlight. A daypack should include adequate water and snacks, insect repellent, and a small first aid kit. If you must be on your own, be sure to let someone know where you intend to search for plants—but better yet, convince a friend (one with patience) to join you.

Photographers quickly learn to bring *more* film than they anticipate using; for digital cameras, this means having enough flash cards and batteries. It always seems that the batteries run out just when you have come upon a rare species in perfect flower. A back-up camera body can save the day if you've traveled far to find and photograph a special plant only to find that your camera is not working.

It's important to carry a good reference in the field so that you can determine on site which species you have just photographed, Gleason and Cronquist's *Manual of Vascular Plants of Northeastern United States and Adjacent Canada*

(Gleason and Cronquist, 2004) and the *Illustrated Companion to Gleason and Cronquist's Manual* by Noel H. Holmgren (1998) were constant, although very heavy, companions on photographic field trips. If photographing in protected areas, where removing flowers or leaves is prohibited, field identification is essential. A notebook or PDA is critical for recording the species photographed and their exact locations. This diary will allow correct labeling of the images. Seeing the small details of a plant is greatly facilitated by the use of a 10× hand lens. This small item can open a new world to the observer much as binoculars do.

When photographing wildflowers, it is important to get down to their level; if you see the plant clearly you will be able to portray it accurately. Sometimes this means lying on damp or muddy ground, so a large, sheet of plastic can come in handy. As in any hiking or fieldwork, it is important to be aware of surrounding vegetation, being sure not to disturb the habitat. When branches from neighboring plants intrude on the photographer's field of view, it is useful to have clothespins, binder clips, or pipe cleaners to keep them out of the way.

Wind is usually an enemy of the photographer. The use of a portable screen to shield the subject specimen is an option, but carrying such equipment may be a burden. The use of flash to capture the image without blur is more sensible. Bright sunshine may also be an enemy, causing distracting shadows or glare. Bring an umbrella—or ask your patient friend to carry one, and to use it to provide shade when needed. An umbrella can also shelter your subject from the wind when held close to the ground—and it will protect you if it rains.

Most important, take the time to enjoy the surroundings and to carefully observe the plants and any insect or bird visitors. Then leave them for others to enjoy.

About the Species Accounts

As has become clear by now, the plants in this book are arranged first by color, following the spectrum but beginning with Purple and Violet, through Red-Pink, Orange, Yellow, Green, Brown, and ending in White, and then by leaf arrangement. Although numerous color variants can be found in nature, very few species have been duplicated in the different color sections of this book. The most common variants are white-flowered versions of flowers of virtually any color. A white flower that doesn't appear in the white section may be presented in one of its other colors.

Quick Guide to Leaves

Within each color section the plants are presented according to leaf arrangement. Leaves are attached to a stem in a variety of ways. Most often single leaves are attached at various points (nodes) along a stem; this is the *alternate* arrangement. If two leaves are attached at a single point, they are *opposite*; if more than two are grouped at one point, they are *whorled*; and if the leaves arise from the base of the plant they are *basal*. In addition to these four leaf arrangements, two additional categories are used, *floating or submersed* and *absent*. Aquatic plants often have leaves that float or are submersed; or there may be leaf arrangements that are difficult to discern. A few plants have either no leaves or leaves that are very small and scale-like. Some species lack leaves at flowering time but produce leaves later; these come under the *leaves absent* category. In the upper left corner of each text page in the Species Accounts is a set of "quick guide" characters to reference the leaf arrangement(s) of the species on that page. These follow a particular sequence: alternate, opposite, whorled, basal, submersed (often difficult to determine arrangement), and absent (at flowering time).

A simple leaf

A compound leaf

The second quick guide character shows whether the leaves are *simple* or *compound*. A simple leaf has only a single blade (broad green portion), whereas a compound leaf has several or many *leaflets*. It is occasionally difficult to determine if a leaf is simple or compound; you may have to look for additional clues to determine this. The most useful approach is to look for the presence of *stipules* (small leaf-like bracts on the stem at the point of attachment of the leaf) or stipule scars. Lobed leaves have deep projections (or lobes) along the margin. A simple leaf or a leaflet of a compound leaf may be lobed.

The third quick guide character is petal number. In most cases this will be straightforward, but there are a few exceptions. The "flowers" of the composite family are actually clusters of flowers in dense heads; the outer flowers are often enlarged and showy and appear to be petals. Such flowers are called *rays* and are referred to by that term in this book. Flowers of a few families, such as some mints, are highly irregular and may have atypical petals. These are found under *petals irregular*.

Headings and General Information

Each page begins with a general heading that indicates which species are presented on that page. This is often followed by a general description of the group. Critical characters are the most important clues to distinguishing species in the group. Occasionally the critical characters are in the fruit, a signal that there may be problems identifying those species when a plant is in flower. Species are usually arranged under bold, green outline headings indicating important characters for distinguishing groups of species.

Plant Families

The name of the plant family is given following the introductory heading if it applies to all the species on the page, at the end of the outline heading if it applies to the outlined group, or at the end of the individual species description when it is particular to that plant. Families are indexed by both common name and scientific name in the Family Index at the back of the book. All family information was brought up-to-date as of January 1, 2005. Classification reflects that found on the Angiosperm Phylogeny Web site (APG, 2005). This may confuse readers who are familiar with plant families, such as the Scroph family, which has been separated into several families, including the Broom-rape and Plantain families.

Names

Some species have many vernacular (common or English) names. We use only one such name for each species. The name usually follows Gleason and Cronquist (1990) or the USDA PLANTS Database; occasionally a name is from another source.

The scientific, or Latin, names match those used on the USDA PLANTS database as of January 1, 2005. Some of the more commonly used synonyms for plant names will be found in the index. Anyone wanting further information should check the USDA PLANTS database. While we have used this database for the names, we have not always used their concept of species. For example, USDA PLANTS database recognizes many more species of stemless blue violets than we do here.

Names with a green wash over them refer to species endemic to the region shown on the map.

Species Accounts

The species accounts are tied to the photographic plates by number. An asterisk (*) preceding the identifying number indicates a non-native species. A species that is labeled with a letter instead of a number is not illustrated.

In the plant descriptions we have avoided the use of scientific terminology in favor of easily understood words. For instance we use "oval" instead of *ovate*, "triangular" instead of *deltate*, and "narrowly oval" instead of *lanceolate*. All measurements are given in English units of inches and feet. We have tried to provide the following information for all species: plant height, leaf shape and length, inflorescence type, and flower width or height. Additional information relevant to the specific group is added where appropriate. Parts of the description in boldface, including habitat, are particularly important for differentiating a species from related ones.

Habitat

The species description is followed by a listing of habitats where the plant is commonly found. This is often an extremely brief listing and doesn't do service to the tremendous amount of ecological information available for any species. Again, we refer you to the Bibliography for comprehensive sourcebooks.

Photographic Plates

In addition to a large, general photograph of the species, there are often smaller inset images to assist in identification. Many of the larger images, for instance, lack leaves. Therefore, insets of leaf images have been added. Many of these leaf images are photographed from herbarium specimens. This means that the plant has been pressed, dried, and stored in a cabinet, often for over one hundred years. The color has often changed, but the shape and vein patterns remain true.

The size bar in each photograph represents the average width of the flower (if the bar is horizontal) or the height of the flower (if the bar is vertical). For Composite family members, the bars represent the averages for the flower heads. The few other instances where the bars indicate size of the inflorescences or some other dimension are indicated.

Name
The common name is given first, followed by the scientific name. Plants with an asterisk (*) before them are not native to the region. Names generally follow the PLANTS Database (USDA, 2005).
Numbers (1) before the name indicate there is an accompanying photo on the opposite page.
Letters (A) indicate there is no photo.
Plants with a green wash are endemic (grow native nowhere other than where indicated on the map.)

Key characters
Key characters for separating species on a page are presented in green.

Description
A brief description is given to assist in identification. For all species, the height, size, and shape of leaves, type of inflorescence, and size of flowers are given. Additional information is given depending on the species. This is followed by habitat information.

Quick Guide
Three characters to quickly place the plant. These characters are:
Leaf arrangement:
alternate, opposite, whorled, basal, aquatic (floating or submersed) and absent
Leaf lobing:
Simple, lobed or compound
Flower type:
Petal number or irregular flowers

Maps
Maps showing the distribution of the species in the Northeast.

Color indicates season of bloom.

Blue	=	**Spring**
Green	=	**Spring–Summer**
Yellow	=	**Summer**
Orange	=	**Summer–Fall**
Red	=	**Fall**
Purple	=	**All Seasons**

Page 4
LEAVES alternate
LEAVES simple
PETALS 4–5

Gentians
Gentian family
Critical characters include petal number (5 unless noted), orientation, sepal shape (see upper inserts of nos. 3-7) and leaf shape. The closed gentians are pollinated by large bumblebees that are strong enough to push the bottle-shaped flowers open and collect the pollen

a. Flowers solitary, stalked
1. Greater fringed gentian *Gentianopsis crinita*
Plants 4–36". Leaves narrowly oval, 1¼–2¼". Flowers solitary, 1½–2½"; the **4 lobes** are spreading, with a **fringe of long hairs**. Low woods, wet meadows, banks.

2. Pine-barren gentian *Gentiana autumnalis*
Plants 8–20". Leaves linear or narrowly oval, 1–3". Flowers solitary, blue speckled with green, 1½–2"; the **5 lobes** are fused with pleats below, the tips are loosely spreading, but **without a fringe**. Moist pine barrens, bogs.

a. Flowers clustered, stalkless
  b. Leaves narrowly oval or oval
A. Prairie gentian *Gentiana puberulenta* (not illus.)
Much like no. 2 in having **flowers that are open**, but they are clustered at the tip of the stem.

3. Meadow closed gentian *Gentiana clausa*
Plants ≤28". Leaves narrowly oval or oval, 2–4". Flowers ½"; the lobes are fused with pleats to near the tip, forming a **closed "bottle"**; **sepal-lobes round or oval**. Moist meadows, woods, thickets.

4. Prairie closed gentian *Gentiana andrewsii*
Plants 12–36". Leaves narrowly oval or oval, 1¼–6". Flowers blue or white, ½"; the lobes are fused with pleats to the tip, forming a **closed "bottle"**; **sepal-lobes narrowly oval**. Moist prairies, open woods, swamps, meadows, thickets.

5. Coastal plain gentian *Gentiana catesbaei*
Plants 4–24". Leaves **narrowly oval or oval**, 1–2½". Flowers ½"; the lobes fused with pleats below, forming a **tube or often flaring above**; **sepal-lobes narrowly oval**. Wet woods, sands, swamps.

  b. Leaves linear or elliptic
6. Soapwort gentian *Gentiana saponaria*
Plants 8–30". Leaves linear or elliptic, 2–4". Flowers ½"; the lobes are fused with pleats to near the tip, forming a **closed "bottle"**; **sepal-lobes narrowly oval**. Woods, thickets, swamps.

7. Narrow-leaved gentian *Gentiana linearis*
Plants 8–32". Leaves linear or nearly so, 1½–3¾". Flowers blue or white, ½"; the lobes are fused with pleats to near the tip, forming a **tube or a closed "bottle"**; **sepal-lobes linear or narrowly oblong**. Wet woods, meadows.

Key to the Species Accounts

Name
The common name is given.
Plants with an asterisk (*) before the
name are not native to the region.

Page 5

1. Greater fringed gentian
2. Pine-barren gentian
3. Meadow closed gentian
4. Prairie closed gentian
5. Coastal plain gentian
6. Soapwort gentian
7. Narrow-leaved gentian

Insets
Photos of leaves,
often from dried
herbarium speci-
mens, or photos of
flowers showing floral
details or color
variations, are shown
in insets.

Size bars
White bars on the
plate signify the
average size of the
flower. Horizontal
bars indicate width
and vertical bars
indicate height
unless otherwise
noted.

For members of the
composite family, the
size bar refers to the
width of the head.

Abbreviations and Symbols

X   times (as in 2× compound or × hybrid, used in the scientific name of a plant to indicate the plant is a hybrid cross

<   less than

≤   less than or equal to

>   greater than

≥   greater than or equal to

'   feet (measurement)

"   inches

*   introduced (non-native) species (found preceding the plant number)

illus.   illustrated

no.   number

spp.   species (plural)

KEY TO THE MAPS

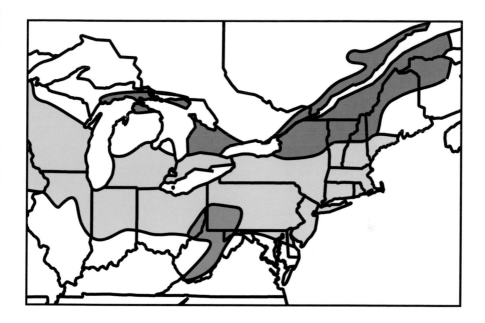

A map like the one above accompanies each entry in the Species Accounts. The colored area shows the range of the species within the geographic area covered by this book. As shown in the legend, different colors indicate flowering times in various parts of the range.

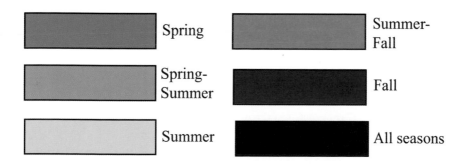

Spring

Summer-Fall

Spring-Summer

Fall

Summer

All seasons

LEAVES alternate
LEAVES simple or lobed
PETALS 2-4

Dayflowers, Spiderworts, and Rockets
Spiderwort and Mustard families
a. Petals 2 large, blue, and 1 small, white (Spiderwort family)
Dayflowers open for a few hours in the morning before they wilt.

*1. Common dayflower *Commelina communis*
Plants ≤32″. Leaves narrowly oval, 2–4¾″. Flowers solitary, **subtended by a round bract ½–1¼″, margins not fused;** flowers ½–1″; the third (lower) petal small and white. Weedy places.

2. Erect dayflower *Commelina erecta*
Plants <36″. Leaves linear or narrowly oval, 1½–6″. Flowers solitary, **subtended by a broadly triangular bract, ⅜–1″; margins fused for lower ⅓;** flowers ¾–2″; the third (lower) petal small and white. Rocky woods, hillsides, sand dunes, fields.

a. Petals 3 (Spiderwort family)
3. Smooth spiderwort *Tradescantia ohiensis*
Plants 16–40″, hairless. Leaves linear, 2–18″, hairless or hairy at tip. Flowers in terminal and axillary clusters, blue or rose, ¾–1½″; **sepals hairless, red-margined.** Meadows, thickets, prairies.

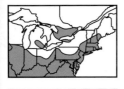

4. Virginia spiderwort *Tradescantia virginiana*
Plants 2–16″. Leaves linear, 5–15″, hairless. Flowers in a terminal cluster, blue, purple, or occasionally white, ¾–1¼″; **sepals hairy.** Moist woods, prairies. Differs from no. 3 in having hairy flower-stalks and sepals. The northern populations are likely garden escapes. Hairs on the filaments change color in response to radiation.

5. Sticky spiderwort *Tradescantia bracteata*
Plants 8–16″, hairless or finely hairy. Leaves linear, 6–12″, hairless or hairy at base. Flowers in a terminal cluster, rose or blue, 1¼–1½″; **sepals densely sticky-hairy.** Prairies.

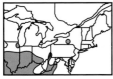

A. Wide-leaved spiderwort *Tradescantia subaspera* (not illus.)
Much like no. 5, but the leaves are wider (narrowly oval), often >¾″ wide, and the **sepals are scarcely sticky-hairy.** Rich woods.

a. Petals 4 (Mustard family)
Mustards are easily recognized by their 4 petals, 6 stamens, and the pod, which retains a frame with a parchment middle after the sides and seeds fall. Critical characters are fruit shape and size.

6. Sea-rocket *Cakile edentula*
Plants ½–2′, succulent. Leaves narrowly oval, broader at tip, 1¼–3″, toothed or pinnately lobed. Flowers ¼″. **Fruit cylindrical, constricted in the middle, ¼–½″.** Beaches, sand dunes.

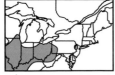

7. Purple rocket *Iodanthus pinnatifidus*
Plants ≤3′. Leaves narrowly oval, broader at base, elliptic or oblong, ≤8″, toothed, often the lower pinnately lobed. Flowers ¾–1″. **Fruit slender, ¾–1½″.** River bottom woods.

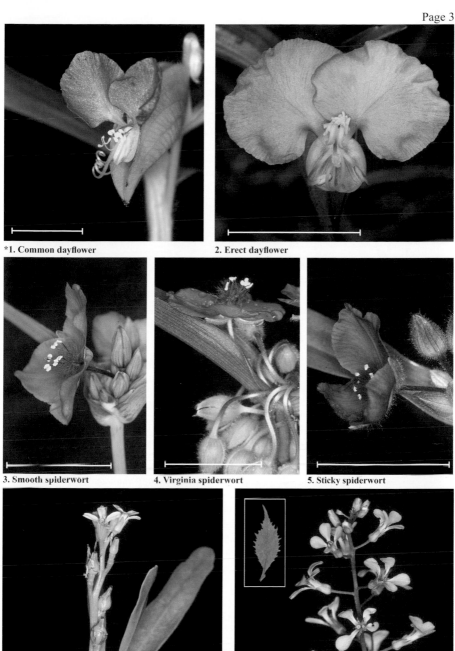

*1. Common dayflower

2. Erect dayflower

3. Smooth spiderwort

4. Virginia spiderwort

5. Sticky spiderwort

6. Sea-rocket

7. Purple rocket

LEAVES alternate
LEAVES simple
PETALS 4–5

Gentians
Gentian family

Critical characters include petal number (5 unless noted), orientation, sepal shape (see upper inserts of nos. 3-7) and leaf shape. The closed gentians are pollinated by large bumblebees that are strong enough to push the bottle-shaped flowers open and collect the pollen.

a. Flowers solitary, stalked

1. Greater fringed gentian *Gentianopsis crinita*
Plants 4–36″. Leaves narrowly oval, 1¼–2¼″. Flowers solitary, 1½–2½″; the **4 lobes** are spreading, with a **fringe of long hairs**. Low woods, wet meadows, banks.

2. Pine-barren gentian *Gentiana autumnalis*
Plants 8–20″. Leaves linear or narrowly oval, 1–3″. Flowers solitary, blue speckled with green, 1½–2″; the **5 lobes** are fused with pleats below; the tips are loosely spreading, but **without a fringe**. Moist pine barrens, bogs.

a. Flowers clustered, stalkless
　b. Leaves narrowly oval or oval
A. Prairie gentian *Gentiana puberulenta* (not illus.)
Much like no. 2 in having **flowers that are open,** but they are clustered at the tip of the stem.

3. Meadow closed gentian *Gentiana clausa*
Plants ≤28″. Leaves narrowly oval or oval, 2–4″. Flowers ½″; the lobes are fused with pleats to near the tip, forming a **closed "bottle"; sepal-lobes round or oval**. Moist meadows, woods, thickets.

4. Prairie closed gentian *Gentiana andrewsii*
Plants 12–36″. Leaves narrowly oval or oval, 1¼–6″. Flowers blue or white, ½″; the lobes are fused with whitish pleats to the tip, forming a **closed "bottle"; sepal-lobes narrowly oval**. Moist prairies, open woods, swamps, meadows, thickets.

5. Coastal plain gentian *Gentiana catesbaei*
Plants 4–24″. Leaves **narrowly oval or oval**, 1–2½″. Flowers ½″; the lobes fused with pleats below, forming a **tube or often flaring above; sepal-lobes narrowly oval**. Wet woods, sands, swamps.

**　b. Leaves linear or elliptic**

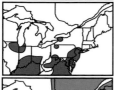

6. Soapwort gentian *Gentiana saponaria*
Plants 8–30″ Leaves linear or elliptic, 2–4″. Flowers ½″; the lobes are fused with pleats to near the tip, forming a **closed "bottle"; sepal-lobes narrowly oval**. Woods, thickets, swamps.

7. Narrow-leaved gentian *Gentiana linearis*
Plants 8–32″. Leaves linear or nearly so, 1½–3¾″. Flowers blue or white, ½″; the lobes are fused with pleats to near the tip, forming a **tube or a closed "bottle"; sepal-lobes linear or narrowly oblong**. Wet woods, meadows.

1. Greater fringed gentian

2. Pine-barren gentian

3. Meadow closed gentian

4. Prairie closed gentian

5. Coastal plain gentian

6. Soapwort gentian

7. Narrow-leaved gentian

**LEAVES alternate, oc-
casionally basal
LEAVES simple
PETALS 5**

Bellflowers and Venus's looking-glass
Bellflower family

Critical characters: flower shape, leaf shape, and inflorescence type.

a. Flowers bell-shaped
b. Flowers ≤½"

1. Marsh bellflower *Campanula aparinoides*
Plants 12–14". Leaves linear or narrowly oval; lower leaves ≤3¾",
upper shorter and narrower. Flowers **solitary on long stalks**, pale
blue or white, ⅛–½"; sepals triangular or narrowly oval. Wet, sunny
meadows. Similar to no. 3, but with smaller flowers.

2. Appalachian bellflower *Campanula divaricata*
Plants 12–24". Leaves linear or narrowly oval, ¾–3", coarsely
toothed. Flowers **dangling from widely branched panicles**, ¼";
sepals narrowly triangular. Rocky woods. The dangling, small
flowers are distinctive.

b. Flowers ≥½"

3. Harebell *Campanula rotundifolia*
Plants 4–32". Leaves **of two sorts**: basal leaves oval or round, ≤¾",
toothed, stalked; stem leaves distinctly different, hair-like, ½–3",
toothless; basal leaves are often withered and gone by flowering-
time. Flowers **nodding, solitary or on lax racemes**, ½–1¼"; sepals
very narrowly triangular. Dry woods, meadows, cliffs, beaches.

*4. Creeping bellflower *Campanula rapunculoides*
Plants 15–36". Leaves oval or narrowly oval, 2½–4", irregularly
toothed. Flowers **on one-sided racemes**, ¾–1¼"; sepals narrowly
oval. Lawns, roadsides, waste ground.

*5 . Clustered bellflower *Campanula glomerata*
Plants 12–28". Leaves oval or narrowly triangular, 3–3½" toothed,
clasping. Flowers in **terminal, stalkless clusters**, ¾–1¼"; sepals
narrowly oval. Roadsides, weedy areas. Similar to no. 4, except for
the arrangement of the flowers.

*6. Nettle-leaved bellflower *Campanula trachelium*
Plants 12–40". The lower leaves heart-shaped or triangular, upper
leaves narrowly oval, 2–2½", coarsely toothed. Flowers **in loose,
open clusters**, 1–1½"; sepals narrowly oval. Roadsides, waste
places. Note the distinctive bristly-hairy sepals.

a. Flowers open with widely spreading petals
7. Tall bellflower *Campanulastrum americanum*
Plants 18–84". Leaves narrowly oval or oval, 2¾–6", toothed.
Flowers **in terminal or axillary clusters**, ¾–1¼"; sepals linear.
Moist borders, open woods.

8. Venus's looking-glass *Triodanis perfoliata*
Plants 4–40". Leaves oval or broadly oval, ¼–1¼", toothed. Flowers
1-few in leaf axils, ¼–⅝"; sepals triangular. Various sites, often
disturbed. The tiny, shiny, black seeds resemble a hand mirror.

1. Marsh bellflower

2. Appalachian bellflower

3. Harebell

*4. Creeping bellflower

*5. Clustered bellflower

*6. Nettle-leaved bellflower

7. Tall bellflower

8. Venus's looking-glass

LEAVES alternate and basal
LEAVES simple
PETALS 5

Forget-me-nots
Borage family

Borages have a characteristic coiled inflorescence that looks somewhat like a scorpion's tail. The lowest flowers usually bloom first, followed by the next lowest, and so on.

Forget-me-nots
Forget-me-nots get their name from the story of a lover collecting these flowers along the edge of a deep pool. He fell into the pool and tossed a cluster of the flowers to the bank, telling his lover, "forget me not," before he disappeared forever beneath the surface. Critical characters are flower size, shape and color, fruits, and seeds. For white-flowered forget-me-nots, see page 284.

a. Flowers with a flat face, flower stalk ≥⅛″
*1. Garden forget-me-not *Myosotis sylvatica*
Plants ≤20″, hairy. Leaves oblong, narrowly oval or spoon-shaped, 1¼–2¾″, hairy, stalkless or with a winged stalk. Flowers on racemes ≤4″, stalked; **flowers ¼–⅜″**. Escaped from gardens.

*2. True forget-me-not *Myosotis scorpioides*
Plants 6–24″, slightly hairy. Leaves narrowly oval or elliptic, 1–3″, rough hairy, stalkless. Flowers on terminal, open panicles, stalked; **flowers ⅜–¾″**. Shallow water, wet soil. Similar to no. 1, but without hooked or gland-tipped hairs on the calyx.

3. Smaller forget-me-not *Myosotis laxa*
Plants 6–20″, slightly hairy. Leaves narrowly oval, oblong or elliptic, 1–3″, hairy, stalkless or sometimes stalked. Flowers on terminal, open panicles, stalked; **flowers ⅛–¼″**. Moist soil, shallow water.

a. Flowers funnel-shaped, flower stalk ≤⅛″
*4. Field forget-me-not *Myosotis arvensis*
Plants 4–16″, hairy. Leaves oblong or narrowly oval, broader toward either end, ⅜–2½″, hairy, stalkless. Flowers stalked, on **terminal open panicles occupying about ½ the height of the plant;** flowers ⅛″. Fields, roadsides, weedy sites.

*5. Strict forget-me-not *Myosotis stricta*
Plants 2–8″, hairy. Basal leaves narrowly oval, broader toward tip; stem leaves oblong or narrowly oval, ≤¾″, hairy, stalkless. Flowers stalkless or nearly so, on **terminal racemes that occupy nearly the whole plant;** flowers <⅛″. Dry, weedy areas; roadsides; old fields.

*A. Changing forget-me-not *Myosotis discolor* (not illus.)
Similar to no. 5, but flowers open yellow and become blue; **the inflorescences have no bracts and occupy only about ½ the plant.** Fields, roadsides.

*2. True forget-me-not

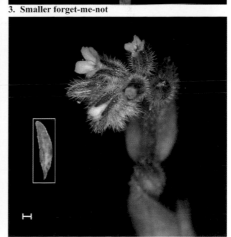

*1. Garden forget-me-not

3. Smaller forget-me-not

*4. Field forget-me-not

*5. Strict forget-me-not

LEAVES alternate and basal
LEAVES simple or lobed
PETALS 5

Bluebells and other blue Borages
Borage family

a. Plants hairless
1. Virginia bluebells *Mertensia virginica*
Plants erect or ascending, 1–2′, hairless. Leaves elliptic, oval or narrowly oval, 2–6″, rounded or pointed. Flowers on scorpion-tail-like racemes; **flowers ½–1″;** sepals oval, blunt. Wet woods, clearings. Look for small holes on the side of the flower near its base. These are indications of nectar-robbing bees.

A. Northern bluebells *Mertensia paniculata* (not illus.)
Similar to no. 1, but the leaves are hairy and the **flowers are only ⅜–½″.** Damp woods.

2. Seaside bluebells *Mertensia maritima*
Plants sprawling or the tips curling upwards, ≤3′, hairless. Leaves oval, ¾–2½″, blunt or pointed, blue-green. Flowers on numerous, widely branched racemes; bracts leaf-like; **flowers ¼–⅜″;** sepals broadly triangular. Shingle, gravelly and sandy beaches.

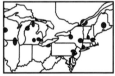

a. Plants hairy
***3. Viper's bugloss** *Echium vulgare*
Plants erect, 1–2½′, bristly. Basal leaves oval, broader toward the tip, 2½–10″, pointed; stem leaves progressively smaller, linear and stalkless. Flowers on a scorpion-tail-like raceme, often pyramid-shaped; flowers ½–¾″, **with long, red, projecting stamens.** Weedy places, roadsides, meadows.

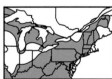

***B. Common bugloss** *Anchusa officinalis* (not illus.)
Somewhat similar to no. 3, but the **flowers are ¼–⅜″; stamens are not projecting,** and the flowers have constricted throats. Weedy areas.

4. Blue hound's-tongue *Cynoglossum virginianum*
Plants 1–2′, rough-hairy. Basal leaves elliptic, 4–8″, stalked; stem **leaves narrowly oval, hairy, clasping.** Flowers on racemes 4–8″ tall, on recurved stalks; flowers ¼–½″. Upland woods.

*5. Stickseed *Lappula squarrosa*
Plants ½–2′, hairy. Leaves linear, ½–1″, rough-hairy, stalkless. Flowers stalkless or nearly so on numerous racemes, 2–4″ tall; flowers ⅛″. Grain fields, weedy areas, railroad sidings. **Note the sticky-hairy fruit.** See the somewhat similar white-flowered Virginia stickseed on page 284.

1. Virginia bluebells

2. Seaside bluebells

4. Blue hound's-tongue

*3. Viper's bugloss

*5. Stickseed

LEAVES alternate and basal
LEAVES simple
PETALS 5

Alternate-leaved Violets and Flax
Several families

Violets are extremely variable. Recent botanists have split some of the following species into several additional species. The spur is a horn-like projection from the back of the flower that contains nectar. Violets are pollinated by bees. See other violets pp. 58, 166, 216, 296 and 378.

a. Leaves broad (Violet family)
b. Stipules small

1. American dog violet *Viola conspersa*
Plants ½–8″, hairless. Leaves heart-shaped or round, ½–1¼″, tip rounded, base notched. Flowers stalked above the leaves, ¼–½″; **spur ¼″;** lateral petals hairy. Woods, meadows, stream banks.

A. Walter 's violet *Viola walteri* (not illus.)
Similar to no. 1, but **prostrate and the flower-stalks are densely hairy.** Dry to moist woods, ledges.

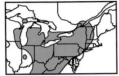

2. Long-spurred violet *Viola rostrata*
Plants 2–5″, hairless. Leaves oval, 1–1½″, pointed, base notched. Flowers stalked above the leaves, ½–1″; **spur ½–¾″;** petals hairless. Shady slopes, woodlands, usually in deep humus.

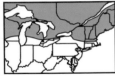

3. Hook-spurred violet *Viola adunca*
Plants 1–3½″, hairy or hairless. Leaves leathery, oval to nearly round, ½–1″, blunt, truncate or notched at base. Flowers long-stalked, ¼–½″; **spur ¼″,** hooked at tip; lateral petals hairy. Dry woods, sandy and gravelly pastures, slopes, alpine ravines.

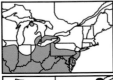

b. Stipules large
4. Wild pansy *Viola bicolor*
Plants 2–16″, hairless. Leaves round, spoon-shaped, or narrowly oval, ½–1″, blunt, tapering to base. **Middle lobe of stipules toothless.** Flowers long-stalked, ¾″. Fields, woods, roadsides, lawns.

*5. Johnny-jump-up *Viola tricolor*
Plants 4–12″, sparsely hairy. Basal leaves round or heart-shaped; stem leaves narrower, ½–2½″, blunt, tapering to base. **Middle lobe of stipules toothed.** Flowers long-stalked, ½–1″. Fields, other cultivated areas.

a. Leaves linear (Flax and Plantain families)
*6. Blue flax *Linum perenne*
Plants 12–28″. **Leaves linear, ⅜–1¼″, 1-nerved or obscurely 3-nerved.** Flowers ⅜–⅝″; **sepals without a fringe of hairs.** Disturbed areas. See also yellow flax on page 148. Flax family.

*7. Lesser toadflax *Chaenorhinum minus*
Plants 2–16″, sticky-hairy. Leaves narrowly oval or linear, ⅜–1″. **Flowers on loose, leafy racemes, ¼″;** spur <⅛″. Weedy places, roadsides, railroad cinders. Plantain family.

1. American dog violet

2. Long-spurred violet

3. Hook-spurred violet

4. Wild pansy

*5. Johnny-jump-up

*6. Blue flax

*7. Lesser toadflax

LEAVES alternate
LEAVES simple or lobed
PETALS 5 or 10

Miscellaneous alternate-leaved species
Several families

a. Vines
  b. Petal-lobes 5 (Morning-glory and Nightshade families)

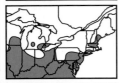

*1. Common morning-glory *Ipomoea purpurea*
Plants climbing, ≤15′, hairy. Leaves round or oval, **rarely 3-lobed**, 1½–3¾″, toothless. Flowers 1–5 in axils, blue, purple, or white, 1½–2½″; **sepals** ⅜–½″. Fields, roadsides, weedy areas. See red-flowered morning-glories on pp. 70, 90 and white-flowered morning-glories on p. 270. Morning-glory family.

*2. Ivy-leaved morning-glory *Ipomoea hederacea*
Plants scrambling, 3–7′, hairy. Leaves round, **usually 3-lobed**, 2–5¾″, toothless. Flowers 1–3 in axils, purple or blue, 1¼–2″; **sepals** ½–1″. Roadsides, fields, weedy areas. Morning-glory family.

*3. Bittersweet nightshade *Solanum dulcamara*
Plants scrambling, 3–10″. Leaves broadly oval or heart-shaped, 1–3″, often lobed. Flowers 10–25 in a cluster, ¼–⅜″. Thickets, clearings, open woods. The fruit changes from green to yellow to orange, and finally to red when fully ripe. See also p. 282. Nightshade family.

  b. "Petals" 10 (Passion-flower family)
4. Purple passion-flower *Passiflora incarnata*
Plants climbing or trailing 26′, hairy. Leaves somewhat triangular, deeply 3–5-lobed, toothed. Flowers on long stalks, white with a purple fringe, 1½–3″. Fields, roadsides, thickets, open woods. Pollinated by large-bodied bees and wasps.

a. Herbs
  b. Petal-lobes 5 (several families)
5. Horse-nettle *Solanum carolinense*
Plants ≤3′. Leaves oval 2¾–4¾″, 2–5 large teeth or lobes on each side, very spiny. **Flowers several on racemes, ¾″**. Fields, weedy areas. See also p. 282. Nightshade family.

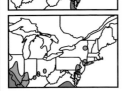

6. Sea-lavender *Limonium carolinianum*
Plants 8–28″. Basal leaves elliptic or oval, 2½–7″. **Flowers on widely branched panicles, ¼″**. Salt marshes. The masses of flowering plants look like sea-mist from a distance. Leadwort family.

7. Common bluestar *Amsonia tabernaemontana*
Plants 16–40″. Leaves narrowly oval, oval, or elliptic, 3–6″. **Flowers on flat-topped or pyramidal panicles, ⅜″**. Moist or wet woods. Dogbane family.

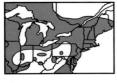

  b. Petal-lobes 6 (Pickerel-weed family)
8. Pickerel-weed *Pontederia cordata*
Plants erect, ≤40″. Basal and lower stem leaves heart-shaped or narrowly oval, ≤7″. Flowers on a crowded raceme 2–6″ tall, subtended by a bract 1¼–2½″; flowers violet-blue to white, ¼″. Marshes, shallow water.

*1. Common morning-glory

*2. Ivy-leaved morning-glory

*3. Bittersweet nightshade

4. Purple passion-flower

5. Horse-nettle

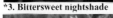

6. Sea-lavender

7. Common bluestar

8. Pickerel-weed

LEAVES alternate
LEAVES simple or
lobed
RAYS many or
PETALS absent

Miscellaneous Composites and Eryngos
Composite and Umbel families

a. Flowers in hemispherical heads (Composite family)
 b. Heads single at branch tips or 1–3 in leaf axils
*1. Cornflower *Centaurea cyanus*
Plants 1–4'. **Leaves narrow, often linear**, ≤5", toothless, toothed or lobed. **Heads** terminating the branches, mostly blue but sometimes pink, purple, or white, ⅜–½" Roadsides, fields, weedy areas.

*2. Chicory *Cichorium intybus*
Plants 1–5'. Lower **leaves narrowly oval**, broader near the tip, 3–10", toothed or lobed; upper leaves reduced. **Heads** 1–3 in axils of reduced leaves, **≤2½"**. Roadsides, fields, weedy places. The flowers open early on clear days and close by noon but later on cloudy days.

 b. Heads 11-34 in panicles
3. Tall blue lettuce *Lactuca biennis*
Plants 2–7'. Leaves elliptic, 4–12", pinnately lobed or only toothed. Heads 15–34 on narrow, **elongate often crowded panicles**; heads ⅜–¾". Seeds with long, **light-brown hairs (pappus)**. Moist places. See yellow lettuces p. 180.

4. Woodland lettuce *Lactuca floridana*
Plants 1½–7'. Leaves elliptic, 3–12", pinnately lobed or only toothed. Heads 11–17 on **open panicles**; heads ⅜–½". **Seeds with long white hairs (pappus)**. Thickets, woods, moist, open places.

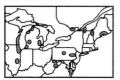

a. Flowers in spherical heads (Composite and Umbel families)
*5. Globe-thistle *Echinops sphaerocephalus*
Plants ≤8'. Leaves elliptic, ≤14", **pinnately lobed**, white-hairy beneath. Heads at the tips of branches, 1½–2½". Weedy places. Composite family.

Eryngos are unusual Umbels because the flowers are in condensed heads. Critical characters are head size and leaf size and shape. See white-flowered eryngium on p. 374. Umbel family.

6. Marsh eryngo *Eryngium aquaticum*
Plants erect, ≤3'. **Leaves linear or narrowly oval, 4–12"**, toothless, toothed, or lobed. Heads in a single, terminal cluster; bracts spreading or drooping; heads oval or cylindrical, ⅜–½". Bogs, marshes.

*7. Plains eryngo *Eryngium planum*
Plants erect, ≤3'. **Leaves elliptic or oval, 2¾–4"**, toothed. Heads terminal, solitary or in clusters; bracts spreading; heads oval, ½–¾". Fields, weedy areas. The insert shows the basal leaf.

A. Spreading eryngo *Eryngium prostratum* (not illust.)
Similar to no. 7, but the heads are smaller and the bracts minute.

Bars for eryngos represent the width of the heads.

*1. Cornflower

*2. Chicory

3. Tall blue lettuce

4. Woodland lettuce

*5. Globe-thistle

6. Marsh eryngo

*7. Plains eryngo

LEAVES basal and alternate
LEAVES simple
RAYS many

Blue Asters with well-developed basal leaves
Composite family

The "flowers" are actually tight clusters (heads) of flowers, the outer, enlarged flowers are called rays, which look like petals. Critical characters include leaf-shape, inflorescence-shape, and particularly the shape and tip of the involucral bracts (the bracts clasping the base of the head). The asters have recently been separated into several different genera. See other asters pp. 288, 357.

a. Basal leaves heart-shaped, stalked
  b. Upper leaves clasping
1. Clasping heart-leaved aster *Symphyotrichum undulatum*

Plants 1–4′. Leaves narrowly heart-shaped, 1½–5½″, pointed, toothed or toothless, hairy; **stalk flared at base and clasping.** Heads on panicles, ½–¾″; involucral bracts sharp; rays 10–20. Open woods, clearings.

  b. Upper leaves not clasping
    c. Inflorescence somewhat flat-topped

2. Big-leaved aster *Eurybia macrophylla*

Plants ½–4′. Leaves heart-shaped, oval, or elliptic, 1½–12″, pointed or blunt, toothed, hairy or not, stalked or not. **Heads on flat-topped panicles,** ½–1″; involucral bracts rounded or pointed; rays 9–20. Woods.

    c. Inflorescence rounded

3. Common heart-leaved aster *Symphyotrichum cordifolium*

Plants ½–4′. Leaves heart-shaped, 1½–6″, narrowly pointed, toothed, **rough above, hairy below,** stalked. **Heads on rounded panicles,** ½–1″; involucral bracts blunt; rays 8–20, blue or white. Woods, clearings. Several similar species are not treated here.

4. Smooth heart-leaved aster *Symphyotrichum lowrieanum*

Plants ½–4′. Leaves heart-shaped or oval, 2–7″, narrowly pointed, toothed, **hairless;** stalk often winged. **Heads on rounded panicles,** ½–1″; involucral bracts blunt; rays 8–20, blue or white. Woods, clearings. Thought to be a hybrid involving no. 3.

A. Midwestern heart-leaved aster *Symphyotrichum shortii* (not illus.).
Similar to no. 4, but the **leaves are toothless.**

a. Basal leaves elliptic
5. Slender aster *Eurybia compacta*

Plants 6–20″, **hairy or hairless.** Basal leaves **elliptic,** ¾–2½″, pointed, toothless or nearly so, hairy or hairless, stalked; stem leaves narrower, ½–3¾″, stalkless. Heads on narrow, flat-topped panicles, ⅜–½″; involucral bracts sharp, spreading; **rays 8–14.** Dry, sandy places.

6. Low showy aster *Eurybia spectabilis*

Plants 4–36″, **sticky-hairy.** Leaves **elliptic,** ¾–6″, pointed, toothless or nearly so, rough or hairless, tapering to stalk, reduced above. Heads on open, flat-topped panicles, ⅜–1″; involucral bracts firm with spreading tips; **rays 15–35.** Dry, sandy soil; pinelands.

1. Clasping heart-leaved aster

2. Big-leaved aster

3. Common heart-leaved aster

4. Smooth heart-leaved aster

5. Slender aster

6. Low showy aster

LEAVES alternate
LEAVES simple
RAYS many

Blue Asters with clasping leaves or densely hairy leaves
Composite family

a. Leaves clasping the stem
  b. Bracts hairy or sticky-hairy

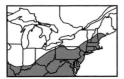

1. Clasping aster *Symphyotrichum patens*
Plants ½–5′. Leaves oval or oblong, 1–6″, pointed, toothless, **hairy,** deeply notched at base, clasping. Heads on open panicles, ½–1¼″; involucral bracts narrowly pointed; rays 15–25. **Woods; dry, open places.**

2. New York aster *Symphyotrichum novi-belgii*
Plants ½–4½′. Leaves narrowly oval, elliptic, or linear, 2–6¾″, pointed, **usually hairless,** usually clasping with basal lobes. Heads several to many on panicles, ⅜–1″; involucral bracts with long, narrow spreading tips; rays 20–50. **Moist places, often in salt marshes.**

  b. Bracts hairless

3. Bristly aster *Symphyotrichum puniceum*
Plants 1½–8′, **bristly hairy.** Leaves narrowly oval, oblong, or elliptic, 2¾–6¼″, pointed, toothless, **hairy below,** clasping with basal lobes. Heads on a panicle, sometimes white, ½–1″; involucral bracts slender, loose, narrowly pointed; **rays 30–60.** Swamps, moist places.

4. Smooth aster *Symphyotrichum laeve*
Plants 1–3′, **hairless.** Leaves narrowly oval, 2–6″, pointed, **usually hairless,** usually toothless, clasping with basal lobes. Heads several to many on open panicles, ½–1¼″; involucral bracts with short, diamond-shaped green tips; **rays 15–30.** Open, dry places.

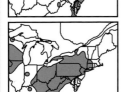

5. New England aster *Symphyotrichum novae-angliae*
Plants 1–6½′, **sparsely hairy.** Leaves narrowly oval, 1¼–4¾″, pointed, toothless, **hairy,** clasping with basal lobes. Heads several to many on short panicles, ¾–1½″; involucral bracts blunt or broadly pointed; **rays 45–100.** Moist, open or wooded places.

A. Zigzag aster *Symphyotrichum prenanthoides* (not illus.)
Similar to no. 5, but the **leaves have a winged stalk which clasps the stem,** and it grows in wetter areas.

a. Leaves not clasping the stem
6. Western silvery aster *Symphyotrichum sericeum*
Plants 1–2½′. Basal leaves soon dying, narrowly oval, oblong, or elliptic, ⅜–1½″, broadly pointed, toothless, silvery-hairy, truncate at the base, Heads **several to many on flat-topped or rounded panicles,** ½–1¼″; involucral bracts broadly pointed; rays 15–25. **Seeds silvery-hairy.** Dry prairies, open places.

7. Eastern silvery aster *Symphyotrichum concolor*
Plants 1–3′. Lower leaves soon dying, narrowly oval, oblong, or elliptic, ¼–1½″, broadly pointed, toothless, silvery-hairy, broad-based. Heads **on narrow racemes,** ½–1″; involucral bracts pointed, rays 8–16. **Seeds hairless.** Dry, sandy areas, especially pinelands.

1. Clasping aster

2. New York aster

3. Bristly aster

4. Smooth aster

5. New England aster

6. Western silvery aster

7. Eastern silvery aster

LEAVES alternate
LEAVES simple
RAYS many

Blue Asters with usually linear leaves
Composite family

a. Leaves usually linear, toothless
b. Plants of dry, upland areas

1. Stiff aster *Ionactis linariifolius*
Plants 4–20″. **Leaves linear, ½–1½″**, pointed, toothless but fringed along margin, minutely roughened, tapering to a stalkless base. Heads solitary or several heads on flat-topped panicles, ½–1″; involucral bracts broadly pointed or rounded; rays 10–20. Dry ground, open woods, especially sandy. Note small, leafy bracts of flower stalks.

2. Long-stalked aster *Symphyotrichum dumosum*
Plants 12–40″. **Leaves linear, narrowly oval, or elliptic, 1¼–4¼″**, broadly pointed, toothless, roughened above, hairless below, blunt based, stalkless. Heads numerous on panicles, ⅜–⅝″; involucral bracts blunt; rays 13–30. Dry or moist, often sandy, places. Note small, leafy bracts of flower stalks.

b. Plants of wetlands
3. Leafy bog aster *Oclemena nemoralis*
Plants 4–24″. Leaves linear, narrowly oval, elliptic, or oblong, ½–2″, blunt, toothless or nearly so, **margins rolled under**, hairy below, blunt-based, stalkless. Heads solitary or several on flat-topped panicles, ½–1¼″; involucral bracts sharp-pointed; rays 13–27. **Sphagnum bogs.**

4. Annual salt-marsh aster *Symphyotrichum subulatum*
Plants 4–40″, fleshy. Leaves linear, 1–8″, pointed, toothless, hairless, narrowed to stalkless base. **Heads** 1-several on open panicles, bluish, purple or white, ⅛–¼″; involucral bracts narrowly pointed; rays <20. **Mostly salt-marshes.**

5. Perennial salt-marsh aster *Symphyotrichum tenuifolium*
Plants 8–28″, often zigzag. Leaves linear, 1½–6″, pointed, toothless, hairless, narrowed to stalkless base. **Heads** 1–several on open panicles; blue, pink or white, ¼–½″; involucral bracts firm, pointed, greenish upward; rays 15–25. **Salt-marshes.**

a. Leaves usually not linear, usually toothed
6. Panicled aster *Symphyotrichum lanceolatum*
Plants 2–8′. **Leaves** narrowly oval, elliptic, or linear, 3¼–6″, pointed, usually toothed, stalkless or stalked. Heads numerous on elongate leafy panicles, ¾–1″; involucral bracts pointed with slender green band; rays 20–40, white, purple or blue. Moist, low places.

1. Stiff aster

2. Long-stalked aster

3. Leafy bog aster

4. Annual salt-marsh aster

5. Perennial salt-marsh aster

6. Panicled aster

LEAVES alternate
LEAVES simple or
lobed
RAYS many

Knapweeds and Pink Thistles
Composite family

Knapweeds and thistles are common roadside weeds, sometimes invasive. Critical characters are in the involucral bracts and the leaf shape. See other thistles on pp. 94, 180, and 318.

a. Heads spiny

*1. Purple star-thistle *Centaurea calcitrapa*
Plants 4–32″, hairy or hairless. Leaves pinnately lobed, ¾–1½″. Heads numerous, ¾–1″; **involucral bracts spiny**. Roadside weed. Note the cobweb-like hairs covering the stems.

a. Heads not spiny
  b. Leaves pinnately lobed

*2. Canada thistle *Cirsium arvense*
Plants 12–60″, hairless or white-hairy. Leaves **pinnately lobed, spiny,** 2–5″. Heads, numerous **in flat-topped panicles**, ⅜–¾″; involucral bracts spiny. Fields, weedy places.

*3. Spotted knapweed *Centaurea biebersteinii*
Plants 12–60″. Leaves pinnately compound or lobed, with linear leaflets, becoming less dissected upward, 4–8″. Heads solitary, ½–1″; **outer flowers of the head not significantly enlarged; involucral bracts without a widened tip, dark, fringed with hairs.** Fields, roadsides, weedy places. Distinctive because of the deeply lobed leaves and lack of a broadened tip to the bracts.

  b. Leaves unlobed or shallowly lobed

*4. Brown knapweed *Centaurea jacea*
Plants 8–40″. Basal leaves narrowly oval, sometimes lobed; stem leaves narrowly oval, 4–6″. Heads solitary, ¾–1¾″; **outer flowers of the heads larger than the others; involucral bracts with a widened, oval tip, brown, without a fringe.** Fields, roadsides, weedy places.

*5. Meadow knapweed *Centaurea × moncktonii*
A hybrid between brown and black knapweeds. Note the intermediate form of the involucral bract tips. See insert on lower right of no. 4. Recent research has shown that what is called *meadow knapweed* in North America is in fact no. 6. Meadow knapweed is rare in the region.

*6. Black knapweed *Centaurea nigra*
Plants 8–32″. Basal leaves oval or narrowly oval, lobed; stem leaves narrowly oval, wider toward the tip, ≤10″. Heads solitary, ¾–1¾″; **outer flowers of the heads not enlarged; involucral bracts with a narrowly oval black tip, with a fringe.** Fields, roadsides, weedy places.

*7. Tyrol knapweed *Centaurea nigrescens*
Plants ≤48″. Basal leaves oblong or narrowly oval, often shallowly lobed; stem leaves smaller and unlobed, ≤6″. Heads solitary, ¾–1¾″, **outer flowers of the heads larger than others; involucral bracts with a triangular blackish tip, with a fringe.** Fields, roadsides, weedy places, Similar to no. 6, but the tips of the bracts are broader and the outer flowers of the heads larger.

*1. Purple star-thistle

*2. Canada thistle

*3. Spotted knapweed

*4. Brown knapweed

*5. Meadow knapweed

*6. Black knapweed

*7. Tyrol knapweed

LEAVES alternate and basal
LEAVES simple
"PETALS" or RAYS many

Ironweeds, Fleabanes, and Sheep's bit
Bellflower and Composite families
Critical characters for ironweeds are head size, number of flowers per head and in the involucral bracts.

a. Bracts with long, narrow tip (Composite family)
1. New York ironweed *Vernonia noveboracensis*
Plants 3–6½', hairy or hairless. Leaves narrowly oval or linear, 4–11", hairy, without black spots. Heads in open, flat-topped or concave panicles, ¼–½", **30–55-flowered. Low, wet woods; marshes.** Note the brownish-purple hairs (pappus) at tip of seeds.

2. Appalachian ironweed *Vernonia glauca*
Plants 2–5', hairless or nearly so. Leaves narrowly oval or oval, 4–10", hairy, without black spots. Heads in loose and irregular panicles, ¼–⅜", **32–48-flowered. Upland woods.** Similar to no. 1, but in dryer sites and with tan or white hairs (pappus) on the seeds.

***3. Ozark ironweed** *Vernonia arkansana*
Plants 2–5', hairless. Leaves linear or narrowly oval, 4–8", hairless or nearly so, with black spots. Heads in very irregular clusters, ⅜–½", **55–100-flowered. Woods, stream banks.** Similar to nos. 1 and 2, but with larger heads and black spots on lower leaf surface.

a. Bracts blunt or broadly pointed
b. Heads a uniform color (Composite and Bellflower families)
4. Smooth ironweed *Vernonia fasciculata*
Plants 1½–4½', hairless or nearly so. Leaves narrowly oval, 3–7", hairless, with black spots, tapered to base. Heads in dense flat-topped panicles, ½–¾", **10–26-flowered. Damp prairies, marshes.** Composite family.

A. Tall ironweed *Vernonia gigantea* (not illus.)
Similar to no. 4, but often taller with leaves hairy below and no black spots. Moist and wet woods, pastures. Composite family.

***5. Sheep's bit** *Jasione montana*
Plants 4–8". Leaves linear, ⅜–¾". Heads blue, ¾–1¼". Sandy fields, weedy areas. Bellflower family. Bar indicates width of head.

b. Heads with a yellow center (Composite family)
See additional fleabanes on pp. 64 and 290.
6. Rough fleabane *Erigeron strigosus*
Plants 12–28", appressed-hairy. Basal leaves narrowly oval or elliptic, toothless or toothed, ≤6"; stem leaves linear or narrowly oval, mostly hairless. Heads several to very many, ≤½"; rays 50–100; involucral bracts slightly sticky and hairy. Disturbed places, weedy.

7. Annual fleabane *Erigeron annuus*
Plants 24–60", spreading-hairy. Basal leaves elliptic or round, ≤4", toothed, stalked; stem leaves narrowly oval or oval, toothed. Heads several to many, ¼–¾"; rays 80–125, white or purple. Disturbed sites. Similar to no. 6, but generally with more leaves.

1. New York ironweed

2. Appalachian ironweed

*3. Ozark ironweed

4. Smooth ironweed

*5. Sheep's bit

6. Rough fleabane

7. Annual fleabane

LEAVES alternate, basal, or submersed
LEAVES simple
PETALS irregular

Lobelias
Bellflower family

In the lobelia flower the top of the corolla is split to below the middle and the stamens are fused. The stigma must push through the fused stamens. In nos. 1 and 2 there are "windows" on either side of the base of the corolla. Pollinated by bees. See also red-flowered lobelias p. 86.

a. Plants terrestrial
   b. Flowers ½–1½"

1. Great blue lobelia *Lobelia siphilitica*
Plants 20–60". Leaves narrowly oval or oblong, 3–4¾", toothless or nearly so. Flowers on a crowded raceme, 4–12" tall; bracts in middle of flower-stalk; **flowers blue or white, ¾–1¼".** Swamps, wet ground.

2. Downy lobelia *Lobelia puberula*
Plants 12–50". Leaves oblong or narrowly oval, 2–4", toothed. Flowers on one-sided racemes, 1–12" tall; bracts at base or middle of flower-stalk; **flowers blue, ½–¾".** Wet woods, clearings.

   b. Flowers <½"

3. Indian tobacco *Lobelia inflata*
Plants 6–40". **Leaves oval,** 2–3", slightly toothed. Flowers on racemes 4–8" tall; **bracts at base of flower-stalk;** flowers ¼–⅜". Open woods, fields, roadsides, weedy places. **Note the inflated calyx.**

4. Spiked lobelia *Lobelia spicata*
Plants 12–40". **Basal leaves oval or round, stem leaves oval or narrowly oval,** usually wider toward tip, 2–4", toothed, upper leaves gradually reduced. Flowers on a slender spike; **bracts at base or middle of flower-stalk;** flowers ¼–⅜". Various habitats.

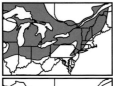

5. Brook lobelia *Lobelia kalmii*
Plants 4–16". **Basal leaves spoon-shaped,** ⅜–1¼", hairy, shallowly toothed; stem leaves linear, ⅜–2". Flowers on one-sided racemes; **bracts on middle of flower-stalk;** flowers ¼–½". Swamps, bogs.

6. Nuttall's lobelia *Lobelia nuttallii*
Plants 8–30". **Lower leaves narrowly oval,** wider toward the tip, ≤1½", toothless; **upper leaves smaller and linear.** Flowers on loose racemes; **bracts at base of flower-stalk;** flowers ¼–⅜". Sandy swamps. Very similar to no. 5, but note the position of the bracts.

7. Canby's lobelia *Lobelia canbyi*
Plants 12–40". **Leaves linear,** ⅜–2", toothless. Flowers on racemes, 4–12"; **bracts at base of flower-stalk;** flowers ⅜–1". Swamps.

a. Plants aquatic

8. Water lobelia *Lobelia dortmanna*
Plants 18–36". **Basal leaves linear, fleshy, with 2 hollow tubes,** ¾–3½", toothless; stem leaves minute. Flowers on racemes; bracts at base of flower-stalk; flowers blue or white, ½–1". Sandy or gravelly pond margins, usually submersed except for flower-stalk.

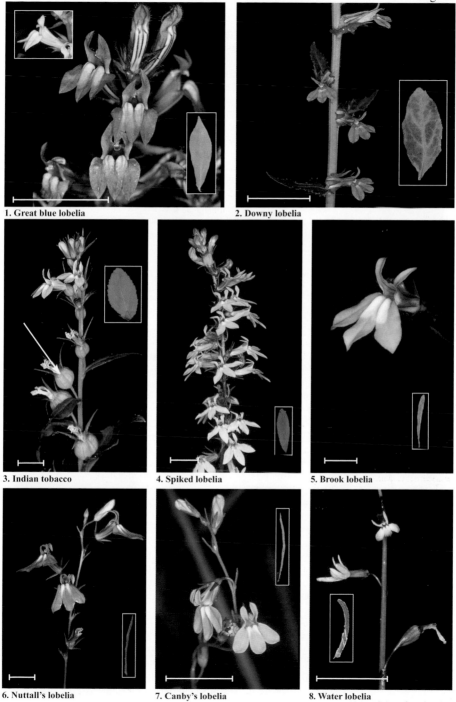

1. Great blue lobelia

2. Downy lobelia

3. Indian tobacco

4. Spiked lobelia

5. Brook lobelia

6. Nuttall's lobelia

7. Canby's lobelia

8. Water lobelia

LEAVES alternate and basal
LEAVES compound
PETALS 5

Waterleafs, Phacelias, and Jacob's-ladders
Borage and Phlox families

Critical characters for phacelias are the flower shape and margin and the leaves; for Jacob's-ladders it is the stamens and inflorescence type.

a. Stigmas 2 (Borage family)

1. Eastern waterleaf *Hydrophyllum virginianum*
Plants 12–32″, hairy. **Leaves** oval or triangular, deeply 5-lobed, **4–8″**; lobes toothed or lobed. Flowers on coiled cymes, ¼–⅜″; **stamens longer than the corolla**. Moist or wet woods; open, wet areas. See white-flowered waterleaves p. 314.

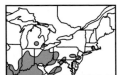

2. Miami-mist *Phacelia purshii*
Plants lax, 8–16″, hairy. Lower **leaves** compound, ¾–2″, **stalked; upper** leaves 3–11-lobed, **clasping;** lobes triangular or oblong, pointed. Flowers 6–20 on one-sided racemes, ¼–½″; **petals fringed**. Moist woods. See white-flowered phacelia p. 314.

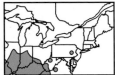

3. Forest phacelia *Phacelia bipinnatifida*
Plants erect, 8–20″, hairy. Leaves broadly triangular or oval, compound, stalked; leaflets 3, oval, toothed, pointed; **upper leaves** lobed, **stalked. Flowers** on curved racemes, ⅜–¾″. Moist woods.

4. Appalachian phacelia *Phacelia dubia*
Plants lax, 4–16″, hairy. Basal leaves stalked, ½–2½″, compound or lobed; lobes 2–10, oval or round, toothed or toothless; **upper leaves** deeply 2–8-lobed; lobes narrowly oval or oval, **stalkless or nearly so. Flowers** on curved racemes, ¼–⅜″. Woods, fields, barrens.

5. Oceanblue phacelia *Phacelia ranunculacea*
Plants erect, 2–8″, hairy. **Leaves** ½–1¼″, **stalked**, 3–7-lobed; lobes broadly pointed, toothed. **Flowers** on small racemes, ⅛″. Moist, river bottom woods.

a. Stigmas 3 (Phlox family)

6. Spreading Jacob's-ladder *Polemonium reptans*
Plants erect or spreading, ½–2′, sticky-hairy or hairless. Leaves compound; leaflets 7–17, narrowly oval, oblong or elliptic, ¾–1½″. **Flowers on loose, open panicles**, ¼–½″; **stamens not exserted** from flower. Rich, moist woods.

7. Appalachian Jacob's-ladder *Polemonium vanbruntiae*
Plants erect, 1–3′, hairless or scarcely hairy in upper parts. Leaves compound; leaflets 15–21, oval or narrowly oval, ¾–1½″, sharp. **Flowers in compact clusters**, ½–¾″; **stamens exserted from flower.** Swamps, stream banks.

8. Western Jacob's-ladder *Polemonium occidentale*
Plants erect, ≤3′, hairy. Leaves compound; leaflets 11–27, narrowly oval, ⅜–½″, pointed. **Flowers on elongate dense panicles**, ⅜–½″; **stamens not exserted**. Rich conifer swamps. Differs from no. 6 in having denser inflorescences and later flowers.

1. Eastern waterleaf

2. Miami-mist

3. Forest phacelia

4. Appalachian phacelia

5. Oceanblue phacelia

6. Spreading Jacob's-ladder

7. Appalachian Jacob's-ladder

8. Western Jacob's-ladder

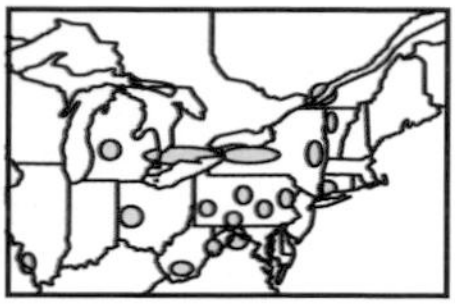

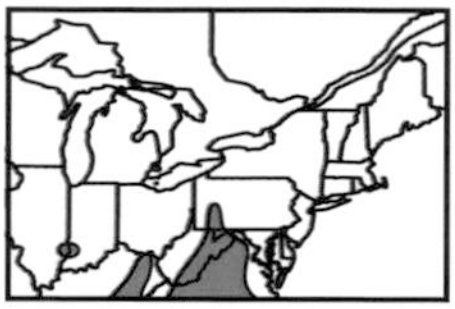

LEAVES alternate and basal
LEAVES compound
PETALS 5 or irregular

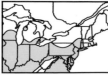

Love-in-a-mist, Monkshoods, and Larkspurs
Crowfoot family

a. Petals not spurred
  b. Petals regular

*1. Love-in-a-mist *Nigella damascena*
Plants 1–2'. **Leaves** ¾–2¼", **finely dissected;** basal leaves stalked. Flowers solitary, ¾–2". Weedy places, gardens.

  b. Petals irregular
The flowers of monkshood have an upper petal that looks like a helmet or monk's hood. Critical characters are in the helmet shape and leaf lobing.

2. Southern monkshood *Aconitum uncinatum*
Plants weakly erect or often leaning on other plants, 1–8'. Leaves ½–4" wide, **3–5 lobed;** lobes toothed. Flowers few on terminal, short racemes; flowers 1–2" high; **helmet mostly higher than broad.** Rich woods.

3. New York monkshood *Aconitum noveboracense*
Plants erect, ≤3'. **Leaves** 2–6" wide, **5–7 lobed;** lobes tapering to base, toothed. Flowers few; flowers ⅝–2" high; **helmet dome-like, as broad as high.** Stream banks, wet rocks.

*4. Garden monkshood *Aconitum napellus*
Plants erect, ≤4'. Leaves 2–4 " wide, **divided into numerous linear leaflets.** Flowers on elongate, terminal racemes; flowers ¾–1" high; **helmet broader than high.** Garden escape.

a. Petals spurred
Larkspurs have a horn-like spur extending from the back of the flower. Critical characters are flower and spur size, and in the fruit.

5. Dwarf larkspur *Delphinium tricorne*
Plants ¼–2'. Leaves mostly near the base, round, ¾–4¾" wide, divided into oblong or linear lobes. Flowers on 3–8" tall racemes; **flowers 1–1¼"; spur ½–⅝".** Rich, moist woods.

6. Tall larkspur *Delphinium exaltatum*
Plants ≤7'. Leaves pentagonal, 1–3¾" wide, divided into narrowly oval lobes. Flowers on elongate racemes ≤12"; **flowers ⅝–1"; spur ⅜–½".** Rich woods.

7. Carolina larkspur *Delphinium carolinianum*
Plants 2–4'. Leaves round or pentagonal, ¾–4¾" wide, dissected into linear lobes; basal leaves dying early. Flowers on elongate, often branched, racemes, 2-16" tall; **flowers ⅝–1¼"; spur ⅜–⅝".** Prairies, woods, sandhills.

*A. Rocket larkspur *Consolida ajacis* (not illus.)
Similar to the other larkspurs, but this species is annual and has highly dissected leaves with linear lobes. Weedy places, roadsides, old homesteads.

*1. Love-in-a-mist

2. Southern monkshood

3. New York monkshood

*4. Garden monkshood

5. Dwarf larkspur

6. Tall larkspur

7. Carolina larkspur

Blue Vetches and other Beans
Bean family

Bean flowers have 5 petals, the uppermost, usually largest, is called a "banner", the two lateral are called "wings," and the lower two are usually fused into a boat-shape called the "keel," which encloses the stamens and pistil.

a. Leaflets >3
b. Leaves with tendrils
Critical characters for vetches are flower number, color and size. See other vetches on pp. 102, 188 and 320.

1. America vetch *Vicia americana*
Plants trailing or climbing, 2–3', **hairless**. Stipules ≤¼", toothed; leaflets 16–32, elliptic, ½–1¼", blunt, spine-tipped. **Flowers 2–9 on racemes**, ½–¾". Fruit 1–1½". Moist woods, thickets, meadows.

***2. Cow vetch** *Vicia cracca*
Plants trailing or climbing, 2–3', **hairy**. Stipules <¼", toothless; leaflets 10–24, linear or oblong, ½–1¼". **Flowers 20–50 on 1-sided racemes**, ¼–½", curved downward. Fruit ¾–1¼". Fields, roadsides, meadows, shores.

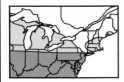

***3. Hairy vetch** *Vicia villosa*
Plants ≤3', **hairy**. Stipules <¼"; leaflets 10–20, narrowly oval, or linear, ½–1". **Flowers 5–40 on dense 1-sided racemes**, ½–¾". Fruit 1–1½". Fields, roadsides, waste places. Similar to no. 2, but the flower-stalk appears to be inserted on the side of the calyx. **Note the distinctive bi-colored flowers**.

b. Leaves without tendrils
c. Leaves pinnately compound (cont. on next page)
Amorpha spp. have a single petal that wraps around the stamens.
4. False indigo *Amorpha fruticosa*
Shrubs ≤13'. **Leaflets 8–20**, ¾–1½", short-hairy below. Flowers on terminal and axillary spikes, ¼"; sepals triangular, ¼". Moist woods, stream banks.

5. Lead-plant *Amorpha canescens*
Shrubs 20–40", densely hairy. **Leaflets 26–40**, ⅜–¾", leaf-stalks <⅛". Flowers on axillary and terminal spikes, ¼"; sepals narrowly oval, <⅛". Dry prairies; sandy, open woods.

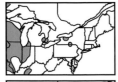

6. Purple prairie-clover *Dalea purpurea*
Plants 8–40". **Leaflets 5,** narrow, ¼–1". Flowers on dense, cone-like, spikes ¾–2¾" tall; flowers ¼". Dry prairies, open glades. See white-flowered prairie-clover p. 320.

7. Field locoweed *Oxytropis campestris*
Plants forming mats, densely hairy becoming less so with age. Leaves nearly basal; **leaflets 15–31**, linear or narrowly oval, ¼–1". Flowers in dense spikes ¾–1¼"; flowers ½–¾". Rocks, cliffs, shores.

1. America vetch

*2. Cow vetch

*3. Hairy vetch

4. False indigo

5. Lead-plant

6. Purple prairie-clover

7. Field locoweed

LEAVES alternate and basal
LEAVES simple or compound
PETALS various

Legumes, Hepatica, and Purple Orchids
Several families
a. Leaflets more than 3 (Bean family) (cont. from previous page)
c. Leaves palmately compound

1. Sundial lupine *Lupinus perennis*
Plants erect, 8–24″, thinly hairy. **Leaflets 7–11**, narrowly oval, wider toward the tip, ¾–2″; leaf-stalk ¾–2½″. Flowers on terminal racemes, 4–8″ tall; flowers blue, white, or pink, ½–⅝″. Dry, open woods; clearings. Caterpillars of the 'Karner' Mellisa Blue butterfly eat the leaves of this plant.

***2. Big-leaved lupine** *Lupinus polyphyllus*
Plants erect, 24–48″, hairless or nearly so. **Leaflets 7–11**, elliptic, 2½–5″. Flowers on terminal racemes, 12–28″ tall; flowers blue, white, pink or purple, ½″. Roadsides, banks. Similar to no. 1, but taller and coarser.

a. Leaflets 3
b. Petals 5–12, regular (Crowfoot family)

3. Round-lobed hepatica *Hepatica nobilis* var. *obtusa*
Plants 3–6″. Leaves round, 3-lobed, ½–2″, notched, **lobes oval, blunt, or rounded**. Flowers solitary, blue, pink, or white, ½–1″; petals 5–12, often 6. Dry or moist upland woods. See the other hepatica p. 382.

b. Petals 5, irregular (Bean family)

4. Blue false indigo *Baptisia australis*
Plants erect, ≤5′, hairless. Leaflets 3, **narrowly oval or oval, wider toward the tip**, 1½–3″; stipules narrowly oval, ¼–¾″. **Flowers 20–27 on racemes**, ≤1¼′ tall; flowers ¾–1″. Moist, rocky soil. See white and yellow baptisias pp. 188 and 320.

***5. Alfalfa** *Medicago sativa*
Plants prostrate, 1–3′. Leaflets 3, **narrowly oval or oblong, ½–1¼″, toothed at summit;** stalk of middle leaflet turned upward. **Flowers in round or short cylindrical heads**, ¼–½″; stalks shorter than calyx. Roadsides, weedy places, old fields. See also p. 184.

a. Leaves simple, flowers irregular (Orchid family)

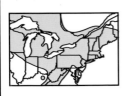

6. Lesser purple fringed orchid *Platanthera psycodes*
Plants 1–5′. Lower leaves narrowly oval or oval; upper leaves reduced. Flowers in dense-cylindric racemes, ⅜–¾″ tall; sepals oval; lateral petals oblong, finely toothed; **lip very broad, deeply 3-parted, each lobe fringed; column with short, rounded lobes**. Wet meadows, wet woods. Pollinated by butterflies. In the process of obtaining nectar from the spur, pollen masses stick to the butterflies' proboscises.

7. Greater purple fringed orchid *Platanthera grandiflora*
Very similar to no. 6, but the column lobes are spreading and angular. Swamps, marshes, floodplains and seeping slopes. Pollinated by long-tongued moths. As a moth obtains nectar from the spur, pollen masses stick to the moth's eye. See other platantheras on pp. 138, 248, and 298.

1. Sundial lupine

*2. Big-leaved lupine

3. Round-lobed hepatica

4. Blue false indigo

*5. Alfalfa

6. Lesser purple fringed orchid

7. Greater purple fringed orchid

LEAVES opposite or whorled
LEAVES simple
PETALS 4–5

Speedwells (1)
Plantain family

Critical characters for speedwells are the inflorescence type and leaf shape. Petals 4 unless noted, leaves opposite except in nos. 1 and 2. Pollinated by bees. See white speedwell p. 348.

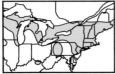

a. Flowers many, on terminal racemes or spikes
*1. Long-leaved speedwell *Veronica longifolia*
Plants erect, 12–60″, hairy. **Leaves opposite or in whorls of 3,** narrowly oval, 2–6″, pointed, sharp-toothed, short-stalked. Flowers on a terminal spike, ¼″; petals 4–5. Fields, roadsides, disturbed places.

*2. Thyme-leaved speedwell *Veronica serpyllifolia*
Plants creeping at the base with ascending branches, 2–12″, hairy. **Leaves mostly opposite,** elliptic to broadly oval, ⅜–1″, rounded to pointed, slightly toothed or toothless, stalked or not. Flowers on elongating, terminal racemes; flowers ⅛–¼″. Fields, meadows, lawns, woods.

A. Northern speedwell *Veronica wormskjoldii* (not illus.)
Similar to no. 2, but erect or curved and flowers ¼–⅜″. Moist meadows, stream banks, bogs.

a. Flowers many on axillary racemes (cont. on next page)
b. Plants hairy
*3. Common speedwell *Veronica officinalis*
Plants creeping at base with ascending tips, 2–10″, hairy. **Leaves elliptic,** ½–2″, blunt, finely toothed, **stalked.** Flowers on axillary racemes, ⅛–¼″; petals 4–5. Dry fields, woods.

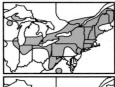

*4. Germander speedwell *Veronica chamaedrys*
Plants prostrate or loosely ascending, 4–12″, hairy. **Leaves oval** ½–1¼″, blunt, coarsely toothed, **stalkless.** Flowers on axillary racemes, ¼–½″. Lawns, disturbed sites.

*B. Broad-leaved speedwell *Veronica austriaca* (not illus.)
Similar to no. 4, but the plants are erect, 13–32″, and leaves are broadly pointed. Garden escape.

b. Plants hairless
5. American speedwell *Veronica americana*
Plants erect or ascending, 4–40″, hairless. **Leaves** narrowly oval or oval, **broader toward the base,** ½–3″, broadly pointed, toothed, short-stalked. Flowers 4–30 on axillary racemes, ¼–⅜″. Swamps, stream banks.

*6. European speedwell *Veronica beccabunga*
Plants erect or ascending, 4–40″, hairless. **Leaves** elliptic or oval, **broader toward the tip,** ¾–1½″, rounded at tip, toothed, short-stalked. Flowers 4–30 on axillary racemes, ¼″. Muddy shores, stream banks.

*1. Long-leaved speedwell

*2. Thyme-leaved speedwell

*3. Common speedwell

*4. Germander speedwell

5. American speedwell

*6. European speedwell

LEAVES opposite
LEAVES simple or
lobed
PETALS 4

Speedwells (2)
Plantain family

a. Flowers many, on axillary racemes (cont. from previous page)
  b. Plants hairless
1. Water speedwell *Veronica anagallis-aquatica*
Plants erect, 1–3′, hairless or slightly sticky-hairy. **Leaves** elliptic, narrowly oval or oval, broader toward the tip, ¾–4″, broadly
pointed, toothed or toothless, stalkless and **clasping**. Flowers **20–65
on axillary racemes**, ⅛–¼″. Ditches, streams, shallow water.

A. Narrow-leaved speedwell *Veronica scutellata* (not illus.)
Similar to no. 1, but shorter with **linear or narrowly oval, non-
clasping leaves** and 5-20-flowered racemes. Swamps, bogs.

a. Flowers solitary in axils of leaves
  c. Plants erect or ascending
*2. Corn speedwell *Veronica arvensis*
Plants erect or ascending, 2–16″, hairy. **Leaves** oval or elliptic,
¼–½″, broadly pointed, **toothed**, short-stalked. **Flowers stalkless** in
axils, <⅛″. Gardens, lawns, fields.

*B. Spring speedwell *Veronica verna* (not illus.)
Very much like no. 2, but with **deeply lobed leaves**. Disturbed areas,
roadsides.

*3. Field speedwell *Veronica agrestis*
Plants prostrate or ascending, 4–12″, hairy. **Leaves** oblong, oval, or
round, ⅜–¾″, blunt, **toothed**, short-stalked. **Flowers on long stalks**
from the axils, ⅛–¼″. Fields, lawns, roadsides.

*4. Bird's-eye speedwell *Veronica persica*
Plants loosely ascending, 4–16″, hairy. **Leaves** broadly elliptic or
oval, ⅜–¾″, rounded, **toothed**, short-stalked. **Flowers long-stalked**
from axils, ⅜–½″. Gardens, lawns, roadsides.

  c. Plants prostrate or creeping
*5. Creeping speedwell *Veronica filiformis*
Plants creeping, ≤10″, with upturned tips, hairy. **Leaves** round or
kidney-shaped, <⅜″, rounded at tip, **slightly toothed**, short-stalked.
Flowers long-stalked from axils, ¼–⅜″. Gardens, lawns.

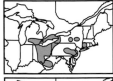

*6. Ivy-leaved speedwell *Veronica hederifolia*
Plants prostrate or loosely ascending, 2–16″, hairy. **Leaves** sometimes alternate, kidney-shaped or rounded, **3-5 lobed**, ¼–½″,
broadly pointed, toothed, long-stalked. Flowers long-stalked from
axils, ⅛–¼″. Fields, disturbed areas.

1. Water speedwell

*2. Corn speedwell

*3. Field speedwell

*5. Creeping speedwell

*4. Bird's-eye speedwell

*6. Ivy-leaved speedwell

LEAVES opposite and basal
LEAVES simple or lobed
PETALS 4–5

Phloxes, Verbenas, and Lopseed
Phlox and Verbena families

a. Flowers on panicles (Phlox family)
Often confused with pinks (p. 106), phloxes have petals fused into a tube with spreading lobes. See pink-flowered phlox p. 116.

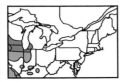

1. Sand phlox *Phlox bifida*
Plants prostrate with ascending stems 4–12″, hairy. Basal leaves linear ≤½″; stem leaves linear to narrowly oval, ½–2½″. Flowers few on panicles, blue-violet or white, ½–¾″; **petals deeply notched**. Dry cliffs, bluffs, sandhills, dunes.

2. Forest phlox *Phlox divaricata*
Plants usually erect, 6–20″, sticky-hairy. Leaves narrowly oval or oblong, 1¼–2″. Flowers on panicles, blue, purple, or white, ¾–1¼″; **petals not notched**. Wet woods, thickets, swamps, prairies.

3. Creeping phlox *Phlox stolonifera*
Plants diffusely spreading, 4–20″, hairy. Leaves narrowly oval, oval, or oblong, ¼–1″. Flowers few on loose panicles, purple, 1–1¼″; **petals not notched**. Moist woods, bottom lands. Similar to no. 2, but the stamens protrude from the flower.

a. Flowers on spikes (Verbena family)
    b. Flowers 5-lobed
Critical characters for verbenas are number of spikes and leaf shape, lobing, and size. See other verbenas p. 332.

4. Common verbena *Verbena hastata*
Plants erect, 12–60″. Leaves narrowly oval, **often 3-lobed**, 1½–7″, toothed. Flowers **on many terminal spikes**, blue, ⅛″. Moist fields, meadows, swamps. Pollinated primarily by Skippers (butterflies).

*5. Prostrate verbena *Verbena bracteata*
Plants prostrate or ascending, 4–20″. Leaves **3-lobed**, ⅜–2½″. Flowers **on 1–3 terminal spikes**, blue or purple, ≤⅛″. Prairies, fields, roadsides.

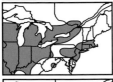

6. Narrow-leaved verbena *Verbena simplex*
Plants erect, 4–28″. Leaves **linear, narrowly oval, or oval**, 1¼–4″, **unlobed**. Flowers on **solitary slender spikes**, lavender or purple, ⅛–¼″. Dry woods, fields, roadsides, rocks.

7. Hoary verbena *Verbena stricta*
Plants erect, 8–48″. Leaves **oval, elliptic, or round**, 1¼–4″, **unlobed**. Flowers **on 1–3 spikes**, blue or purple, ¼–⅜″. Prairies, fields, roadsides. Similar to no. 6, but longer-hairy, with stouter spikes and rounder leaves.

    b. Flowers 4-lobed
8. Lopseed *Phryma leptostachya*
Plants erect, 12–36″. Leaves oval, 2–6″. Flowers on a terminal spike, pale purple or white, ¼″. Rich thickets, moist woods. Note the downward-pointed and appressed fruit.

1. Sand phlox

2. Forest phlox

3. Creeping phlox

4. Common verbena

*5. Prostrate verbena

6. Narrow-leaved verbena

7. Hoary verbena

8. Lopseed

LEAVES opposite and basal

LEAVES simple or lobed PETALS 4–5 or irregular

Miscellaneous opposite-leaved species
Several families

a. Petals 4 (Plantain family)

1. Water-hyssop *Bacopa caroliniana*
Plants creeping or floating, 4–8″. Leaves oval, ⅜–1″, blunt, toothed, broadly rounded to base. **Flowers solitary in upper axils,** ⅜″. Shores, shallow water. See pink-flowered herb-of-grace p. 122.

2. Eastern blue-eyed Mary *Collinsia verna*
Plants often reclining, 8–16″. Leaves triangular or oval, ¾–2″. **Flowers in 1–3 whorls of 4–6 flowers,** ⅜″. Rich, moist woods. Note the distinctive bi-colored flower.

a. Petals 5
  b. Flowers relatively small, ≤1¼″ (Dogbane and Pink families)
*3. Bouncing Bet *Saponaria officinalis*
Plants erect, 16–32″, hairless. Leaves elliptic or narrowly oval, 2¾–4″, pointed. **Flowers on congested or open panicles**, bluish-white, white, or pink, ½–1″. Roadsides, weedy places. Flowers probably pollinated by nocturnal moths, such as hawk moths, but visited by a variety of insects seeking nectar. Pink family.

*4. Periwinkle *Vinca minor*
Plants trailing or scrambling, ≤3′. Leaves narrowly oval, 1¼–2″, toothless, hairless. **Flowers solitary in axils,** ¾–1¼″; tube ¼–½″ long. Roadsides, woods. Dogbane family.

  b. Flowers relatively large ≥1¼″ (Acanthus family)
5. Hairy wild petunia *Ruellia humilis*
Plants 8–24″. **Leaves** oval or elliptic, ¾–3″, **stalkless. Flowers in clusters,** 1¼–2¾″. Prairies; dry, upland woods; fields; roadsides. Primarily pollinated by nocturnal hawk moths. What remains of the pollen in the morning is collected by bees. If all else fails this species also has self-fertilizing, closed flowers.

6. Stalked wild petunia *Ruellia pedunculata*
Plants 12–28″. **Leaves** narrowly oval or oval, 2–4″; **leaf-stalks** ⅛–⅜″. **Flowers solitary in axils,** 1¼-2″. Dry or rocky, upland woods.

a. Petals irregular (Teasel family)
*7. Fuller's teasel *Dipsacus fullonum*
Plants 1½–7′. Basal leaves narrowly oval, broader toward the tip, toothed; **stem leaves** narrowly oval, ≤12″, **fused with the opposing leaf around the stem (see insert). Flowers in thimble-shaped heads 1¼–4″ wide**. Roadsides, weedy ground. Pollinated by bumblebees. Bar indicates width of head.

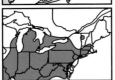

*8. Blue buttons *Knautia arvensis*
Plants 1–3′. Lower leaves narrowly oval, broader toward the tip, simple or lobed, coarsely toothed; upper leaves deeply lobed; terminal lobe ≤1″. **Flowers in heads ½–1½″ wide**. Weedy areas. Bar indicates width of head.

1. Water-hyssop

2. Eastern blue-eyed Mary

*3. Bouncing Bet

*4. Periwinkle

5. Hairy wild petunia

6. Stalked wild petunia

*7. Fuller's teasel

*8. Blue buttons

LEAVES opposite
LEAVES simple
PETALS irregular

Mints (1)
Mint family

The flowers of mints usually have the petals fused into two irregular lobes that look like lips. They can be distinguished from similar plants by the square stems and four nutlets in the fruits.

a. Flowers on panicles (cont. on next page)

1. Blue curls *Trichostema dichotomum*
Plants much branched, ≤2½', hairy. Leaves oblong, elliptic, or oval, ½–2½", toothed, lobed, or entire, leaf-stalk ≤½". Flowers on a panicle, blue changing to purple, ¼–½"; **stamens arching out of the flower**. Upland woods, fields, usually sandy areas.

a. Flowers on spikes
b. Upper lip very short

*2. Carpet-bugle *Ajuga reptans*
Plants forming loose mats, stems erect, 4–12", hairy. Leaves coppery or purplish, oval or round, 1–2½". Flowers in loose spikes with 6 flowers per node, ½–¾". Lawns. **Note the very small upper lip.**

b. Upper lip well developed

3. Self-heal *Prunella vulgaris*
Plants prostrate to erect, 4–20". Leaves narrowly oval or elliptic, ¾–4". Flowers on spikes ¾–2" tall; flowers purple, violet, or white, ⅜–¾"; **stamens 4, not longer than the corolla**. Fields, roadsides, weedy areas. Both native and non-native populations. Pollinated by bumblebees, which can visit all the flowers of an inflorescence in ½ minute.

4. Downy woodmint *Blephilia ciliata*
Plants 16–32", hairy. Leaves narrowly oval or oval, 1¼–2½", toothless or nearly so, stalkless or stalked ≤⅜". Flowers crowded on a **terminal, head-like spike** ¾–2" tall; flowers ½" long; **stamens 2, longer than the corolla**. Woods. See also p. 338.

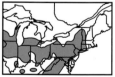

5. Purple giant hyssop *Agastache scrophulariifolia*
Plants 3–5', hairless or nearly so. Leaves oval or narrowly oval, ≤6", toothed, hairless or hairy. Flowers on spikes ≤6" tall; **flowers ½" long; stamens 4, longer than corolla**. Upland woods. Similar to no. 6, but spikes without separated lower clusters. See also p. 206.

*6. Spearmint *Mentha spicata*
Plants 1–3', hairless. Leaves narrowly oval or elliptic, 1¼–2¾", toothed, stalkless. Flowers crowded on spikes, 1¼–4¾" tall; **flowers ≤⅛" long; stamens 4, longer than corolla**. Streams, ditches.

7. Nettle-leaved salvia *Salvia urticifolia*
Plants 12–20", sparsely sticky-hairy above. Leaves triangular or oval, 2–3", pointed, toothed, narrowed to stalked base. **Flowers on interrupted spikes;** flowers ¼", ⅜–½" long; **stamens 2, longer than corolla**. Woods, thickets. See other salvia on p. 50.

1. Blue curls

*2. Carpet-bugle

3. Self-heal

4. Downy woodmint

5. Purple giant-hyssop

*6. Spearmint

7. Nettle-leaved salvia

LEAVES opposite
LEAVES simple or compound
PETALS irregular

Mints (2)
Mint family

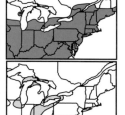

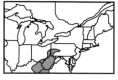

a. **Flowers in leaf axils (cont. from previous page)**
  b. **Lobes of the flower the same size and shape or nearly so**
1. **Field-mint** *Mentha arvensis*
Plants erect, 8–32″. Leaves narrowly oval, ¾–3″, toothed. **Flowers numerous in clusters at the nodes**, white, purple, or pink, ¼″. Moist places, streams, shores. Both native and non-native populations occur.

2. **False pennyroyal** *Isanthus brachiatus*
Plants branching, 8–16″. Leaves elliptic or narrowly oval, ⅜–1½″, toothless. **Flowers 1–3 in axils, bluish; petals equal, spreading**, <⅛″. Dry soil.

  b. **Lobes of the flower differing in size and shape**
*3. **Ground ivy** *Glechoma hederacea*
Plants lax, 4–12″. Leaves round or kidney-shaped, ⅜–1¼″, toothed. **Flowers 3 per axil**, blue to purple, **2-lipped**, ½–1″. Moist woods, disturbed sites. Many flowers are male the first day with the pistil becoming receptive only after this. Strong scented when stepped on.

4. **American pennyroyal** *Hedeoma pulegioides*
Plants erect, 4–16″, finely hairy on upper parts. Leaves narrowly oval or oval, broader toward base or tip, ⅜–1¼″, toothed or not. **Flowers few in axillary clusters**, blue, ¼″. Upland woods, usually rocky.

5. **Ozark calamint** *Clinopodium arkansanum*
Plants erect, 4–24″, hairless except at nodes. Leaves linear or narrowly oval, broader toward the tip, ⅜–1½″, toothless. **Flowers solitary or 4 or 6 per cluster at well-separated upper nodes**, ¼–½″. Limestone barrens, beaches. See also p. 118.

*6. **Mother-of-thyme** *Acinos arvensis*
Plants erect, 4–8″, hairy. Leaves oval, broader toward tip or elliptic, ¼–½″, toothed or toothless. **Flowers 1–3 in axils**, purple, ¼–⅜″. Roadsides, waste places.

7. **Lyre-leaved sage** *Salvia lyrata*
Plants erect, 12–24″. **Basal leaves pinnately compound**, oblong or oval, broader toward tip, 4–8″. **Flowers in widely separated 6-flowered clusters**, blue or violet, ¾–1″ long. Upland woods, thickets.

8. **Meehan's mint** *Meehania cordata*
Plants creeping, forming carpets, ≤2′, hairy. Leaves oval, 1¼–2½″, **notched at base; leaf-stalk** ½–2½″. Flowers few on leafy terminal clusters; flowers 1–1¼″ long. Rocky woods.

1. Field-mint

2. False pennyroyal

*3. Ground ivy

4. American pennyroyal

5. Ozark calamint

*6. Mother-of-thyme

7. Lyre-leaved sage

8. Meehan's mint

LEAVES opposite
LEAVES simple
PETALS irregular

Skullcaps
Mint family

Skullcaps have a distinctive, dish- or hat-like appendage at the top of the calyx (see insert in image 6). Critical characters are inflorescence type, flower size, and leaf shape.

a. Flowers 1-2 in leaf axils

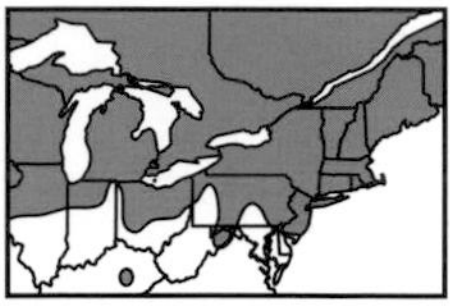

1. Marsh skullcap *Scutellaria galericulata*
Plants 6–30″, hairy. Leaves narrowly oblong or oval, ½–3″, narrowly pointed, mostly toothed, hairless, tapering to stalkless base. Flowers 1–2 in axils, ¾–1″ **long**. Shores, meadows, swamps.

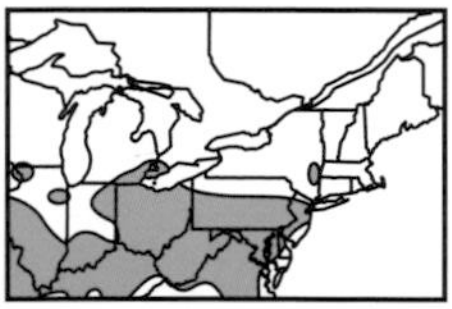

2. Veiny skullcap *Scutellaria nervosa*
Plants erect, 6–24″, mostly hairless. Leaves oval or round, 1–1¼″, pointed, **toothed**, rounded at base; stalk ≤¼″. Flowers solitary in axils, ⅛–¼″ **long**. Prairies, upland woods, rocks.

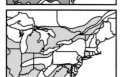

3. Little skullcap *Scutellaria parvula*
Plants 3–12″, hairy, sometimes sticky. Leaves narrowly oval or triangular, ½–1″, **toothless or with a few teeth,** hairy or hairless, rounded or notched, stalkless. Flowers solitary in axils, ¼–⅜″ **long**. Dry, upland woods; prairies.

a. Flowers on racemes
   b. Racemes axillary

4. Mad-dog skullcap *Scutellaria lateriflora*
Plants 12–36″, hairy. Leaves oval or narrowly oval, ½–5″, sharppointed, toothed, hairless or nearly so, rounded or notched at base; stalks ¼–1″. **Flowers on axillary, one-sided racemes**, 1–4″ tall; flowers ⅛–¼″ long. Rich thickets, meadows, swampy woods.

   b. Racemes terminal

A. Forest skullcap *Scutellaria ovata* (not illus.)
Plants 12–28″, hairy. **Leaves heart-shaped**, 2–4¾″, blunt or broadly pointed, toothed. Flowers on racemes ≤4″ tall; **flowers ½–1″ long**. Woods.

5. Showy skullcap *Scutellaria serrata*
Plants 12–24″, **hairless**. Leaves oval or elliptic, 2–4½″, **sharppointed**, toothed, hairless, broadly rounded or tapered at base. Flowers on solitary, terminal racemes, 2¼–6″ tall; **flowers 1–1¼″ long**. Rich woods, bluffs.

6. Hairy skullcap *Scutellaria elliptica*
Plants 6–36″, **hairy**. Leaves oval or triangular, 1¼–2¾″, **blunt**, toothed, hairy, broadly tapered at base. Flowers on panicles 1–4″ tall; **flowers ½–¾″ long**. Dry, upland woods; fields; thickets.

7. Downy skullcap *Scutellaria incana*
Plants 12–48″, **downy**. Leaves narrowly oval or oval, 2–4″, **blunt**, toothed, broadly rounded or notched at base. Flowers on numerous terminal racemes, <½″ tall; **flowers ¾–1″ long**. Dry, upland woods; thickets; clearings.

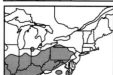

1. Marsh skullcap

2. Veiny skullcap

3. Little skullcap

4. Mad-dog skullcap

5. Showy skullcap

6. Hairy skullcap

7. Downy skullcap

LEAVES basal and alternate
LEAVES simple
PETALS 3

Blue irises
Iris family

Iris flowers are peculiar: sepals are spreading to reflexed and very showy, petals are smaller and often erect, and the style is a broad, colorful flap over the lower part of the sepals (see arrow in no. 4). See yellow irises p. 214 and orange iris p. 136.

a. Stems well-developed, leaves usually >6″

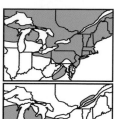

1. Northern blue iris *Iris versicolor*
Stems 4–32″. Leaves 4–32″. Flowers 2–4, 2½–3″; sepals spreading, **unspotted or with a green-yellow blotch at base surrounded by heavy blue veins on a white background; petals erect.** Marshes, swamps, meadows, shores.

A. Southern blue iris *Iris virginica* (not illus.)
Very similar to no. 1, but this species has a **yellow blotch at the base of the sepals without surrounding heavy blue veins.**

2. Slender blue iris *Iris prismatica*
Stems 16–48″. Leaves 20–28″. Flowers 1–3, 2½–3″; sepals spreading or reflexed, **veined with dark purple, mottled with white; petals erect.** Marshes, swamps, damp meadows. Similar to the previous two, but leaves are <½″ wide (vs. often >½″), and the capsule is sharply angled.

3. Zigzag iris *Iris brevicaulis*
Stems 8–16″. Leaves 14–28″. Flowers 2 on the terminal branch, 1 on side branches, 3–4″; sepals with a **white mottled area at base and yellow hairy crest; petals erect or spreading.** Swamps, wet woods, shores.

4. Arctic blue iris *Iris setosa*
Stems 4–20″. Leaves ¾–1¾″. Flowers 2–3, 2½–3″; sepals with **light streaks and a white blotch with purple veins; petals very small.** Shores, beaches, rocks. Note the small leaves and petals.

a. Stems very small or absent, leaves <6″

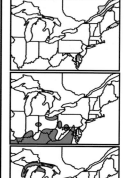

5. Dwarf crested iris *Iris cristata*
Stems 1–1¾″. Leaves 4–8″. Flowers 1–2, 2½–3″; **flower base green;** sepals with a **strong, ridged, yellow or white crest; petals spreading.** Rich woods, banks, cliffs.

6. Dwarf lake iris *Iris lacustris*
Stems ¼–1½″. Leaves 1½–2½″, later in the season becoming ≤7″. Flower solitary, 2–2½″; **flower base yellow;** sepals with **3 yellow crests; petals spreading.** Gravel shores, cliffs. Similar to no. 5, but smaller and with a yellow flower base.

7. Dwarf iris *Iris verna*
Stems 2–6″. Leaves at base 1–6″, on stem oval, broader toward tip, 1¼–1½″, later elongate. Flowers 1–2, 2–2½″; **flower base dull green; sepals with a broad yellow band; petals erect.** Sandy woods, pinelands.

1. Northern blue iris

2. Slender blue iris

3. Zigzag iris

Style

4. Arctic blue iris

5. Dwarf crested iris

6. Dwarf lake iris

7. Dwarf iris

Violets with only basal leaves
Violet family
Violets are extremely variable. Recently they have been split into several additional species. The lower petal is spurred (having a horn-like projection containing nectar). Most violets are pollinated by small bees. Caterpillars of several Fritillary butterflies feed on violets. See other violets pp. 12, 166, 216, 296, and 378.

a. Leaves compound or deeply lobed

1. Bird's-foot violet *Viola pedata*

Plants 4–10″. Leaves compound, ¾–1″; leaflets 3, **highly dissected into narrow lobes,** hairy. Flowers stalked above the leaves, violet or sometimes bicolored with upper petals deep purple, ¾–1½″, **with a flat face (like a pansy) and orange center**. Dry, sandy fields; open woods; slopes; rocks.

2. Wood violet *Viola palmata*

Plants 4–8″. Leaves heart-shaped, ≤7″; **summer leaves highly dissected; spring leaves lobed or toothed, hairy.** Flowers stalked above leaves, **violet with a white center,** 1–1½″; lower petals densely hairy at base. Woods, clearings, meadows, thickets.

a. Leaves not deeply lobed
  b. Leaves erect or spreading

3. Arrow-leaved violet *Viola sagittata*

Plants 6–8″. Leaves **arrow-shaped to broadly triangular, with or without basal lobes,** ½–1½″. Flowers long-stalked, violet-purple, ¾–1″, hairy. Dry to moist, open woods; clearings; meadows.

4. Common blue violet *Viola sororia*

Plants 3–8″. Leaves **broadly heart-shaped**, 1–3″, toothed, hairless. **Flowers not stalked above the leaves,** violet or sometimes white with purple markings, 1–1½″; **lower 3 petals hairy (hairs not club-shaped) at base.** Damp woods, meadows, clearings, weedy areas.

5. Marsh blue violet *Viola cucullata*

Plants 5–12″. Leaves **heart-shaped, 1–2″,** broadly toothed, hairless. **Flowers stalked above the leaves,** ½–1″; **lateral petals with club-shaped hairs,** usually with a dark bases (the very base is usually white, however). Bogs, swamps, wet or moist places, not weedy.

6. Northern bog violet *Viola nephrophylla*

Plants 5–6″. Leaves **heart-shaped to kidney-shaped,** 1–2½″, hairless. **Flowers stalked above the leaves,** ½–1¾″. Shores, slopes, bogs, other low grounds. Similar to no. 5, but the flower is narrower with the lateral petals angled forward.

  b. Leaves lying flat on the ground

7. Southern woolly violet *Viola villosa*

Plants 2–3″. **Leaves lying flat on ground,** heart- or kidney-shaped or round, ½–3″, hairy and purple-veined above, purple below. Flowers stalked well above the leaves, ½–1″; **lower 3 petals hairy (hairs not club-shaped).** Dry or moist, often sandy, woods; clearings.

1. Bird's-foot violet

2. Wood violet

3. Arrow-leaved violet

4. Common blue violet

5. Marsh blue violet

6. Northern bog violet

7. Southern woolly violet

Blue-eyed grasses
Iris family

Blue-eyed grasses, sometimes more violet than blue, are difficult to distinguish, often requiring technical characters for conclusive identification. Critical characters are branching, presence of winged stems, inflorescence type, and bracts. Most leaves are basal.

a. Stems branched
   b. Stems winged, ≥⅛″ wide

1. Narrow-leaved blue-eyed grass *Sisyrinchium angustifolium*
Stems 6–20″, broadly winged, ⅛–¼″ wide. Leaves ⅛–¼″ wide. **Outer bracts ¾–1½″, fused at base for ⅛–¼″.** Flowers erect, blue or violet, ½–¾″; petals rounded or notched. Meadows, stream banks, damp woods.

   b. Stems not winged, <⅛″ wide

2. Coastal plain blue-eyed grass *Sisyrinchium fuscatum*
Stems 4–20″, not winged, <⅛″ wide. Leaves ≤⅛″ wide. **Outer bracts ½–¾″, fused at base for ≤⅛″.** Flowers erect, blue or bluish-violet, ½–¾″; petals notched. Pine barrens, marshes. Note the persistent leaf bases forming thick, fibrous tufts.

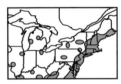

3. Eastern blue-eyed grass *Sisyrinchium atlanticum*
Stems ≤23″, wiry, not winged, <⅛″ wide. Leaves ≤⅛″ wide. **Outer bracts ½–¾″, fused at base for ⅛–¼″.** Flowers on angled stalks, light blue, violet, or sometimes white, ½–1″; petals notched or truncate. Meadows, dunes, open woods.

a. Stems unbranched
   c. Stems winged, usually ⅛″ wide

4. White blue-eyed grass *Sisyrinchium albidum*
Stems 4–16″, winged, ≤⅛″ wide. Leaves ≤⅛″ wide. **Outer bracts often purplish, 1½–2¾″, not fused.** Flowers nodding, white or violet, ¾–1″; petals notched. Prairies, meadows, roadsides, open woods. Note the long, erect leaf immediately subtending the flowers, making them appear almost lateral.

5. Strict blue-eyed grass *Sisyrinchium montanum*
Stems 4–20″, winged, ⅛″ wide. Leaves ≤⅛″ wide. **Outer bracts 1¼–2¾″, fused at base for ⅛–¼″.** Flowers mostly erect, dark blue-violet, ⅝–1″; petals deeply notched. Sandy, open ground; meadows.

   c. Stems not winged, <⅛″ wide

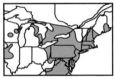

6. Needletip blue-eyed grass *Sisyrinchium mucronatum*
Stems 4–20″, wiry, barely winged or not winged, <⅛″ wide. Leaves <⅛″ wide. **Outer bracts 1¼–2″, fused at base for ≤⅛″.** Flowers mostly bent over, blue or blue-violet, ⅝–1″; petals rounded. Meadows, fields, sandy places, woods. Often difficult to distinguish from narrow examples of no. 5, but can be distinguished by more rounded petals.

1. Narrow-leaved blue-eyed grass

2. Coastal plain blue-eyed grass

3. Eastern blue-eyed grass

4. White blue-eyed grass

5. Strict blue-eyed grass

6. Needletip blue-eyed grass

Page 62
LEAVES opposite or
basal
LEAVES simple or
compound
PETALS 4-7

Miscellaneous opposite or basal-leaved species
Several families

a. Leaves opposite
1. American spurred gentian *Halenia deflexa*
Plants erect, 4–30″. **Leaves spoon-shaped, narrowly oval or oval,**
¾–2″. Flowers 5–9 on loose racemes, greenish or purplish, ⅜–½″;
petals 4, narrowly oval. Moist or wet woods, exposed areas near the
sea. Gentian family.

***2. Meadow cranesbill** *Geranium pratense*
Plants 8–20″. **Leaves 2, deeply cleft** into 5–7 tapered, sharply and
deeply cut leaflets. Flowers few-several, 1½–1¾″; **petals 5**. Fields,
roadsides. See also p. 124. Geranium family.

a. Leaves basal
  b. Petals 4

3. American plantain *Plantago rugelii*
Plants 2–10″. Leaves broadly elliptic or broadly oval, 1½–7″, hairless, reddish at base. Flowers on a terminal spike, 2–12″ tall; flowers
greenish or purplish, ⅛″. Lawns, gardens, roadsides, weedy places.
See other plantains p. 360. Plantain family.

  b. Petals 5

4. Laurentide primrose *Primula laurentiana*
Plants ≤16″. Leaves narrowly oval, wider toward the tip, ≤4″,
toothed or not. Flowers ≤15 clustered at tip of stem, ⅜–½″. Wet
rocks. See the smaller Mistassini primrose p. 128. Primrose family.

  b. Petals 6 (cont. on next page)

5. Pasque-flower *Pulsatilla patens*
Plants 4–16″, hairy. Basal leaves kidney-shaped, **deeply lobed,**
long-stalked; stem leaves similar but stalkless. Flowers blue, purple,
or white, 2–3″; petals 5–7. Dry prairies, barrens. Crowfoot family.

6. Wild hyacinth *Camassia scilloides*
Plants 12–24″. **Leaves linear,** 4–8″. **Flowers many on racemes,**
8–19″ tall; flowers blue or white, ½–1″. Prairies, open woods. The
flowers are visited by an amazing array of insects, most nectaring,
but short-tounged bees also collect pollen. Hyacinth family.

***7. Siberian squill** *Scilla siberica*
Plants 4–8″. **Leaves linear,** 4–6″. **Flowers 1-3 on short racemes,**
⅜–½″, tubular or bell-shaped. Garden escape. One of the first
flowers of spring. Hyacinth family.

1. American spurred gentian

*2. Meadow cranesbill

3. American plantain

4. Laurentide primrose

5. Pasque-flower

2½'

6. Wild hyacinth

*7. Siberian squill

LEAVES basal, floating or submersed
LEAVES simple
PETALS 6 or irregular

Miscellaneous species (2)
Several families

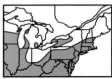

a. Leaves basal (cont. from previous page)
   b. Petals 6 (Arrow-grass and onion families)
1. Wild onion *Allium stellatum*
Plants 12–28″. **Leaves** 3–6, linear, **folded, edge angled**, mostly gone by flowering-time. **Flower on an erect umbel;** flowers ¼–½″. Prairies, rocky hills. See other onions pp. 130 and 396. Onion family.

*2. Field garlic *Allium vineale*
Plants 1–4″. **Leaves** linear, **round in cross section. Flowers 1–50 in an erect, round cluster;** flowers <⅛″. Lawns, fields, meadows. Note the flowers are often replaced by bulblets. Onion family.

3. Common arrow-grass *Triglochin maritimum*
Plants 8–32″. Leaves linear, ≤20″. Flowers many on racemes 4–16″ tall; flowers green, ⅛″; stigmas purple. Marshes, bogs. Common arrow-grass forms elevated rings that develop unique plant communities within nearly bare substrate. Arrow-grasses are wind pollinated. Arrow-grass family.

   b. Rays many (Composite family)
4. Robin's plantain *Erigeron pulchellus*
Plants 6–24″. Basal leaves narrowly oval, broader toward the tip, or round, ¾–5″. Flowers ⅜–¾″; rays 50–100. Woods, stream banks. See other erigerons on pp. 26, 290.

   b. Flowers irregular (Bladderwort family)
5. Butterwort *Pinguicula vulgaris*
Plants 2–6″, hairless or minutely sticky-hairy. **Leaves 3–6,** oval or elliptic, ¾–2″, **narrowed to base,** greasy. **Flowers** solitary, ½–¾″. Wet rocks, bogs, meadows, shores. Insects are trapped by mucilage secreted on the leaves. Leaves then absorb nutrients from the decaying insects. Leaves were once used to coagulate milk.

a. Leaves floating or submersed (Pickerel-weed family)
6. Mud-plantain *Heteranthera multiflora*
Plants submersed, emergent, or creeping on mud. Leaves kidney-shaped, ⅜–2″. Flowers 3–16 on a spike, pale purple or white with yellow spots, ⅛–½″ long. Shallow water or mud. The flowers all open within one or two days of each other. See also p. 228.

*7. Water-hyacinth *Eichhornia crassipes*
Plants floating. Leaves radiating at water level, round or kidney-shaped, 1½–4¾″, leaf-stalks swollen, providing flotation. Flowers on a panicle, 1½–6″ tall; flowers lilac or white, 2–2¾″. Streams, canals, and rivers.

1. Wild onion

*2. Field garlic

3. Common arrow-grass

4. Robin's plantain

5. Butterwort

6. Mud-plantain

*7. Water hyacinth

LEAVES alternate and basal
LEAVES simple or compound
PETALS 3-4

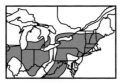

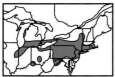

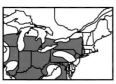

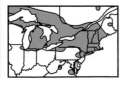

Miscellaneous alternate-leaved species
Several families

a. Petals 3 (Spiderwort and Knotweed families)
1. Sticky spiderwort *Tradescantia bracteata*
Plants 8–16″. Leaves linear, 6–12″. Flowers rose or blue, 1¼–1½″; **sepals densely sticky-hairy.** Prairies. See page 2 for further details and other spiderworts. Spiderwort family.

*2. Red sorrel *Rumex acetosella*
Plants 2–20″. **Leaves** variable, **usually 3-lobed, arrow-shaped,** ¾–2″; terminal lobe narrow elliptic to oblong; lateral lobes divergent. Flowers reddish, on slender racemes, <⅛″, nodding. Fruit oval. Fields, lawns, weedy places. See also p. 232. Knotweed family.

a. Petals 4
b. Petals fused (Gentian family)
3. Stiff gentian *Gentianella quinquefolia*
Plants 8–32″, branched or unbranched, hairless. Leaves narrowly oval, ¾–2¾″, rounded at base. Flowers on dense panicles; flowers ¼″ wide, ½–1″ long; petals 4–5, barely opening at tip. Woods, openings.

b. Petals free (several families)
4. Fireweed *Chamerion angustifolium*
Plants 1–8′, hairless. **Leaves narrowly oval or linear,** 1–8″, stalkless, pale beneath, toothed. Flowers on terminal spikes; flowers ¾–1″. Fruit 1–3″. Burned woodlands, clearings, roadsides. An early arrival after fire or clearing. Evening-primrose family.

*5. Dame's-rocket *Hesperis matronalis*
Plants 1½–3′. **Leaves narrowly oval,** 3–8″, short-stalked, sharply toothed, hairy above. Flowers on racemes; flowers ¾–1″. Fruit linear, 2–4″. Roadsides, woods, bottomlands, thickets. Among the differences with no. 4 are the concealed stamens and styles. Mustard family.

*6. Honesty *Lunaria annua*
Plants ≤3′, scarcely hairy. **Leaves triangular or heart-shaped,** ≤3½″, narrowly pointed, short-stalked or stalkless. Flowers on racemes; flowers ¾″. Fruit broadly elliptic (silver-dollar-like), 1½–2″. Garden escape. Fruit used as a decoration. Mustard family.

7. Pink spring-cress *Cardamine douglassii*
Plants 4–12″, long-hairy. **Basal leaves round or heart-shaped,** ½–1½″, long-stalked; stem leaves 3–5, oblong or oval, slightly toothed, stalkless. Flowers on racemes; flowers ½–1″. Fruit linear, ½–1″. Moist or wet woods, springs. See other spring-cresses on pp. 268, 300. Mustard family.

8. Cuckoo-flower *Cardamine pratensis*
Plants 8–20″. Basal leaves 1½–1¾″, long-stalked, **compound with 3–8 broadly oval leaflets;** stem leaves shorter, with 5–17 oval or oblong leaflets. Flowers on crowded racemes, ¾–1″. Fruit linear, 1–1½″. A variety of wet sites. Mustard family.

1. Sticky spiderwort

*2. Red sorrel

3. Stiff gentian

4 Fireweed

*5. Dame's-rocket

*6. Honesty

7. Pink spring-cress

8. Cuckoo-flower

Bearberries, Bilberries, and Cranberries
Heath family

Members of the Heath family typically have anthers with pores at the tip instead of slits down the side. Most are pollinated by bumblebees, which grasp the flower and "buzz" it, causing it to vibrate so that the pollen falls out of the pores. Most of these species also occur as white flowers.

a. Fruits berries (cont. on next page)
    b. Petals 4

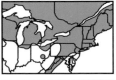

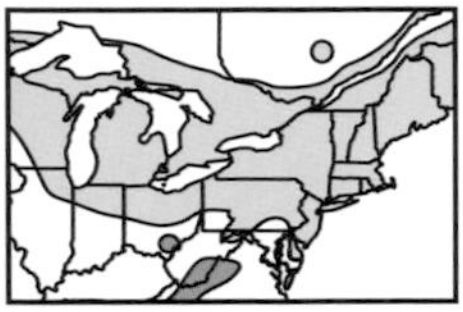

1. Large cranberry *Vaccinium macrocarpon*
Creeping shrubs, 1–6″. **Leaves elliptic, ¼–½″**, rounded, pale beneath, hairless or rarely slightly hairy, toothless, stalkless. Flowers 1–6 from axils, ½″; the petals strongly curled backward. **Fruits red berries, ½–¾″.** Open bogs, swamps, wet shores. Caterpillars of the Bog Copper butterfly feed on large cranberry leaves. This is the source of cultivated cranberries.

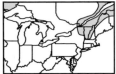

2. Small cranberry *Vaccinium oxycoccos*
Creeping shrubs ⅜–1¼″. **Leaves oval, strongly inrolled, ⅛–½″**, pointed, whitened beneath, hairy or not, toothless, stalkless. Flowers 1–4 in a terminal cluster, ½″; the petals strongly reflexed. **Fruits red berries, ¼–½″.** Cold bogs.

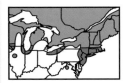

3. Bog bilberry *Vaccinium uliginosum*
Matted shrubs, 6–24″. **Leaves elliptic, ¼–1″**, blunt, densely hairy, stalkless. Flowers 1–4 in axillary clusters, ⅛″. **Fruits dark blue or black berries, ≤¼″.** Alpine rocks, bogs.

    b. Petals 5

4. Common bearberry *Arctostaphylos uva-ursi*
Prostrate shrubs ≤36″ wide, with papery, reddish bark. **Leaves narrowly oval, broader toward the tip, ½–1″**, blunt, becoming hairless, toothless, tapering to short stalk. Flowers 5–12 in a terminal cluster, ¼–⅜″. **Fruits dry, red berries, ¼–⅜″.** Open sands, rocks. Caterpillars of the Hoary Elfin butterfly feed on common bearberry.

5. Dwarf bilberry *Vaccinium caespitosum*
Shrubs, 2–24″. **Leaves narrowly oval, broader toward the tip, ½–1½″**, blunt, hairless above, glandular beneath, finely toothed, tapering to base, stalkless. Flowers solitary in leaf axils, ¼″. **Fruits blue berries, ¼″.** Alpine rocks and shores.

6. Mountain cranberry *Vaccinium vitis-idaea*
Creeping shrubs, 3–8″. **Leaves elliptic, ¼–¾″**, rounded, shiny, with black dots beneath, toothless, tapering to stalkless base. Flowers few in small, terminal clusters, ¼″. **Fruits red berries, ½″.** Rocks, peats.

1. Large cranberry

2. Small cranberry

3. Bog bilberry

4. Common bearberry

5. Dwarf bilberry

6. Mountain cranberry

Heaths, Hardhack, Morning-glory, and Twisted-stalk
Several families

a. Fruits capsules (cont. from previous page)
  b. Petals 5

1. Mountain-heath *Phyllodoce caerulea*
Plants 2–6″. **Leaves needle-like, ¼–½″,** blunt, toothless, stalk <⅛″.
Flowers in small clusters at ends of branches, ½″. Alpine rocks, peat.
Heath family.

2. Sand-myrtle *Leiophyllum buxifolium*
Plants 4–18″. **Leaves alternate or opposite, oval, ⅛–½″,** pointed,
toothless. **Flowers** in umbel-like clusters, ⅛″. Sandy pine barrens,
rock outcrops. Note the petals are separate and sometimes spreading.
Heath family.

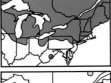

3. Bog-rosemary *Andromeda polifolia*
Plants ≤20″. **Leaves linear or narrowly elliptic, ¾–2″,** toothless.
Flowers ¼″. Open sands, rocks. Note the inrolled leaf margins and
white hairs on the lower leaf surface. Heath family.

4. Lapland rose-bay *Rhododendron lapponicum*
Plants forming mats, aromatic shrubs, 4–20″. **Leaves elliptic,
½–¾″,** densely covered with brown scales. **Flowers** few in terminal
clusters, ½–¾″. Barrens, cliffs, mountain summits. Heath family.

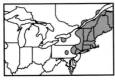

5. Rhodora *Rhododendron canadense*
Shrubs ≤3′. **Leaves narrowly oblong to elliptic, ¾–2″.** Flowers in
terminal umbel-like clusters, ¾″; **flowers divided into 3 segments,
the upper segment shallow-lobed.** Bogs, wet woods. Heath family.

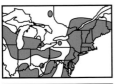

6. Hardhack *Spiraea tomentosa*
Plants shrubby, ≤4′. **Leaves oval, narrowly oval, or oblong, 1¼–
2″,** white-hairy below. Flowers on terminal panicles, ⅛″. Swamps,
wet meadows. Produces little nectar but is visited by a variety of
flies, beetles, and bees seeking pollen. See another spirea p. 282.
Rose family.

*7. Red morning-glory *Ipomoea coccinea*
Plants twining, 3–10′. Leaves broadly oval, 2–4″, pointed, heart-
shaped, toothed, toothless, or lobed. Flowers on short racemes or
panicles at ends of axillary stalks, ¾″; tube ¾–1¼″ long, often
yellow inside. Thickets, roadsides, waste places. See other morning-
glories on pp. 14, 90, 270. Morning-glory family.

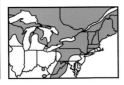

  b. Petals 6
8. Twisted stalk *Streptopus lanceolatus*
Plants 12–32″, finely hairy. Leaves narrowly oval, 2–3½″, pointed,
rounded at base. **Flowers solitary, axillary, ¾″.** Rich woods. See
also p. 286. Lily family.

1. Mountain-heath

2. Sand-myrtle

3. Bog-rosemary

4. Lapland rose-bay

5. Rhodora

6. Hardhack

*7. Red morning-glory

8. Twisted-stalk

LEAVES alternate
LEAVES simple
PETALS 5

Gerardias
Broom-rape family
Gerardias are partially parasitic on the roots of other plants. Flowers open for only a day. Bees enter the flower upside down and hang from the stamens. Critical characters are flower-stalks, flower hairs, and petal shape.

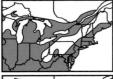

a. Flowers stalkless or nearly so
1. Smooth gerardia *Agalinis purpurea*
Plants 4–40″, **hairless**. **Leaves** linear, ¾–1¾″. **Flowers** few to many, short-stalked (<⅛″) flowers on racemes, purple, pink, or white, ⅜–1¼″; sepals pointed; petals downy inside. Usually wet, acid sands.

A. Fascicled gerardia *Agalinis fasciculata* (not illus.)
Similar to no. 1, but with **rough hairs on stem**, and numerous small leaves in axils of other leaves. Sandy areas, often weedy.

B. Eared gerardia *Agalinis auriculata* (not illus.)
A distinctive species with lobes at base of narrowly oval leaves and **downward-pointing, long hairs** on stems. Prairies, open woods.

a. Flowers stalked
  b. Flowers hairy inside
2. Thread gerardia *Agalinis setacea*
Plants 6–28″. **Leaves** thread-like, ⅜–1¼″. **Flowers** divergent, long stalked (⅜–1¼″), on racemes, purple, ½–1″; sepals pointed; **petals** hairy inside, **rounded.** Dry, sandy openings.

3. Sand-plain gerardia *Agalinis acuta*
Plants 4–16″. **Leaves** linear, ⅜–1″. **Flower** stalks diverging at 45°, on lateral racemes, ¼–¾″ long; flowers pink, ⅜–½″; sepals broadly triangular; **petals** hairy inside, **notched.** Dry, sandy soils.

C. Midwestern gerardia *Agalinis skinneriana* (not illus.)
Similar to no. 3, but with **rounded petals.** Prairies, open woods, barrens. Rare, endangered in Canada.

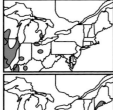
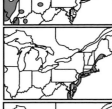

  b. Flowers hairless inside
4. Salt-marsh gerardia *Agalinis maritima*
Plants 2–14″. **Leaves** linear, succulent, ½–1¼″, blunt. **Flowers** 2–5 pairs on short stalks (⅛–⅜″) on short racemes, purple, ½–¾″; **sepals blunt or rounded;** petals hairless. **Salt-marshes.**

5. Slender gerardia *Agalinis tenuifolia*
Plants 6–24″. **Leaves** linear, 1″, pointed. **Flowers** divergent, stalked ⅜–¾″, on racemes; flowers purple, ⅜–¾″; **sepals broadly triangular;** petals hairless. **Dry woods, thickets, fields.** Note how the upper petals arch over the stamens.

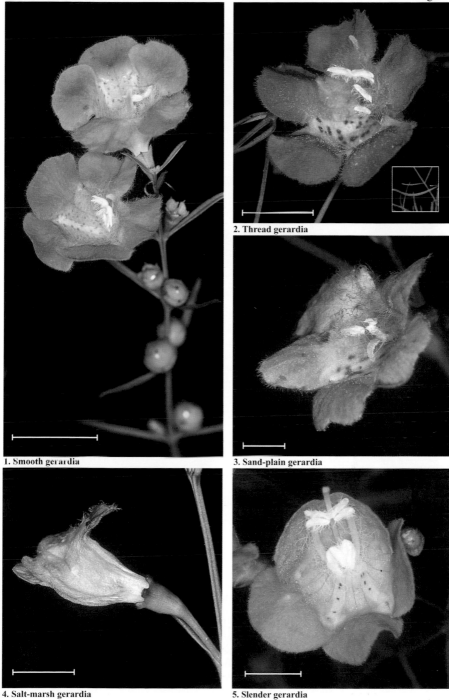

1. Smooth gerardia

2. Thread gerardia

3. Sand-plain gerardia

4. Salt-marsh gerardia

5. Slender gerardia

LEAVES alternate
LEAVES simple or lobed
PETALS 5

Mallows
Mallow family

Mallows are recognized by the fused stamens forming a tube around the style. Critical characters are petal shape, leaf shape, and shape of bractlets just outside the sepals (see inserts in nos. 4 and 5).

a. Petals rounded or truncate, native

1. Finger poppy-mallow *Callirhoe digitata*
Plants erect, 1½–7', hairless or nearly so. Leaves triangular, deeply 5–10-lobed; lobes finger-like, toothless. Flowers in open panicles, deep pink or purple with a basal white spot, 1½–3"; **bractlets absent.** Limestone glades, rocky places.

2. Purple poppy-mallow *Callirhoe involucrata*
Plants **creeping on ground,** ≤3', rough-hairy. Leaves round, deeply 5–7-lobed; lobes linear, toothed. Flowers axillary, stalked, deep pink or purple with a basal white spot, 1½–3"; **bractlets 3, narrowly oval or linear.** Open places, upland prairies. Pollinated by bees.

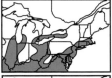

3. Clustered poppy-mallow *Callirhoe triangulata*
Plants **ascending,** 1–3', rough-hairy. Leaves triangular, toothed, and sometimes lobed. Flowers on panicles, deep pink with a basal white spot, 1½–3"; **bractlets 3, oval.** Sandy prairies, dry woods.

4. Rose-mallow *Hibiscus moscheutos*
Plants erect, 3–7', white-downy. Leaves narrowly oval or oval, sometimes shallowly 3-lobed, toothed. Flowers solitary in axils, pink or white, often with a red center, **4–8"; bractlets 12, linear.** Brackish or freshwater marshes.

5. Seashore-mallow *Kosteletzkya virginica*
Plants erect, 1–5', minutely rough-hairy. Leaves triangular-oval, 2½–6", often with basal lobes, toothed. Flowers on axillary or terminal panicles, **1–3"; bractlets 6–10, linear.** Brackish or salt-marshes.

a. Petals notched, non-native

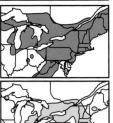

*6. Common mallow *Malva neglecta*
Plants trailing on ground, 4–24", densely hairy. Leaves round or kidney-shaped, ¾–2½", **shallowly 5–9-lobed;** lobes toothed. Flowers in axillary clusters, white, light pink, or light purple, ½–1"; **bractlets 3, linear.** Barnyards, gardens, weedy places.

*7. Musk mallow *Malva moschata*
Plants erect, 8–40", rough-hairy. Leaves round, 1¼–2½", **deeply 5–7-lobed or basal leaves unlobed;** lobes parted again. Flowers on terminal panicles, white or pink, 1½–3"; **bractlets 3, linear or narrowly oval.** Roadsides, weedy places.

*8. Vervain mallow *Malva alcea*
Plants erect, 12–40", hairy. Leaves round or kidney-shaped, 1½–3", **deeply 3–7-lobed;** lobes toothed. Flowers in axillary clusters, 1–2"; **bractlets 3, oval.** Gardens, weedy places.

1. Finger poppy-mallow

2. Purple poppy-mallow

3. Clustered poppy-mallow

4. Rose-mallow

5. Seashore-mallow

*6. Common mallow

*7. Musk mallow

*8. Vervain mallow

Pink Smartweeds
Smartweed family

Critical characters are inflorescence shape and orientation, leaf shape, size and stalk, and sheath margin. The sheath surrounds the stem above the node. See white smartweeds pp. 272–76.

a. Spikes erect
b. Wet areas

1. Water smartweed *Polygonum amphibium*
Plants ≤3′. Leaves often narrowly oval, 2–4″; **stalk ¾–3″;** sheaths with or without bristles. Flowers on 1–2 dense spikes, **1–7″ tall;** flowers ¼″. Wet soil, ponds. Terrestrial form is erect, with sharp-pointed, hairy leaves; aquatic form has floating branches with rounded, hairless leaves.

2. Mild water-pepper *Polygonum hydropiperoides*
Plants sprawling to erect, ½–3′. Leaves narrowly oval or linear, 1½–8″; **stalk ≤¼″;** sheaths short-bristly. Flowers on spikes, 1–3″ tall; flowers white or pink, <¼″. Beaches, marshes, shallow water.

b. Weedy areas
3. Pennsylvania smartweed *Polygonum pensylvanicum*
Plants erect, 1–4′, sticky upward. Leaves narrowly oval or oval, 4–6″; stalk ⅜–1½″; **sheaths torn and hairless.** Flowers numerous on dense spikes, ½–2″ tall; flowers ≤¼″. Fields, gardens, weedy areas.

***4. Cespitose smartweed** *Polygonum caespitosum*
Plants sprawling or ascending, ≤3′. Leaves narrowly oval, 1–3″, stalk ≤¼″; **sheaths long-bristly.** Flowers on 1–a few dense spikes, ¾–1½″ tall; flowers <¼″. Roadsides.

***5. Lady's thumb** *Polygonum persicaria*
Plants sprawling to erect, ½–2′. Leaves narrowly oval, 3–6″; stalk ¼–½″; **sheaths short-fringed.** Flowers on 1–several dense spikes, ½–1¾″ tall; flowers <¼″. Weedy areas. **Note the dark-purple blotch in the middle of the leaf.**

a. Spikes nodding
***6. Dock-leaved smartweed** *Polygonum lapathifolium*
Plants erect or sprawling, 1–6′. Leaves variable, commonly narrowly oval, 4–8″; stalk ⅜–¾″; **sheaths entire or torn at top.** Flowers on numerous spikes ½–3″ tall; flowers <¼″. Moist or wet fields, weedy areas. Note the "anchor-shaped" nerves of the outer petals.

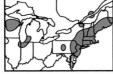

7. Carey's smartweed *Polygonum careyi*
Plants erect, 1–4′, sticky upward. Leaves narrowly oval, 2½–7″, stalked; **sheaths spreading-hairy, bristly at top. Flowers on loosely flowered spikes, 1-2″ tall;** flowers <¼″. Wet areas, swamps.

***8. Prince's-feather** *Polygonum orientale*
Plants erect, 1–8′. Leaves oval, 4–12″; stalk 1¼–2½″; **sheaths with flaring collar, hairy and fringed at top. Flowers on dense, drooping spikes, 1–4″ tall;** flowers <¼″. Near gardens.

Aquatic form   Terrestrial form

1. Water smartweed   2. Mild water-pepper

3. Pennsylvania smartweed   *4. Cespitose smartweed   *5. Lady's thumb

*6. Dock-leaved smartweed   7. Carey's smartweed   *8. Prince's-feather

LEAVES alternate
LEAVES simple
"PETALS" many

Blazing stars
Composite family
Critical characters are head shape and size and involucral bracts. The "petals" are actually flowers but not differentiated into rays and disk.
a. Heads several to many
   b. Heads cylindrical with 3–14 flowers

1. Dense blazing star *Liatris spicata*
Plants 1–5′, usually hairless. Leaves linear, 4–16″. Heads on dense spikes, stalkless, ⅛–¼″; **involucral bracts appressed, rounded or blunt, outer sharper.** Wet meadows; prairies; moist, open places.

2. Thick-spike blazing star *Liatris pycnostachya*
Plants 1–4′, usually hairy. Leaves linear, 4–20″. Heads on dense spikes, stalkless, ¼–⅜″; **involucral bracts reflexed, sharp.** Mostly dry prairie sites, open woods, occasionally moist sites.

3. Grass-leaved blazing star *Liatris pilosa*
Plants 1–3′, hairy or hairless. Leaves linear, 2½–12″. Heads on spikes (often one-sided spikes), short-stalked or rarely stalkless, ¼″; **involucral bracts appressed, rounded.** Dry, open woods; especially sandy soil, often among pines.

4. Small-headed blazing star *Liatris microcephala*
Plants 12–32″, hairless. Leaves linear, 1¼–6½″. Heads on racemes, stalked, ⅛″; **involucral bracts appressed, sharp or rounded.** Exposed rocky places, glades, sandy shores, meadows, clearings, old fields.

   b. Heads button-shaped with 14–many flowers
5. Rough blazing star *Liatris aspera*
Plants 1–4′, short-hairy or hairless. Leaves narrowly elliptic, 2–16″. Heads on spikes, stalkless or nearly so, ⅜–¾″; **involucral bracts spreading to reflexed, rounded.** Dry, open places; thin woods, especially sandy soil.

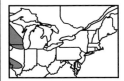

6. Northern plains blazing star *Liatris ligulistylis*
Plants ½–2′, usually hairless. Leaves linear, 3–10″. Heads on racemes, stalked, ½–¾″, the terminal head the largest; **involucral bracts seldom reflexed, blunt.** Mostly in damp, low places; occasionally in drier soil.

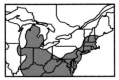

7. Northern blazing star *Liatris scariosa*
Plants 1–4′, hairless or slightly hairy. Leaves elliptic, linear, or oval, 4–14″. Heads on spikes, stalkless or stalked, ⅜–¾″; **involucral bracts appressed or usually loose, occasionally reflexed, broadly rounded.** Prairies; open woods; dry, open places.

a. Heads 1–a few
8. Few-headed blazing star *Liatris cylindracea*
Plants 8–24″, hairless or slightly hairy. Leaves linear, 4–10″. Heads 1–a few, stalked or stalkless, ⅜–¾″; **involucral bracts appressed or slightly loose, broadly rounded.** Dry, open places.

1. Dense blazing star

2. Thick-spike blazing star

3. Grass-leaved blazing star

4. Small-headed blazing star

5. Rough blazing star

6. Northern plains blazing star

7. Northern blazing star

8. Few-headed blazing star

LEAVES alternate
LEAVES simple
RAYS many or absent

Miscellaneous composites
Composite family

Composite "flowers" are actually tight clusters (heads) of flowers, often of two types, disk (in the center) and ray (around the outside).

a. Rays present

1. Purple coneflower *Echinacea purpurea*
Plants 2–5′, **hairless. Leaves elliptic or oval to broadly oval**, ≤6″. Heads solitary, terminal, 2½–4″; disk ½–1¼″; rays 1¼–3″, drooping. Woods, prairies. Generally in moister sites than no. 2.

2. Prairie coneflower *Echinacea pallida*
Plants ¼–3′, **hairy. Leaves** mostly toward the base, **narrowly oval**, ≤8″. Heads solitary, terminal; disk ½–1¼″; rays ¾–3″, spreading or drooping. Dry, open prairies; plains. Note: petals are paler than no. 1.

3. Rose-ring blanket-flower *Gaillardia pulchella*
Plants 8–24″, hairy. Leaves narrowly oval, broader toward the tip, toothed or lobed, densely hairy below. Heads long-stalked, purple, or the tip yellow-orange, 1¼–2¾″; disk ½–1¼″; rays ⅜–¾″. Sandy places, sea beaches. See another blanket-flower p. 164.

a. Rays absent
  b. Involucral bracts densely bristly

*4. Common burdock *Arctium minus*
Plants 2–5′. Leaves narrowly to very broadly oval, ≤2′. **Heads** stalkless or short-stalked, **on elongate racemes; heads ½–1″**. Roadsides, weedy places. The bristly heads were the inspiration for Velcro.

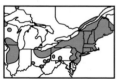

*5. Great burdock *Arctium lappa*
Plants 3–8′. Leaves oval or broadly oval, ≤2′. **Heads in flat-topped clusters, 1¼–1½″**. Roadsides, weedy places. The major differences between this and no. 4 are the inflorescence type and the size of the heads.

  b. Involucral bracts not bristly

6. Appalachian Barbara's buttons *Marshallia grandiflora*
Plants 8–40″. Leaves narrowly oval or elliptic, 3-nerved, 2–6″. **Heads** long-stalked, ¾–1½″. Moist to wet woods, meadows, along streams.

7. Marsh-fleabane *Pluchea odorata*
Plants ≤3′, **sticky-hairy**. Leaves narrowly oval, elliptic, or oval, 1½–6″, pointed, toothed, short-stalked. **Heads** on flat-topped panicles, ⅛–⅜″. Salt-marshes, brackish marshes.

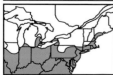

8. Purple cudweed *Gamochaeta purpurea*
Plants 4–16″, **thinly woolly**. Leaves narrowly oval, broader toward the tip, ≤4″. **Heads** numerous on terminal clusters, ⅛–¼″. Sandy areas, weedy areas.

1. Purple coneflower

2. Prairie coneflower

3. Rose-ring blanket-flower

*4. Common burdock

*5. Great burdock

6. Appalachian Barbara's buttons

7. Marsh-fleabane

8. Purple cudweed

LEAVES alternate or whorled
LEAVES simple
PETALS 3

Milkworts
Milkwort family

Milkworts have 3 outer sepals, 2 inner sepals forming wings, and 3 petals united around the filament tube. Critical characters include flower shape and wing shape and size, inflorescence type, and leaf size. See other milkworts on pp. 136 and 282.

a. Leaves alternate

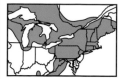

1. Fringed polygala *Polygala paucifolia*
Plants erect, 3–6″. Lower leaves scale-like, ⅛–¼″; upper leaves elliptic to oval, ½–6″. Flowers occasionally white, 1–4; **long-stalked from leaf axils; flowers ½–¾″; wings narrowly oval, ½–1″; lower lip fringed at tip**. Moist woods.

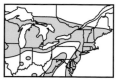

2. Bitter milkwort *Polygala polygama*
Plants prostrate, 4–16″. **Basal leaves spoon-shaped**, ½–1″; stem leaves narrowly oval, ½–1½″. **Flowers on loose, 1-sided racemes, 1–6 × ½–⅝″; flowers ¼″; wings narrowly oval, ⅛–¼″**. Sandy woods, clearings.

3. Blood milkwort *Polygala sanguinea*
Plants 3–16″. Leaves linear or narrowly elliptic, ½–1½″. Flowers on very **dense, egg-shaped racemes, ≤1½ × ½″; flowers ⅛″; wings oval, ¼″**. Fields, meadows, open woods. Similar to nos. 4 and 5, but the **wings are twice the size of the corolla.**

4. Maryland milkwort *Polygala mariana*
Plants 4–16″. Leaves linear, ½–¾″. Flowers on egg-shaped or cylindrical **racemes, ≤1¼ × ¼–½″; flowers ⅛″; wings oval and wider toward the tip, ⅛″**. Dry or moist sands, clays, peats.

5. Nuttall's milkwort *Polygala nuttallii*
Plants 2–12″. Leaves linear, ¼–½″. Flowers on cylindrical **racemes, ⅜–¾ × ⅛–¼″; flowers ≤⅛″; wings oval, broader toward the tip, ≤⅛″**. Dry, sandy soil; barrens.

a. Leaves whorled

6. Whorled milkwort *Polygala verticillata*
Plants 4–16″. Leaves linear, ⅜–¾″; lower leaves in whorls of 2–5. Flowers on **tapering, pointed racemes, ¼–2 × 1¼–2½″; flowers white, green, or pink, ½″; wings elliptic or oval, <⅛″**. Moist grasslands, woods.

7. Short-leaved milkwort *Polygala brevifolia*
Plants 2–12″. Leaves linear or narrowly oval, wider toward the tip, ½–1½″, in whorls of 3–5. Flowers on **blunt racemes, ⅜–1¼ × ⅜–½″; flowers ⅛–¼″; wings ovate or oblong, ⅛″**. Sandy swamps. Similar to no. 8, but the racemes are stalked ¾–4″ (vs. ≤⅜″).

8. Cross-leaved milkwort *Polygala cruciata*
Plants 4–12″. Leaves linear or narrowly oval, broader toward the tip, ½–1½″, in whorls of 3–4. Flowers on dense, cylindrical, **blunt racemes, ½–2½ × ⅜–½″; flowers ⅛–¼″; wings triangular, ⅛–¼″**. A variety of damp or wet sites. Note, leaves appear to form crosses.

1. Fringed polygala

2. Bitter milkwort

3. Blood milkwort

4. Maryland milkwort

5. Nuttall's milkwort

6. Whorled milkwort

7. Short-leaved milkwort

8. Cross-leaved milkwort

LEAVES alternate and basal
LEAVES simple
PETALS absent, 3 or 5

Miscellaneous alternate-leaved species
Various families

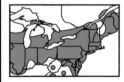

a. Flowers minute, petals absent
*1. Halberd-leaved orache *Atriplex patula*
Plants erect, ≤3′. Leaves narrowly oval, 1½–4¾″. Flowers numerous in spikes; flowers ⅛″. **Fruit enclosed in 2 bracts; bracts oval or triangular, ⅛–¼″.** Weedy areas. See also p. 234. Amaranth family.

a. Flowers minute, petals 3
Pinweeds are a small group of eight species in our area found in dry, often sandy areas. They have 3 petals and 5 sepals; the outer 2 sepals are narrowly oval; the inner sepals, oval. Critical characters are the relative lengths of the outer sepals, inner sepals, and capsule. Rock-rose family.

2. Maritime pinweed *Lechea maritima*
Plants erect or ascending, 4–16″, **densely hairy.** Leaves on basal shoots elliptic, ¼–½″; stem leaves narrowly oval. Flowers on a panicle forming ½–⅔ the entire height of the plant; flower ⅛″; the sepals about equal the fruit. **Fruit about the same length as width.** Sandy areas near the sea.

3. Pinweed *Lechea racemulosa*
Plants spreading, 8–16″, **thinly hairy.** Leaves on basal shoots oblong or elliptic, ¼″, often whorled; stem leaves narrowly oval. Flowers on a panicle ½ the entire height of the plant; flowers ⅛″; the outer sepals shorter or equaling the inner. Dry soil. **Differs from no. 2 in that the fruit is twice as long as wide.**

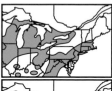

a. Flowers large, petals fused
4. Indian paintbrush *Castilleja coccinea*
Plants 8–24″, hairy. Basal leaves elliptic, unlobed; stem leaves 3–5 lobed; lobes linear or oblong, ≤1½″. **Flowers on dense spikes 1½–2½″; bracts red; flowers greenish-yellow; 1″ long.** Meadows; moist prairies; damp, sandy soil. See also p. 168. Broom-rape family.

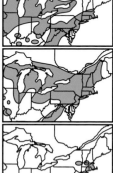

*5. Common comfrey *Symphytum officinale*
Plants 1–4′, hairy. Leaves oval or narrowly oval, 6–12″. **Flowers on small, curled racemes, nodding, pink, blue, purplish or cream, ½–⅝″.** Weedy places. Borage family.

6. Chaffseed *Schwalbea americana*
Plants 12–24″, finely hairy. Leaves narrowly oval, ¾-2″. **Flowers on a leafy spike, 1¼–1½″ long.** Moist, sandy areas. Broom-rape family

*7. Tobacco *Nicotiana tabacum*
Plants ≤7′. Leaves narrowly oval, oval, or elliptic; lower ones ≤20″; upper reduced. **Flowers on a panicle; flowers pink to red and white, 2-2½″ long.** Escape from cultivation in tobacco-growing regions. Nightshade family.

*1. Halberd-leaved orache

2. Maritime pinweed

3. Pinweed

4. Indian paintbrush

*5. Common comfrey

6. Chaffseed

*7. Tobacco

LEAVES alternate
LEAVES simple
PETALS irregular

Miscellaneous irregular-flowered species
Bellflower, Plantain, and Orchid families

a. Flowers with petals more or less the same size and shape.
1. Cardinal-flower *Lobelia cardinalis*
Plants erect, 2–6′. Leaves narrowly oval to oblong, ≤6¼″, pointed, toothed; lower leaves stalked; upper leaves stalkless. Flowers on racemes 4–20″ tall; flowers red, 1¼–1¾″; **the petals spreading, three somewhat fused.** Damp shores, meadows, swamps. Note the stamens exserted from slit in petals. Pollinated by hummingbirds. See also p. 28. Bellflower family.

*2. Chinese lobelia *Lobelia chinensis*
Plants prostrate, branches 2–8″. Leaves narrowly oval, ¼–1″. Flowers solitary at the tip, pink, 1½″; **the 5 petals are all directed to one side.** Escape, potentially naturalizing. Bellflower family.

*3. Common foxglove *Digitalis purpurea*
Plants erect, 2–6′. Leaves narrowly oval, 5–10″, toothed, hairy below. Flowers on one-sided racemes; flowers deep pink or white spotted with purple inside, 1½–2″, **tubular with lower portion slightly longer than upper.** Escape near gardens. See also p. 168. Plantain family.

a. Flowers with a lip noticeably different from the other petals.
    b. Lip forming a pouch (Orchid family)
4. Showy lady-slipper *Cypripedium reginae*
Plants 1–3′, hairy. Leaves elliptic or oval, 4–8″, clasping, strongly ribbed, hairy. **Flowers 1–3, terminal, 1–2″;** sepals round-oval, blunt, 1–1½″, lower 2 united; lateral petals oblong, equaling sepals but narrower; lip 1¼–1½″. See also pp. 132, 166, 298. Swamps, bogs, wet woods.

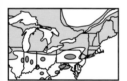

5. Ram's head lady-slipper *Cypripedium arietinum*
Plants 4–16″, thinly hairy. Leaves narrowly oval or elliptic, 2–4″, often folded, hairy only on margin. **Flowers solitary, ½–1″;** sepals and lateral petals ½–1″; lip ½–1″, with white hairs around the edge of the pouch. Moist, usually acid soils in conifer woods.

    b. Lip not forming a pouch (Orchid family)
6. Showy orchis *Galearis spectabilis*
Plants 4–8″. Leaves 2, narrowly oval and wider toward the tip, or elliptic, 3–6″. **Flowers up to 5 on terminal racemes, pink or white; the hood usually pink; lip white, ≤1¼″;** sepals and lateral petals fused to form a hood, ≤½″; lip oval, ½–¾″. Rich woods.

7. Rose pogonia *Pogonia ophioglossoides*
Plants 2–24″. Leaf solitary, narrowly oval or narrow elliptic, 1¼–3¾″. **Flowers solitary,** pink, ¾″; sepals and lateral petals ½–1¼″; sepals narrowly oval; lateral petals elliptic; lip spoon-shaped, ½–¾″; beset with yellow hairs. Open, wet meadows; sphagnum bogs.

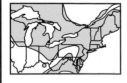

1. Cardinal-flower   *2. Chinese lobelia   *3. Common foxglove

4. Showy lady-slipper   5. Ram's head lady-slipper

6. Showy orchis   7. Rose pogonia

LEAVES alternate
LEAVES compound or lobed
PETALS 5

Roses and Raspberries
Rose family

Roses and raspberries are difficult to properly identify. In roses the problems are caused by escaped, cultivated species and hybrids. Critical characters are leaflet number, stipule shape, flower size, and sepal orientation. In raspberries, problems arise from hybrids and from asexual reproduction that fixes variation. Critical characters are in the habit and leaflet shape. See other roses and rubus spp. p. 304

a. Stems densely prickly

1. Bristly rose *Rosa acicularis*
Plants ≤3′, densely prickly; **twigs hairless.** Leaves with 5 or 7 leaflets; leaflets elliptic or oval, ½–1½″, toothed, often sticky-hairy. **Flowers** solitary, 1¼–2¼″; **sepals erect, persistent.** Upland woods, hills, rocky areas.

2. Pasture rose *Rosa carolina*
Plants ≤3′, very prickly; **twigs hairless.** Leaves with 3–7 leaflets; leaflets oblong, oval, or round, coarsely toothed. **Flowers** usually solitary, 1½–2½″; **sepals spreading or reflexed and soon falling.** Upland woods, dunes, prairies.

***3. Beach rose** *Rosa rugosa*
Plants 3–7′, densely prickly; twigs hairy. Leaves with 7 or 9 leaflets; leaflets 1–2″, wrinkled above. **Flowers** solitary, sometimes white, 2½–4″; **sepals reflexed and soon falling.** Beaches, garden escape.

a. Stems without prickles or with a few prickles
    b. Leaves compound

4. Smooth rose *Rosa blanda*
Plants ≤7′, with few or no prickles; twigs hairless. **Leaves with 5 or 7 leaflets;** leaflets narrow elliptic or oval, ≤2″, coarsely toothed. **Flowers** solitary or in lateral clusters, 1½–2½″; **sepals erect and persistent.** Dry woods, hills, prairies, dunes.

5. Climbing rose *Rosa setigera*
Plants climbing, 7–15′, hairless, with few stout, downward-curved prickles; twigs hairless. **Leaves with 3 or 5 leaflets,** leaflets narrowly oval or oval, 1¼–4″, veins impressed above, paler and hairless or hairy below. **Flowers** numerous, 1½–2½″; **sepals reflexed, falling early.** Thickets, fence-rows.

6. Dwarf raspberry *Rubus arcticus*
Plants 1½–9″, without prickles. **Leaves trifoliate;** terminal leaflet short-stalked, broadly oval, broader toward tip, ⅜–1½″, blunt and toothed, tapering to base; lateral leaflets asymmetric. **Flowers** solitary, ¾–1¼″; sepals narrowly triangular. Cold bogs, wet meadows.

    b. Leaves lobed
7. Flowering raspberry *Rubus odoratus*
Plants 3–6′, without prickles. Leaves round, triangular or angled, 4–8″, 5-lobed; lobes triangular. Flowers on open cymes, 1–2″, sepals densely sticky-hairy. Moist, shady areas; wood edges.

1. Bristly rose

2. Pasture rose

*3. Beach rose

4. Smooth rose

5. Climbing rose

6. Dwarf raspberry

7. Flowering raspberry

LEAVES alternate
LEAVES compound
PETALS 4–5

Miscellaneous alternate, compound leaved species
Several families

a. Petals 5
  b. Leaflets 3

1. Violet wood-sorrel *Oxalis violacea*
Plants 4–10″. Leaves trifoliate, "clover-like", 1–2″; leaflets heart-shaped, hairless. **Flowers several on umbels, ½″.** Dry, upland woods; prairies. See also p. 176. Wood-sorrel family.

2. Northern wood-sorrel *Oxalis montana*
Plants 2½–6″. Leaves trifoliate; leaflets heart-shaped, 1–1¾″, sparsely hairy; leaflets fold at night and on overcast days. **Flowers solitary, ¾–1¼″.** Rich, moist woods. Wood-sorrel family.

  b. Leaflets >3 (cont. on next page)

3. Water avens *Geum rivale*
Plants 12–24″, sparsely hairy. Basal leaves ≤12″; terminal leaflet oval, broader toward tip, or round, toothed, 3-lobed; lateral leaflets not lobed; stem leaves much smaller. **Flowers several, nodding, ½–¾″.** Swamps, wet meadows. See also pp. 174, 302. Rose family.

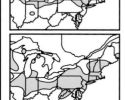

4. Prairie smoke *Geum triflorum*
Plants 6–16″, hairy. Basal leaves narrowly oval, broader toward the tip, 4–8″, pinnately compound; **leaflets 7–17, the lateral progressively smaller toward tip,** ≤2″, "torn" or lobed; terminal leaflet similar but wider; stem leaves few, small, lacerate. Flowers in clusters of 3, ½–1″, nodding. Calcareous woods, prairies. Rose family.

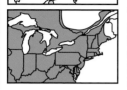

5. Marsh-potentilla *Comarum palustre*
Plants prostrate or ascending, 4–24″. Leaves long-stalked, pinnately compound; **leaflets 5–7,** narrow oblong to elliptic, 2–4″, toothed. Flowers few, ¾–1″. Swamps, bogs, stream banks. Rose family.

6. Queen-of-the-prairie *Filipendula rubra*
Plants 2–10′, hairless. Leaves pinnately compound, **leaflets 5–11,** ≤4″, shallowly to deeply lobed; terminal leaflet kidney-shaped, ≤8″, deeply lobed. Flowers on cone-shaped panicles, ¼–½″. Low woods, wet prairies. See also p. 302.3. Rose family.

7. Wild columbine *Aquilegia canadensis*
Plants ½–3′. **Basal leaves, 2–3 times compound;** leaflets oval, broader toward the tip, or round, toothed or lobed; stem leaves reduced. Flowers solitary, nodding, 1–2″. Dry woods, rocky cliffs, ledges. Pollinated by hummingbirds. Caterpillars of the Columbine Duskywing butterfly feed on wild columbine. Crowfoot family.

*8. Cypress-vine *Ipomoea quamoclit*
Plants climbing, 3–16′, hairless. Leaves oval, pinnately divided into linear lobes, 1½–3″. Flowers 1–a few in axils, 1–1½″; sepals blunt. Fields, weedy places. See also pp. 14, 70, 270. Morning-glory family.

1. Violet wood-sorrel

2. Northern wood-sorrel

3. Water avens

4. Prairie smoke

5. Marsh-potentilla

6. Queen-of-the-prairie

7. Wild columbine

*8. Cypress-vine

LEAVES alternate
LEAVES compound or lobed
PETALS absent, 4, or 5

Miscellaneous alternate, compound-leaved species
Several families

a. Petals 5
  b. Leaflets >3 (cont. from previous page)
*1. Standing cypress *Ipomopsis rubra*
Plants erect, ≤4'. Leaves pinnately lobed; lobes linear. Flowers on elongate panicle, 8–20″ tall; flowers red outside, yellow inside, ½–¾″; tube ¾–1½″ long. Pastures, roadsides, gardens. Phlox family

a. Petals 4
  b. Flowers irregular (Cleome and Poppy families)
*2. Spider-flower *Cleome hassleriana*
Plants ≤5', sticky hairy. **Leaves with short spines at base of stalk,** compound; leaflets 5–7, narrowly oval, broader toward the tip, toothed, fine hairy; central leaflet 2½–4″. Flowers on terminal racemes, ¾″. Escape from cultivation. Cleome family.

3. Tall corydalis *Corydalis sempervirens*
Plants erect, 12–32″. Basal leaves 3–5-lobed, stalked, becoming smaller and stalkless upward. Flowers on small panicles, ½–⅝″ including ⅛–¼″ spur; sepals broadly oval, ≤⅛″. Dry or rocky woods. Poppy family.

4. Allegheny-vine *Adlumia fungosa*
Plants climbing, ≤10'. Leaves pinnately compound; leaflets lobed. Flowers on axillary racemes; flowers ½–¾″. Woods. Poppy family.

  b. Flowers regular, petals spreading (Poppy family)
*5. Long-headed poppy *Papaver dubium*
Plants 1–2', hairy. Leaves narrowly oval and broader toward the tip or elliptic, 1¼–3″, pinnately compound; leaflets 8–16, toothed or lobed. **Flowers** red, pink, or orange with a central dark spot, **1¼–2½″**. Weedy areas. Similar to no. 7, but **flower stalks have appressed hairs and fruit is generally longer than wide.**

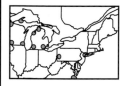

*6. Oriental poppy *Papaver orientalis*
Plants ≤3', hairy. Leaves elliptic, ≤12″, pinnately compound or lobed, hairy; leaflets elliptic. **Flowers** orange or red, **>4″**. Fields, clearings, roadsides, weedy areas.

*7. Shirley poppy *Papaver rhoeas*
Plants ≤3', hairy. Leaves pinnately compound; leaflets lobed or incised. **Flowers** red, purple, pink, white, or streaked, often with a basal dark spot, **1–3″**. Weedy places, near gardens. **Note flowerstalks have spreading hairs.**

a. Petals absent (Rose family)
*8. Salad burnet *Sanguisorba minor*
Plants ¾–2¼'. Leaves compound; leaflets 7–17, oval or round, ¼–¾″, with 3–7 teeth on each side, smaller upwards. Flowers in dense, **spike-like heads,** flowers ⅛–¼″; **petals absent,** lower flowers male or bisexual, upper female. See also p. 302. Fields, weedy places.

*1. Standing cypress

*2. Spider-flower

3. Tall corydalis

4. Allegheny-vine

*5. Long-headed poppy

*6. Oriental poppy

4½"

*7. Shirley poppy

*8. Salad burnet

LEAVES alternate
LEAVES compound
"PETALS" many

Thistles

Composite family

All these thistles produce a cluster of leaves the first year, the stem and flowers the second year, and then the whole plant dies. See yellow thistles p. 180, white thistles p. 318, and pink thistles p. 24.

a. Stems with spiny wings running down from the leaves

*1. Bull thistle *Cirsium vulgare*
Plants 3–6′, stem spiny winged. **Leaves** strongly spiny, lobed, **often thinly white-hairy beneath**. Heads 1–3 at ends of prickly branches, 1½–2½″; involucral bracts narrowly pointed, rigid, not sticky, yellow spine-tipped. Pastures, fields, roadsides, weedy places.

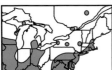

*2. Nodding thistle *Carduus nutans*
Plants 1–6′, spiny stemmed. **Leaves** deeply lobed, ≤ 10″, spiny, **not white-woolly**. Heads mostly solitary, **nodding**, 1½–2½″; involucral bracts ⅛–¼″ wide, purple; middle and outer bracts broad with long, flat, spreading or usually reflexed, spine-pointed tip. Roadsides, weedy places.

*3. Plumeless thistle *Carduus acanthoides*
Plants 1–5′, strongly spiny. **Leaves** deeply lobed or pinnatifid, ≤12″, **not white-woolly**. Heads 1–several at the ends of branches, ½–1″, **erect**; involucral bracts narrow, <⅛″ wide, erect or loosely spreading; middle and outer bracts spine-tipped. Roadsides, pastures, weedy places.

*4. Welted thistle *Carduus crispus*
Plants 2–7′, weakly spiny. **Leaves** cleft, **white-woolly beneath when young**. Heads clustered at ends of branches, ½–1″, **erect**; involucral bracts narrow, <⅛″ wide, erect or loosely spreading; middle and outer bracts spine-tipped. Roadsides, weedy places.

a. Stems without spiny wings
b. Leaves white-woolly beneath

5. Field thistle *Cirsium discolor*
Plants 3–10′. **Leaves** deeply lobed, firm, and spiny; lobes linear, narrowly oval or narrowly oblong, **white-woolly beneath**. Heads solitary; upper leaves embrace the heads. **Heads** 1½–2″; involucral bracts with long, colorless bristles and weakly spreading spines. Fields, open woods, river bottoms, weedy places.

b. Leaves green or thinly hairy beneath

6. Swamp thistle *Cirsium muticum*
Plants 2–10′, stems hollow. Leaves deeply lobed, narrowly oval, or oblong, ≤ 22″, weakly spiny, thinly hairy beneath. Heads clustered, 1½–2½″; base of head cobwebby; **involucral bracts** hairy, generally with a sticky ridge, **spineless or nearly so**. Various wet areas.

7. Pasture thistle *Cirsium pumilum*
Plants 1–3′, very hairy. Leaves lobed, narrowly oval, or oblong, numerous marginal spines and scattered larger ones, 5–1″, not white-woolly. Heads 1-several, 2–3″; **middle and outer involucral bracts tipped by a short erect spine**. Pastures, old fields, open woods.

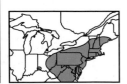

*1. Bull thistle

*2. Nodding thistle

*3. Plumeless thistle

*4. Welted thistle

5. Field thistle

6. Swamp thistle

7. Pasture thistle

LEAVES alternate
LEAVES trifoliate
PETALS irregular

Tick-trefoils
Bean family
A large, difficult group with many more species than portrayed here. The fruit segments are dispersed by animals and often by pant legs and socks. Critical characters are leaflet shape and size and fruit segment shape and number.

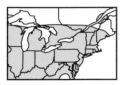

a. Leaves only just below the inflorescence
1. Cluster-leaved tick-trefoil *Desmodium glutinosum*
Plants erect, 1–4'. Stipules narrow; **leaves just below inflorescence only;** terminal leaflet oval or round, 2½–5". Flowers ¼". Fruit joints 2–3, rounded-triangular, ¼–½". Dry or rocky woods.

a. Leaves scattered on stem
2. Sessile-leaved tick-trefoil *Desmodium sessilifolium*
Plants unbranched, 2–4', hairy. Stipules narrowly oval, ¼", falling early; **leaf-stalks <⅛" or absent; leaflets oblong, 1¼–2¾", densely hairy.** Flowers ⅛". Fruit joints 1–3, convex above, rounded below, ¼". Dry, sandy soil.

3. Pine-barren tick-trefoil *Desmodium strictum*
Plants ≤4', unbranched, minutely hairy. Stipules narrow, ⅛", **leaf-stalks ⅜–¾"; leaflets linear, 1¼–2¾", hairless, veiny.** Flowers ⅛". Fruit joints 1–2, flat above, rounded below, ¼". Sandy soil in pine barrens.

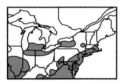

4. Little-leaf tick-trefoil *Desmodium ciliare*
Plants 1–3', branched from base, downy. Stipules narrowly oval, ⅛–¼", falling early; **leaf-stalks ⅛–½"; leaflets oval or elliptic, ½–1¼", hairy.** Flowers ⅛". Fruit joints 1-3, gradually curved, ¼". Dry, sandy woods.

5. Panicled tick-trefoil *Desmodium paniculatum*
Plants 2–4', **hairless.** Stipules linear, ⅛", falling early, leaf-stalks ½–2"; leaflets narrowly oval or oblong; terminal leaflet 1¾–4". **Flowers on panicles, 4–16" tall, with horizontally spreading branches;** flowers ¼". Fruit joints 3–6, triangular, ¼". Dry woods, clearings.

6. Canadian tick-trefoil *Desmodium canadense*
Plants 2–6', **downy.** Stipules linear, ⅛–⅜", leaf-stalks ⅛–1"; leaflets oblong; terminal leaflet 2–4", blunt or broadly pointed, hairy. **Flowers on densely-flowered racemes;** flowers ¼–½". **Fruit joints 3–5,** rounded below, ¼", **very sticky.** Moist woods, thickets, banks.

7. Prairie tick-trefoil *Desmodium illinoense*
Plants 3–6', **hairy.** Stipules oval, ⅜–½"; leaf-stalks long; leaflets narrowly oval; terminal leaflet 2½–4". **Flowers on long, terminal racemes or few-flowered panicles;** flowers ¼–½". Fruit joints 2–7, round, ¼". Rich, dry prairies.

1. Cluster-leaved tick-trefoil
2. Sessile-leaved tick-trefoil
3. Pine-barren tick-trefoil
4. Little-leaf tick-trefoil
5. Panicled tick-trefoil
6. Canadian tick-trefoil
7. Prairie tick-trefoil

LEAVES alternate
LEAVES compound
PETALS irregular

Bush-clovers
Bean family

Bush-clovers are somewhat similar to tick-trefoils, but they do not have the characteristic segmented fruit. Critical characters are habit, leaflet shape and size, and flower number. See also p. 320.

a. Plants trailing

1. Downy trailing bush-clover *Lespedeza procumbens*
Plants creeping or prostrate, ≤ 4′, densely hairy. Leaf-stalks shorter than leaflets; leaflets oval, ½–1″, hairy. Flowers 8–12 on racemes, ¼″. Dry, sandy or rocky woods; clearings. **Note that hairs are spreading.**

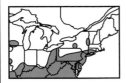

2. Smooth trailing bush-clover *Lespedeza repens*
Plants creeping or prostrate, ≤ 3′, finely hairy or hairless. Leaf-stalks shorter than leaflets; leaflets round or oval and broader toward the tip, ½–1″, silky. Flowers 3–8 on loose racemes, ¼″. Dry, sandy or rocky woods; fields. **Much like no. 1, but hairs are appressed.**

a. Plants erect

3. Violet bush-clover *Lespedeza violacea*
Plants erect or ascending, much branched, ½–3′, hairless or nearly so. Leaf-stalks same length as leaflets; **leaflets elliptic**, ¼–1½″, hairy or hairless. Flowers few on loose **racemes, ¼–½″.** Dry, upland woods; clearings.

A. Velvety bush-clover *Lespedeza stuevei* (not illus.)
Plants erect, unbranched, or branched, ≤3′, spreading-hairy. Leaf-stalks shorter than leaflets; leaflets oval or elliptic, ½–1¼″, velvety beneath. Flowers numerous in axillary clusters, ¼″. Dry, upland woods; barrens.

*4. Japanese bush-clover *Kummerowia striata*
Plants erect, diffusely branched, 4–16″, slightly hairy. Leaf-stalks shorter than leaflets; **leaflets oval**, broader toward the tip, ½–1″. Flowers 1–5, clustered in upper leaf axils, ¼″. Fields, roadsides, upland woods.

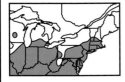

5. Slender bush-clover *Lespedeza virginica*
Plants erect or ascending, simple and wand-like or branching above, 1–3′, hairy. Leaf-stalks shorter than leaflets; leaves erect, very crowded; **leaflets linear**, ½–1½″. Flowers few, clustered in upper axils, very crowded, ¼″. Dry, upland woods; openings.

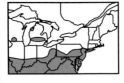

*6. Chinese bush-clover *Lespedeza cuneata*
Plants erect, much branched, ≤ 4½′, very leafy. Leaf-stalks ⅛–¼″; **leaflets** ascending, **linear**, ½–1″, hairless above, hairy below. Flowers mostly solitary in upper axils, ¼″. Roadsides, weedy areas.

1. Downy trailing bush-clover

2. Smooth trailing bush-clover

3. Violet bush-clover

*4. Japanese bush-clover

5. Slender bush-clover

*6. Chinese bush-clover

Miscellaneous beans with 3 leaflets
Bean family

The distinctive legume flower has an upper "showy" petal called the "banner," 2 lateral petals called "wings," and 2 petals fused into a boat-shape called the "keel."

a. Leaflets toothless (some lobed)

1. Trailing milk-pea *Galactia regularis*

Plants prostrate or twining, 1–3′, minutely hairy. Leaflets 3, elliptic or oval, 1–1½″, hairy or hairless below. **Flowers few on racemes, 1¼–2½″ long;** flowers ½–¾″. Dry, sandy woods; barrens.

2. Hog-peanut *Amphicarpaea bracteata*

Plants climbing, ≤5′, slightly hairy. Leaflets 3, oval, ¾–3″, rounded at base. **Flowers nodding, 2–15 on racemes,** pale purple to white, ½–⅝″; also small flowers without petals near base. Sometimes the fruit are underground. Woods, thickets.

3. Trailing wild bean *Strophostyles helvula*

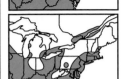

Plants twining, 1–7′, hairless or hairy. Leaflets 3, oval, ¾–2½″, sparsely hairy, often lobed. **Flowers few-several in dense, stalked clusters;** flowers ¼–½″, standard ⅜–¾″. Fruit woolly. Dry, sandy soil; thickets; shores; often in cinders. See also p. 256.

4. Pink wild bean *Strophostyles umbellata*

Plants twining, ≤3′. Leaflets 3, oblong, ¾–2″, hairy below, not lobed. **Flowers few-several in dense, stalked clusters;** flowers ¼–½″. Fruit woolly. Dry, sandy soil; upland woods, clearings, fields. **Similar to no. 3, but flower clusters are stalked above the leaves.**

a. Leaflets toothed. Bars indicate head width.

*5. Alsike clover *Trifolium hybridum*

Plants arching to ascending, 10–24″. Leaflets 3, oval or elliptic, 1–2½″, broadly rounded or notched at the tip. **Flowers in round heads, ½–¾″.** Fields, open woods, roadsides, weedy places. See also pp. 184, 320.

*6. Red clover *Trifolium pratense*

Plants ascending or erect, 6–24″, hairless or hairy. Leaflets 3, oval, ½–1″, blunt, with **pale chevrons** on upper surface. **Flowers in round heads, 1–1½″.** Fields, roadsides, meadows, clearings, lawns.

*7. Rabbit-foot clover *Trifolium arvense*

Plants erect, 4–16″, freely branching, silky-hairy. Leaflets 3, narrowly oval and broader toward the tip, ½–1″, silky-hairy, toothed to the tip. **Flowers in dense egg-shaped or cylindrical heads, ⅜–1¼″.** Roadsides, weedy places. **Note the heads appear furry.**

*8. Crimson clover *Trifolium incarnatum*

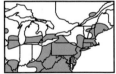

Plants 6–30″, hairy. Leaflets 3, oval, broader toward the tip, ⅜–1¼″, broadly rounded or blunt at tip. **Flowers in egg-shaped or cylindrical heads, ¾–1½″.** Fields, roadsides, weedy places.

1. Trailing milk-pea

2. Hog-peanut

3. Trailing wild bean

4. Pink wild bean

*5. Alsike clover

*6. Red clover

*7. Rabbit-foot clover

*8. Crimson clover

LEAVES alternate
LEAVES compound
PETALS irregular

Miscellaneous trifoliate beans
Bean family

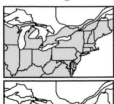

a. Leaves without tendrils
1. Ground-nut *Apios americana*
Plants twining. **Leaflets 5–7,** narrowly oval or oval, 1½–2½″, short-hairy or hairless. Flowers on dense, rounded racemes, ⅜–½″; standard pinched and hood-like. Moist woods.

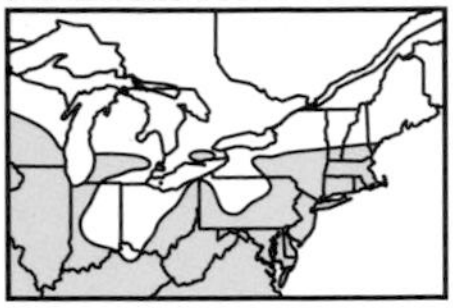

2. Goat's-rue *Tephrosia virginiana*
Plants erect, 8–28″, silky-hairy. **Leaflets 15–29, elliptic or linear,** ⅜–1¼″, dense silky-hairy to partly hairy. Flowers on dense, terminal racemes, 1½–3″ tall; flowers ½–¾″; **standard yellow; keel and wings pink.** Fruit 1½–2¼″, shaggy-hairy. Old fields, open woods, dunes, often in sandy soils.

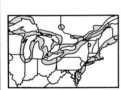

***3. Crown-vetch** *Coronilla varia*
Plants creeping, 12–40″. Leaves unstalked, **leaflets 11–25, oblong or oval and broader toward the tip,** ⅜–¾″. Flowers 10–20 on umbels, ⅜–½″. Roadsides, weedy places, old homesteads.

a. Leaves with tendrils
Vetches and wild peas are climbing, with compound leaves and tendrils at ends. Vetches generally have smaller flowers and more numerous leaflets. See also pp. 34, 188.

b. Stems not winged
4. Beach pea *Lathyrus japonicus*
Plants prostrate or erect, 1–3′, hairless or hairy. Stipules large, arrowhead-shaped, clasping; leaflets 6–20, oblong or oval, broader toward tip, ½–3″. **Flowers 3–10 on racemes, ½–1″.** Fruit 1–2½″. Sea beaches, lake shores.

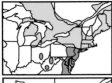

***5. Common vetch** *Vicia sativa*
Plants erect, ascending or climbing, 1–3′, hairy. Stipules <¼″; **leaflets 6–16, oblong or linear, 1–1½″, notched at base. Flowers 1–2 in leaf axils, ½–1¼″.** Fruit 1–2¾″. Roadsides, fields, weedy places.

***6. Slender vetch** *Vicia tetrasperma*
Plants prostrate or climbing, 6–24″. Stipules <¼″; **leaflets 4–10,** linear or narrowly oval, ½–¾″, **rounded at base. Flowers 1–2 in axils, ⅛–¼″.** Fruit ½″. Roadsides, fields, waste places.

b. Stems winged
***7. Everlasting pea** *Lathyrus latifolius*
Plants creeping or climbing, ≤6′, stem winged. Stipules >½″, narrowly oval; **leaflets 2,** narrowly oval or elliptic, 1½–3″. Flowers 4–10 on racemes, ½–1¼″. Fruit 2½–4″. Roadsides, thickets, weedy places.

8. Marsh pea *Lathyrus palustris*
Plants climbing, ½–4′, often winged, hairless or hairy. Stipules >¼″, half arrowhead-shaped; **leaflets 4–20,** linear or elliptic, 1–3″. Flowers 2–9 on racemes, ½–1″. Wet meadows, swamps, shores, wet woods.

1. Ground-nut

2. Goat's-rue

*3. Crown-vetch

4. Beach pea

*5. Common vetch

*6. Slender vetch

*7. Everlasting pea

8. Marsh pea

LEAVES opposite
LEAVES simple
PETALS 4–13

Marsh-pinks
Gentian family

Marsh-pinks attract many insects; the flower's yellow, central marking serves as a guide to the nectar. Individual flowers shed their pollen before the stigma is open and receptive, insuring against self-fertilization. Critical characters are petal number and calyx nerves.

a. Petals 4-7

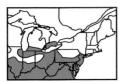

1. Common marsh-pink *Sabatia angularis*
Plants 1–3′, stems strongly 4-angled, branching above the middle. **Leaves oval or narrowly oval,** ½–1½″, rounded or notched and clasping at base; 3–7 nerves. Flowers 1–2″; calyx base and flower-stalk winged; stalks ¼–1¼″; sepals narrowly oblong to narrowly oval, ¼–½″; petals ½–1″; narrowly oval or oblong. Open woods, clearings, prairies, fields, sometimes weedy. **Note the opposite branching.**

2. Slender marsh-pink *Sabatia campanulata*
Plants 6–24″, with 2–several stems, unbranched or branched. Leaves oblong, narrowly oval, or linear, ¾–1½″, blunt, rounded at base, 3-nerved. Flowers 1″; **calyx tube not winged, inconspicuously 10-nerved;** sepals linear; petals ½″; elliptic or oval. Moist or wet places, damp sands, peats. **Note the sepals are about the same size as the petals.**

3. Annual sea-pink *Sabatia stellaris*
Plants 1–20″, unbranched or branched. Leaves narrowly oval and broader toward the tip or elliptical, 1–1½″, pointed at base and apex. Flowers ½–1″; **calyx not winged, conspicuously nerved;** sepals linear; petals 4–7, narrowly oval, ¼–½″. Salt or brackish marshes. **Similar to no. 2, but sepals are shorter than the petals.**

A. Western marsh-pink *Sabatia campestris* (not illus.)
Plants 4–16″, commonly branched below the middle. Leaves narrowly oval, ¾–1½″, rounded or notched at base. Flowers 1–2″; **calyx tube 5-winged;** sepals narrowly oval, 3-nerved, ½–1″; petals mostly ½–1″. Wet or dry, usually open, prairies; fields.

a. Petals 8–13

4. Large marsh-pink *Sabatia dodecandra*
Plants 1–2′, branched. Leaves narrowly oval or narrowly elliptic, 1–2″, blunt or broadly pointed. Flowers 1–2″; **sepals narrowly oval or spoon-shaped**, 3–5 nerved; petals 9–12, narrowly oval, ½–1″. Salt and brackish marshes, meadows, rarely freshwater.

5. Plymouth gentian *Sabatia kennedyana*
Plants 10–30″, branched. Leaves narrowly oval or linear, firm, clasping. Flowers long-stalked, 1–2½″; **sepals linear,** 1–3 obscure nerves; petals 7–13, ½–1¼″. Fresh water marshes, margins of streams or ponds. **Distinguished from no. 4 by the linear sepals with 1–3 obscure nerves.**

1½"

1. Common marsh-pink

2. Slender marsh-pink

3. Annual sea-pink

4. Large marsh-pink

5. Plymouth gentian

LEAVES opposite
LEAVES simple
PETALS 5

Miscellaneous pinks
Pink family

Pinks are a diverse group recognized by the opposite, entire leaves and regular flowers. The first five species have nectaries deep in a tube, an adaptation for the long tongues of butterflies and moths.

a. Petals toothed or notched
  b. Styles 2
*1. Deptford pink *Dianthus armeria*

Plants 8–24″, hairy. Basal leaves narrowly oval, broader toward the tip; stem leaves linear or narrowly oval, 1¼–3″, hairy. Flowers 3–9 in congested clusters; flowers ½″; calyx ½–¾″; hairy; petals elliptic or oval, toothed. Weedy areas.

*2. Maiden pink *Dianthus deltoides*

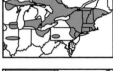

Plants 4–16″, hairless or nearly so. Basal leaves narrowly oval, broader toward the tip, ½–1¼″; stem leaves linear, ¾–1½″. Flowers solitary, various colors, usually pink, ½–¾″; calyx ½–¾″; petals toothed. A garden escape. Similar to no. 1, but with fewer flowers, rounded and broader petals.

*3. Sweet William *Dianthus barbatus*

Plants 12–24″, hairless. Leaves narrowly oval, 2½–4″. Flowers many, ½–¾″; calyx hairless, ½–¾″; petals toothed. A garden escape.

  b. Styles 3-6
*4. Corn-cockle *Agrostemma githago*

Plants ≤3′. Leaves linear or narrowly oval, 3–4½″. Flowers solitary at ends of branches, 1½–2½″, stalked ≤8″; calyx-tube ½–¾″; sepals ¾–1½″, longer than petals, strongly ribbed and widely veined; petals narrowly oval, notched. Grain fields, waste places. Pollination is by butterflies and nocturnal moths.

a. Petals not toothed
  c. Styles 2
*5. Childing pink *Petrorhagia prolifera*

Plants ≤24″, mostly hairless. Leaves linear, ⅜–2½″. Flowers 3–7 in dense clusters ≤¼″; flowers enclosed in bracts, only one flower opening at a time; calyx cylindrical, ⅜–½″; petals narrowly oval. Weedy areas.

  c. Styles 3
*6. Salt-marsh sand-spurry *Spergularia salina*

Plants erect or prostrate, ≤14″, hairless or sticky-hairy. Leaves linear, ¼–1½″, fleshy. Flowers white or pink, ¼″; sepals oval, ⅛–¼″; petals oval, shorter than sepals. Brackish and saline marshes, highways.

*7. Roadside sand-spurry *Spergularia rubra*

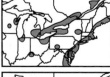

Plants prostrate or ascending, 2–12″. Leaves linear, ⅛–1½″. Flowers ¼″; sepals ⅛–¼″; narrowly oval; petals shorter than sepals. Sandy or gravelly soil.

*1. Deptford pink

*2. Maiden pink

*3. Sweet William

*4. Corn-cockle

*5. Childing pink

*6. Salt-marsh sand-spurry

*7. Roadside sand-spurry

Campions, Catchflies, and related plants
Pink family
See other Silene and Lychnis species pp. 326–28.

a. Plants less than 1′, often forming mats

1. Moss campion *Silene acaulis*
Plants forming mats, 1–3″, hairless. Leaves crowded, narrowly oval or linear, ⅛–½″, stalkless. **Flowers solitary, ½″,** male and female separate; male flower smaller than female flower; calyx tubular, hairless; petals unlobed or 2-lobed. Alpine areas.

2. Wild pink *Silene caroliniana*
Plants usually clumped, 3–10″, sticky-hairy. Basal leaves narrowly oval, broader toward the tip or oblong, 2–5″; stem leaves linear ≤2″, stalkless. **Flowers 4–13 on short, dense and flat-topped cymes; flowers 1″;** calyx cylindrical, sticky; petals wedge-shaped, slightly notched or rounded. Dry, sandy or gravelly woods; openings.

a. Plants usually 1′ or taller, not forming mats
*3. Sweet William catchfly** *Silene armeria*
Plants 4–24″, **hairless or sparsely hairy; stem with sticky areas.** Leaves numerous, elliptic, oval, or narrowly oval, 1–2″, stalkless and clasping. Flowers on open or compact, terminal umbels; **flowers ¼–½″;** calyx tubular to club-shaped; **petals truncate or shallowly notched.** Roadsides, weedy areas.

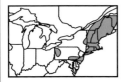

*4. Red campion** *Silene dioica*
Plants 1–3′, hairless. Leaves oval or oblong, 1½–5″; basal leaves stalked. Flowers on a loosely forking panicle; **flowers 1½–3″,** with separate male and female flowers; calyx hairy, inflated; **petals deeply 2-lobed.** Roadsides, weedy places.

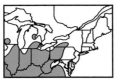

5. Fire-pink *Silene virginica*
Plants 1–2′, **sticky-hairy.** Basal leaves narrowly oval, broader toward the tip, or spoon-shaped, 1½–4″, stalked; stem leaves 2–4 pair, 6″, stalkless. Flowers 7–11 on loose, open panicles; **flowers 1–1½″,** calyx broadly tubular; **petals notched.** Rich woods, clearings, rocky slopes.

*6. Scarlet lychnis** *Lychnis chalcedonica*
Plants 1–2′, **hairy.** Basal leaves spoon-shaped or narrowly oval, short-stalked; stem leaves 10–20 pair, narrowly oval or oval, 2–5″, notched at base. **Flowers on terminal, dense panicles, ¾–1″; petals Y-shaped.** Roadsides, gardens, open woods, thickets.

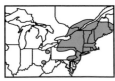

*7. Ragged robin** *Lychnis flos-cuculi*
Plants 1–2′, *thinly hairy*, sticky above. Basal leaves spoon-shaped; stem leaves mostly 4–5 pair, narrowly oval, 2–4″, stalkless. **Flowers** on open, much-branched panicles, ¾–1″; calyx bell-shaped becoming round; **petals deeply 4-cleft, with a ragged look.** Fields, meadows, swales.

1. Moss campion

2. Wild pink

*3. Sweet William catchfly

*4. Red campion

5. Fire-pink

*6. Scarlet lychnis

*7. Ragged robin

LEAVES opposite
LEAVES simple
PETALS 5

Milkweeds
Dogbane family

Milkweed flowers are unique with their colorful reflexed petals and cluster of "hoods" with "horns." Insects put their legs into slots between the hoods, their legs becoming wrapped in a "wishbone"-shaped object with two clumps of pollen at the ends. This wishbone is removed at a slot on the next flower. Monarch butterflies are completely dependent on these plants, as milkweeds are the sole food for their caterpillars. See other milkweeds, pp. 136, 246, 330, 354.

a. Stems erect
    b. Leaves stalked; stems often hairy

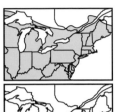

1. Common milkweed *Asclepias syriaca*
Plants 3–6'. Leaves variable, 4–6", **lateral veins spread at right angles.** Flowers in two or more round clusters, 2–4" wide; **flowers dull, dusky pink,** ½–¾". Fields, meadows, roadsides. Note **warty pods.**

2. Purple milkweed *Asclepias purpurascens*
Plants 2–3'. Leaves elliptic, narrowly oval or oblong, ½–6"; **lateral veins in the leaf spread at right angles.** Flowers in 1–3 clusters; clusters 1½–3" wide; **flowers purplish,** ½–¾". Dry fields, thickets, swamp forests, alluvial woods, open woods, roadsides. **Differs from no. 1 in having purplish flowers and smooth pods.**

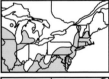

3. Swamp milkweed *Asclepias incarnata*
Plants 1–6'. Leaves narrowly oval, broader toward either end, oblong or linear, 2½–6"; **lateral veins in the leaf spread at an acute angle.** Flowers in 4–20 flat clusters; clusters 1–2" wide; **flowers magenta,** ⅜". Open swamps, ditches, marshes, stream banks, wet prairies.

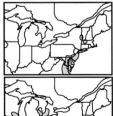

    b. Leaves stalkless or nearly so; stems hairless
4. Blunt-leaved milkweed *Asclepias amplexicaulis*
Plants 2–3'. Leaves oval or broadly oblong, 3–6", **blunt,** base rounded, and **often clasping.** Flowers in a single large cluster, 3–4" wide; flowers greenish-purple to magenta, ½–¾". Dry, sandy fields; prairies; open woods. **Note wavy leaf margins.**

5. Red milkweed *Asclepias rubra*
Plants 1–4'. Leaves narrowly oval, 3½–4½", **sharp-pointed, stalked.** Flowers in 1–4 small clusters, 1–2" wide; flowers reddish, ½–¾". Swamps, bogs, savannahs, pocosins, wet pinelands, wet woods.

6. Prairie milkweed *Asclepias sullivantii*
Plants 2–5'. Leaves oblong or oval, 4½–8", **sharply pointed, somewhat clasping.** Flowers in few to several clusters; flowers purplish-rose, ⅝–¾". Moist prairies, fence rows, roadside banks, dry fields.

a. Stems twining
7. Oblique milkvine *Matelea obliqua*
Plants twining. Leaves oval or round, 4–6", deeply lobed at base. Flowers on axillary racemes; flowers deep red, ⅜–½". Rocky woods, thickets.

1. Common milkweed

2. Purple milkweed

3. Swamp milkweed

4. Blunt-leaved milkweed

5. Red milkweed

6. Prairie milkweed

7. Oblique milkvine

**LEAVES opposite,
sometimes alternate
LEAVES simple or
compound
PETALS 4**

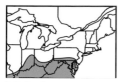

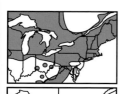

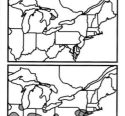

Miscellaneous opposite-leaved species (1)
Several families
a. Petals 4 (cont. on next page)
 b. Herbs
  c. Only "petals" present (Crowfoot family)
See other Clematis pp. 322, 346.
1. Leather-flower *Clematis viorna*
Plants climbing, ≤12′. Leaves compound, leaflets 4–8, narrowly
oval or oval; some leaflets 3-lobed, hairy below. Flowers solitary in
axils; sepals showy, thick, and leathery, ½–1″. Moist woods, thickets.

2. Erect silky leather-flower *Clematis ochroleuca*
Plants erect, ¾–2¾″, hairy. **Leaves oval,** unlobed or shallowly
lobed, 2½–4¾″, rounded at base, stalkless, **hairy beneath.** Flowers
solitary, terminal, yellowish or reddish, ¾–1½″, **silky-hairy outside.**
Woods; thickets; rocky slopes; disturbed, open areas.

3. Millboro leather-flower *Clematis viticaulis*
Plants erect, 8–24″. **Leaves oval,** 1½–3¼″, **hairless or nearly
so.** Flowers solitary, terminal, ¾″, **nearly hairless outside.** Shale
barrens. **This species and no. 4 differ from no. 2 in having leaves
arching over the flowers.**

4. White-haired leather-flower *Clematis albicoma*
Plants erect, 8–24″. Leaves narrowly oval or oval, 1½–3″, hairless
or nearly so. Flowers solitary, terminal, ½–1¼″. Shale-barrens. **Differs from no. 3 in having larger and densely silky-hairy flowers.**

  **c. Petals and sepals present, petals notched
   (Evening-primrose family)**
5. American willow-herb *Epilobium ciliatum*
Plants 1–16″. **Leaves sometimes alternate**, elliptic, narrowly oval,
or oval, ½–5″. **Flowers axillary, ¼–½″.** Springs; wet rocks; weedy,
damp spots. See other willow-herbs p. 322.

6. Annual willow-herb *Epilobium brachycarpum*
Plants 8–36″. **Leaves mostly alternate,** linear, narrowly elliptic,
or narrowly oval, ¾–2½″. **Flowers on racemes terminating the
branches; flowers pink or white, ⅛–¼″.** Dry, open places.

7. White evening-primrose *Oenothera speciosa*
Plants creeping to erect, 4–24″. **Leaves linear or narrowly oval,
broader toward either end,** 1¼–3″. **Flowers stalkless in upper
axils, white or pink, 1½–3″.** Dry, open places.

  b. Shrubs (Heath family)
***8. Heather** *Calluna vulgaris*
Plants shrubs ≤3′. Leaves narrowly oval, ⅛–¼″, stalkless. Flowers
on one-sided racemes, ⅛″. Sandy places.

1. Leather-flower

2. Erect silky leather-flower

3. Millboro leather-flower

4. White-haired leather-flower

5. American willow-herb

6. Annual willow-herb

7. White evening-primrose

*8. Heather

LEAVES opposite, sometimes alternate or basal
LEAVES simple
PETALS 5

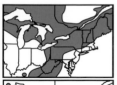

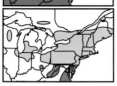

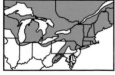

Miscellaneous opposite-leaved species (2)
Several families
a. Petals 5 (cont. from previous page)
  b. Petals not fused
Spring-beauties are among the first flowers of spring. The pink lines on the petals are nectar guides for the pollinating bees.
1. Carolina spring-beauty *Claytonia caroliniana*
Plants 1–12″. Leaves at base and on stem; **stem leaves oval**, narrowly oval, or diamond-shaped, ½–3″. **Flowers 2–11 on loose, terminal racemes,** ½–¾″. Rich woods, thickets, slopes. Purslane family.

2. Virginia spring-beauty *Claytonia virginica*
Plants 4–12″. Leaves at base and on stem; **stem leaves linear, 2–7″. Flowers 5–19 on loose, terminal racemes,** ½–¾″. Rich woods, fields. Purslane family.

3. Marsh St. John's-wort *Triadenum virginicum*
Plants 12–24″. Leaves oblong, oval, or elliptic, ¾–2½″, rounded or notched at base, stalkless, dark-spotted below. **Flowers in axillary and terminal clusters,** ¾″. Bogs, marshes, wet shores. The flowers only open for a short time in early afternoon.

4. Live forever *Hylotelephium telephium*
Plants tufted, 6–24″. Leaves alternate or opposite, elliptic, 1¼–2½″, often coarsely toothed. **Flowers on broadly rounded panicles,** 2–3″ wide; flowers deep pink, ½″. Roadsides, old fields, clearings. See other sedums p. 150. Stonecrop family.

  b. Petals fused
    c. Shrubs
5. Alpine azalea *Loiseleuria procumbens*
Low, mound-like shrubs, ≤ 4″. Leaves narrowly elliptic, ¼″, crowded, hairy beneath, stalks <⅛″. Flowers on umbel-like clusters in leaf axils, 2–5″ wide; flowers ⅛″. Peats, exposed rocks. Heath family.

6. Swamp laurel *Kalmia polifolia*
Shrubs ≤3′. Leaves linear or narrowly oval, ⅜–1½″, inrolled, white below, stalkless. Flowers from upper axils, ⅜–½″. Bogs. Heath family.

    c. Herbs (cont. on next page)
7. Rose verbena *Glandularia canadensis*
Plants prostrate to ascending, **12–24″**. Leaves oval or narrower, lobed, 1¼–3¾″, hairy or hairless. Flowers on stalked spikes, blue, purple, or rose-colored, ⅜–½″. Various dry habitats. Verbena family.

8. Twinflower *Linnaea borealis*
Plants creeping with erect leafy shoots, **≤4″**. Leaves oval, broader toward either end, ⅜–¾″, short-stalked. The pairs of nodding flowers are pink or white, ⅜–½″. Moist or dry woods, cold bogs. Pollinated by bees. Twinflower family.

1. Carolina spring-beauty

2. Virginia spring-beauty

3. Marsh St. John's-wort

4. Live forever

5. Alpine azalea

6. Swamp laurel

7. Rose verbena

8. Twinflower

Miscellaneous opposite-leaved species (3)
Several familes
a. Petals 5, fused, herbs (cont. from previous page)
  b. Flowers spreading open (Phlox family)
Phlox
Critical characters include leaf shape, inflorescence type, and petal shape. See other phlox p. 44.

1. Moss phlox *Phlox subulata*
Plants mat forming, 2–8″. Leaves needle-shaped, ¼–¾″, sharp. Flowers on few-flowered panicles, ½–¾″; **petals notched.** Sands, gravels, rocky areas.

2. Garden phlox *Phlox paniculata*
Plants erect, 2–6′, **hairy or hairless**. Leaves narrowly oblong, narrowly oval or elliptic, 3–6″, nearly stalkless, not clasping. Flowers on large panicles, densely hairy; flowers ½–¾″. Rich woods, thickets, borders, bottom lands.

3. Downy phlox *Phlox pilosa*
Plants erect, 1–2′, **downy**. Leaves linear or narrowly oval, 1–4″; upper leaves clasping. Flowers on loosely branched panicles, sticky or not, ½–¾″. Open woods, sand hills, prairie grasslands.

  b. Flowers tubular (Honeysuckle and Logania families)
These red, tubular flowers are adapted for hummingbird pollination.
4. Horse-gentian *Triosteum aurantiacum*
Plants ≤4′. **Leaves** oval, broader toward either end, 4–12″, **tapering to base**. **Flowers axillary**, ⅜–½″. Rich woods, thickets. See also the yellow-flowered horse-gentian p. 206. Honeysuckle family.

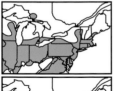

5. Perfoliate horse-gentian *Triosteum perfoliatum*
Plants ≤4½′. **Leaves** oval, broader toward the tip, or oblong, 4–12″; **base fused with the opposing leaf base. Flowers 3-4 in axils,** sometimes green-yellow, ¼–⅝″. Woods, thickets. See also p. 206. Honeysuckle family.

6. Pinkroot *Spigelia marilandica*
Plants 12–28″. Leaves narrowly oval or narrowly triangular, 2–4½″. **Flowers in a solitary terminal cluster;** flower scarlet outside, yellow inside, 1¼–2½″. Moist woods, thickets. Logania family.

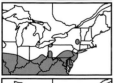

a. Rays many (Composite family)
7. Mist-flower *Conoclinium coelestinum*
Plants 1–3′. Leaves triangular, 1¼–4″, pointed, coarsely toothed. Flowers 35–70 per head; heads ¼″. Woods, stream banks, meadows, fields.

8. Pink tickseed *Coreopsis rosea*
Plants 8–24″. Leaves linear, ¾–2″. Flowers in heads; heads solitary at ends of branches, 1″; disk ¼–⅜″; rays ⅜″. Wet soils, shallow water. See also pp. 200, 208.

1. Moss phlox

2. Garden phlox

3. Downy phlox

4. Horse-gentian

5. Perfoliate horse-gentian

6. Pinkroot

7. Mist-flower

8. Pink tickseed

LEAVES opposite
LEAVES simple
PETALS irregular

Miscellaneous Mints (1)
Mint family
The flowers of mints usually have petals fused into two irregular lobes called lips. They can be distinguished from similar plants by the square stems and 4 nutlets in fruit.
a. Flowers on terminal heads, spikes, racemes, or panicles
  b. Flowers in heads

1. Bee-balm *Monarda didyma*

Plants 2–5′, hairy at nodes. Leaves oval or triangular, 2½–6″, narrowly pointed, toothed. **Flowers in terminal heads ¾–1½″ wide; flowers scarlet or crimson, 1¼–1¾″ long, hairless or nearly so; stamens 2**. Moist woods, thickets, bottomlands. See also pp. 206, 338.

2. Wild bergamot *Monarda fistulosa*

Plants 1½–4′, usually hairy. Leaves narrowly triangular or narrowly oval, 3½–4″, pointed, toothed. Flowers in terminal heads, ⅜–1½″ wide; **flowers pink, ¾–1½″ long**, hairy; **stamens 2**. Upland woods, thickets, prairies. Note strong minty smell of crushed heads.

3. Purple bergamot *Monarda media*

Plants ≤3′, hairless or nearly so. Leaves oval or broadly triangular, 2½–6″. Flowers in terminal heads, ⅜–1½″ wide; **flowers rose-purple, ¾–1¼″ long, hairless or nearly so; stamens 2**. Moist woods, thickets.

4. Wild basil *Clinopodium vulgare*

Plants erect, 8–20″. Leaves oval or oblong, ¾–1½″. Flowers in dense terminal heads; **flowers purple, pink, or white, ½″; stamens 4**. Upland woods. See also p. 50.

  b. Flowers on spikes or racemes (cont. on next page)

5. Rough hedge-nettle *Stachys aspera*

Plants 16–32″, hairless or nearly so. Leaves narrowly oblong, 1¼–2¾″, pointed, **stalkless**. Flowers on terminal spikes, ⅜–½″; sepals triangular. Damp sands, swamps, prairies.

6. Marsh hedge-nettle *Stachys palustris*

Plants 1–3′, downy and often sticky. Leaves narrowly triangular or elliptic, 1½–3½″, broadly rounded, **stalkless or nearly so**. Flowers on terminal spikes, ½–⅝″; sepals narrowly oval, downy. Ditches, roadsides, shores, stream margins, meadows.

7. Smooth hedge-nettle *Stachys tenuifolia*

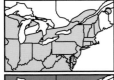

Plants 1–5′, hairless or bristly. Leaves linear, narrowly oval, broader toward either end, or oblong, 2½–3″, pointed; **stalk ¼–1″**. Flowers on terminal spikes, ⅜–½″; sepals narrowly oval. Wet areas.

***8. Wild thyme** *Thymus praecox*

Plants mat-forming, flowering stems erect, ≤3″. Leaves aromatic, linear to round, commonly elliptic or oval, ¼–⅜″. Flowers on a terminal spike, ⅜–1½″ tall; flowers ¼″. Upland woods, fields.

1. Bee-balm

2. Wild bergamot

3. Purple bergamot

4. Wild basil

5. Rough hedge-nettle

6. Marsh hedge-nettle

7. Smooth hedge-nettle

*8. Wild thyme

LEAVES opposite or whorled
LEAVES simple
PETALS irregular

Mints (2)
Mint family

a. **Flowers on terminal heads, spikes, racemes, or panicles**
  b. **Flowers on spikes or racemes (cont. from previous page)**

1. American germander *Teucrium canadense*
Plants 1–4′, hairy. Leaves narrowly oval or oblong, 2–4¾″, toothed, blunt or rounded at base, stalked ¼–½″. **Flowers on crowded spikes,** 2–8″ tall; **flowers ½–⅝″; upper lip apparently absent.** Shores, thickets, woods.

2. Obedience *Physostegia virginiana*
Plants ≤5′, hairless. Leaves narrowly oval, broader toward the tip, or elliptic, 2–6″. **Flowers on terminal racemes; flowers ½–1½″,** sticky-hairy. A wide variety of open habitats. The flowers will stay in position when moved, thus, an obedient plant.

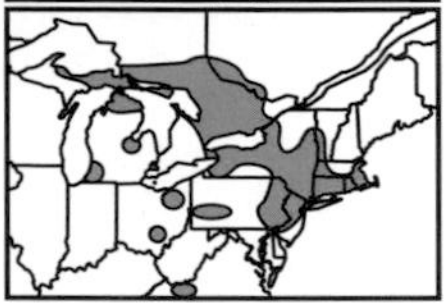

  b. **Flowers on panicles**
*3. **Wild marjoram** *Origanum vulgare*
Plants 16–32″, aromatic. Leaves oval, ½–1¼″, toothless or nearly so; stalk ¼–½″. **Flowers on rounded or pyramidal panicles,** 2–6″ wide; flowers ¼″. Weedy areas.

a. **Flowers in axillary whorls**
  b. **Herbs**

*4. **Henbit** *Lamium amplexicaule*
Plants arching, 2–16″, hairy, or hairless. **Leaves** round or kidney-shaped, ≤½″, notched at base, toothed or lobed; **uppermost leaves clasping.** Flowers ¼–¾″. Fields, weedy places. See also p. 338.

*5. **Red dead nettle** *Lamium purpureum*
Plants prostrate, 4–16″, inconspicuously hairy. **Leaves** round or oval, ¼–1¼″, notched at base, toothed, **stalked,** often tinged with red-violet. Flowers ¼–¾″. Fields, gardens, weedy places.

*6. **Common motherwort** *Leonurus cardiaca*
Plants erect, 1½–5′, short-hairy on the angles. Leaves diamond-shaped; cleft, 2–4″. **Flowers on long, interrupted spikes,** pink or white, ¼–½″. Weedy places.

*7. **Hemp-nettle** *Galeopsis tetrahit*
Plants 8–28″. **Leaves** diamond-shaped or oval, 1¼–4″, toothed, **stalked.** Flowers in dense whorls, ⅜–1″. Fields, roadsides, weedy areas.

  b. **Shrubs**
8. Cumberland false rosemary *Conradina verticillata*
Plants shrubs, ≤20″. Leaves appearing whorled, linear, ⅜–¾″, inrolled, gray-hairy below. Flowers 1–3 in several whorls, ¼″ long. Sandy riverbanks.

1. American germander   2. Obedience   *3. Wild marjoram

*4. Henbit   *5. Red dead nettle

*6. Common motherwort   *7. Hemp-nettle   8. Cumberland false rosemary

LEAVES opposite
LEAVES simple or
lobed
PETALS various

Miscellaneous opposite-leaved species
Several families

a. Flowers with 2 lips (Plantain family)
1. Purple turtlehead *Chelone obliqua*
Plants erect, 1–2½′. Leaves narrowly oval or oblong, 2½–4¾″, pointed, toothed, **rounded, or tapering to stalk; stalk ¼–½″.** Flowers on compact spikes, 1–1½″, with a white or yellow beard. Cypress swamps, wet woods. See also p. 348.

*2. Pink turtlehead *Chelone lyonii*
Plants erect, 16–40″. Leaves oval or narrowly oval, 3–5½″, pointed, toothed, **rounded at base; stalk ½–1¼″.** Flowers on compact spikes, 1¼–1½″, with a yellow beard. Garden escape.

a. Flowers with 4 petals (Melastome family)
Meadow-pitchers have anthers opening by pores at their tips. They are pollinated by bees vibrating (buzzing) the flower to release the pollen.

3. Dull meadow-pitchers *Rhexia mariana*
Plants 8–40″, hairy; stems not winged. Leaves narrowly oval, ¾–2″. Flowers on terminal racemes, 1–1½″; tube ¼–⅜″. Moist, open places.

4. Wing-stem meadow-pitchers *Rhexia virginica*
Plants 8–40″, bristly, especially at the nodes; stems winged. Leaves oval or narrowly oval, ¾–2½″. Flowers on terminal racemes, 1¼–1½″; tube ¼–⅜″. Moist, open places.

a. Flowers with 5 petals (Several families)
*5. Heart-leaved umbrella-wort *Mirabilis nyctaginea*
Plants ≤5′. Leaves oval or triangular, pointed. **Flowers surrounded by saucer-shaped bract,** ⅜″; the calyx is the showy part. Dry, weedy places.

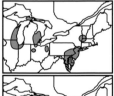

*6. Branching centaury *Centaurium pulchellum*
Plants erect, ≤8″. Leaves narrowly oval, ⅜–¾″, stalkless. Flowers many on dense spikes; flowers ⅜″. Fields, weedy places. Gentian family.

7. Blue-hearts *Buchnera americana*
Plants erect, 1–3′, rough-hairy. Lower leaves narrowly oval or oblong, 2–4″, pointed, coarsely toothed, rough; upper leaves narrower. **Flowers on stalked spikes,** ≤8″ tall; flowers ½″; tube ⅜–¾″; calyx hairy. Sandy or gravelly soil of upland woods, prairies. Broom-rape family.

8. Herb of grace *Bacopa monnieri*
Plants creeping and forming mats. Leaves narrowly oval or oval, broader toward the tip, ¼–1″, toothless. **Flowers solitary at nodes,** white, pink, or blue, ¼–⅜″. Wet sandy shores. Plantain family. See also p. 46.

1. Purple turtlehead

*2. Pink turtlehead

3. Dull meadow-pitchers

4. Wing-stem meadow-pitchers

*5. Heart-leaved umbrella-wort

*6. Branching centaury

7. Blue-hearts

8. Herb of grace

LEAVES opposite and basal
LEAVES compound or lobed
PETALS 5

Cranesbills and Valerian
Geranium and Valerian families

Cranesbill refers to the prolongation of the ovary into a "beak." The hairs inside the flower protect the nectar from washing out with rain. Critical characters are leaf shape, petal shape, and flower-stalk length. See also p. 62.

a. Leaves palmately lobed or compound (Geranium family)
    b. Flowers ≥1″

1. Wild geranium *Geranium maculatum*
Plants 8–20″. **Leaves** 2, **cleft** into 5–7 tapered, deeply cut lobes. **Flowers** few-several, **1–1½″**; petals not notched. Woods, shady roadsides, thickets, meadows.

    b. Flowers ≤1″
*2. Herb-Robert *Geranium robertianum*
Plants 6–18″. **Leaves compound;** terminal leaflet stalked. Flowers 2 per stalk, ½–1″; **petals not notched.** Damp, rich woods; rocky woods; ravines. With disagreeable odor and pollinated by flies.

*3. Dissected cranesbill *Geranium dissectum*
Plants ≤ 24″. **Leaves** broadly heart-shaped, 1–2½″ wide, **5–7 lobes;** lobes shallowly or deeply incised into additional linear lobes. Flowers 2 per stalk; **stalk equalling the sepals;** flowers ≤½″; **petals notched.** Roadsides, weedy places. Note spreading hairs (appressed in no. 4).

4. Carolina cranesbill *Geranium carolinianum*
Plants 4–20″. **Leaves** round to kidney-shaped, 12–36″ wide, **deeply 5–9 cleft;** lobes oblong, deeply toothed. Flowers 2–6 on 2-flowered stalks in compact clusters; **stalk shorter than sepals;** flowers ≤1″; **petals notched.** Dry woods, sandy soil, weedy places.

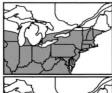

*5. Siberian cranesbill *Geranium sibericum*
Plants ≤3′. **Leaves** kidney-shaped, 1½–3½″ wide, **lobed;** lobes toothed or further lobed. Flowers 1–2 per stalk; **stalk twice the length of sepals;** flowers ½″; **petals jagged-edged.** Roadsides, fields, weedy places.

*6. Dove's-foot cranesbill *Geranium molle*
Plants 8–20″. **Leaves** round or kidney-shaped, ¾–2″ wide, 5–9 **cleft.** **Flowers** ⅜–½″. Weedy areas

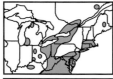

a. Leaves pinnately compound
*7. Storksbill *Erodium cicutarium*
Plants 3–12″. Leaves lying on ground, narrowly oval, wider toward the tip, mostly pinnately compound; leaflets oval or oblong, ½–1″. Flowers 2–8, long-stalked on umbels, purplish, pinkish, or white, ⅜–½″. Roadsides, sandy ground, weedy areas. Geranium family.

*8. Garden valerian *Valeriana officinalis*
Plants 1½–5′, hairy below. Leaves pinnately compound, leaflets 5–21 narrowly oval, toothed. Flowers on large, open panicles, ¼″; tube ⅛–¼″. Escape from gardens. Valerian family. See also p. 332.

1. Wild geranium

*2. Herb-Robert

*3. Dissected cranesbill

4. Carolina cranesbill

*5. Siberian cranesbill

*6. Dove's-foot cranesbill

*7. Storksbill

*8. Garden valerian

Miscellaneous species with whorled leaves (1)
Several families
a. Petals 3 (Bunch-flower family)
See other trilliums pp. 210, 258, 350.

1. Purple trillium *Trillium erectum*
Plants 8–24″. Leaves broadly diamond-shaped, 1¼–3″, tapering to base. **Flowers maroon, 1–3″**, erect or pendant; flower-stalks 1¼–3″; **petals widely spreading from the base.** Moist woods.

2. Large-flowered trillium *Trillium grandiflorum*
Plants 8–18″. Leaves oval, diamond-shaped, or round, 3¼–4¾″, tapering to base. **Flowers white, turning pink with age, 3–5″**, erect; flower-stalks 2–3¼″; **petals spreading from the middle.** Rich, moist woods; thickets.

a. Petals 4 (Madder family)
See other bedstraws pp. 210, 244, 352.

3. Wild licorice *Galium lanceolatum*
Plants 12–28″, hairless. Leaves in whorls of 4, elliptic or narrowly oval, 1¼–3″. Flowers on widely spreading, 1–3-forked inflorescences; flowers yellow, turning purple with age, ⅛″; petals pointed. Dry woods, thickets.

4. Hairy bedstraw *Galium pilosum*
Plants 6–40″, hairy. Leaves in whorls of 4, elliptic or oval, ½–1″. Flowers on terminal and axillary panicles; flowers greenish-white or purple, ⅛″. Dry woods.

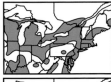

a. Petals 5 (Loosestrife family)
5. Water-willow *Decodon verticillatus*
Plants 3–10′, usually hairy. **Leaves opposite or in whorls of 3–4**, narrowly oval, 2–6″. Flowers in dense clusters in upper axils; flowers ¾–1¼″. Swamps, still water.

a. Petals 6 (Loosestrife family)
6. Blue wax-weed *Cuphea viscosissima*
Plants 6–24″, sticky-hairy; stems not winged. Leaves narrowly oval or oval, ¾–2″, long-stalked. Flowers in upper axils, 1–2″; tube ⅜″. Dry soil. Usually with 6 petals, sometimes 5.

7. Winged loosestrife *Lythrum alatum*
Plants 16–32″, hairless; stems winged. Leaves linear, ≤1½″, stalkless. Flowers solitary in axils, ¼–½″; tube ⅛–¼″, 12-winged. Moist or wet soils, especially in prairies. See also p. 330.

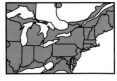

*8. Purple loosestrife *Lythrum salicaria*
Plants 1½–5′, hairless or more often hairy; stems not winged. Leaves narrowly oval or linear, 1¼–4″, stalkless. Flowers on terminal spikes 4–16″ tall; flowers ½–1″. Wet places.

1. Purple trillium

2. Large-flowered trillium

3. Wild licorice

4. Hairy bedstraw

5. Water-willow

6. Blue wax-weed

7. Winged loosestrife

*8. Purple loosestrife

LEAVES whorled or basal
LEAVES simple
PETALS 5 or RAYS many

Miscellaneous species with whorled or basal leaves
Several families
a. Leaves whorled
  b. Rays many (Composite family)

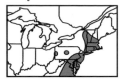

1. Three-nerved Joe-Pye weed *Eupatorium dubium*
Plants 16–40″. **Leaves** in whorls of 3–4, oval or narrowly oval, 2–4¾″, **strongly 3-nerved**. Flowers in heads on large panicles; **heads 5–8-flowered**, ¼–⅜″. Moist places, especially in sandy or gravelly, acid soil. See also pp. 334–36, 356.

2. Spotted Joe-Pye weed *Eupatorium maculatum*
Plants 2–7′. **Leaves** in whorls of 4–5, narrowly elliptic or narrowly oval, 2½–8″, **not 3-nerved**. Flowers in heads on flat-topped panicles; **heads 10–16-flowered**, ¼–⅜″. Moist places, especially calcareous.

3. Purple-node Joe-Pye weed *Eupatorium purpureum*
Plants 2–7′. **Leaves** mostly in whorls of 3 or 4, narrowly oval, oval, or elliptic, 3–12″, **not 3-nerved**. Flowers in heads on rounded panicles; **heads 4–7-flowered**, ¼–⅜″. Thickets, open woods, often in drier habitats than other Joe-Pye weeds.

  b. Petals 5 (Gentian family)
4. American columbo *Frasera caroliniensis*
Plants 3–6′. Leaves in whorls of 4, narrowly oval, broader toward the tip, ≤16″; upper leaves smaller. **Flowers on panicles, green-yellow with purple dots and a large green gland on each petal.** ⅜–⅝″; petals narrowly oval, pointed. Rich woods.

a. Leaves basal
  b. Petals regular in size and shape
    c. Petals 5 (Heath & Primrose families) (cont. on next page)
5. Western shooting-star *Dodecatheon amethystinum*
Plants 8–20″. Leaves oblong or oval, broader at either end, 1½–10″, usually tinged with red at base. Flowers 3–25, nodding, on umbels; flowers pink, rarely white, ⅜–¾″. Prairies, moist hillsides. Primrose family.

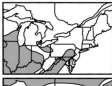

6. Eastern shooting-star *Dodecatheon meadia*
Plants 8–24″. Leaves oblong or oval, broader at either end, 2¼–8″, usually tinged with red at base. Flowers 4–125, nodding on umbels; flowers white or pink, ½–1″. Moist or dry woods, prairies. Primrose family.

7. Mistassini primrose *Primula mistassinica*
Plants ≤10″. Leaves oval, broader toward the tip, or spoon-shaped, ¾–2¾″. Flowers 1–2; tube yellow; lobes white or pink, ⅜–¾″. Rocks, cliffs, gravelly shores. Primrose family.

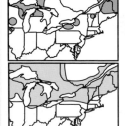

8. Pink shinleaf *Pyrola asarifolia*
Plants 6–12″. Leaves kidney-shaped, 1¼–2½″, notched at base. Flowers on erect terminal racemes, ¼–½″; sepals triangular, ≤⅛″, pointed. Moist woods, bogs. Heath family. See also p. 364.

1. Three-nerved Joe-Pye weed

2. Spotted Joe-Pye weed

3. Purple-node Joe-Pye weed

4. American columbo

5. Western shooting-star

6. Eastern shooting-star

7. Mistassini primrose

8. Pink shinleaf

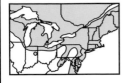

Miscellaneous species with basal leaves
Several families
c. Petals 5 (several families) (cont. from previous page)

1. Purple pitcher-plant *Sarracenia purpurea*
Plants 12–20″. **Leaves forming pitchers, 4–8″ tall,** ⅜–2″ wide.
Flowers solitary, long-stalked, 2–2¾″. Sphagnum bogs, sandy
shores, other wet places. Populations from southwestern Pennsylvania to West Virginia are introduced. The pitcher-leaves capture
insects that fall in, eventually decomposing; the plant absorbs their
nutrients. Pollinated by bumblebees, which enter between the stigma
and petal seeking nectar. Pitcher-plant family. See also p. 214.

2. Thread-leaved sundew *Drosera filiformis*
Plants erect, 3–10″ with long, sticky hairs. **Leaves thread-like,
forming a fiddlehead when uncurling, 5½–12″.** Flowers 4–16 on
a coiled raceme, ¼–½″. Wet areas. See other sundews p. 366. The
long, sticky hairs capture insects, which decompose, thus releasing
nutrients used by the plant. Sundew family.

3. Rockpink fame-flower *Talinum calycinum*
Plants 4–12″. **Leaves cylindrical, 1¼–3″.** Flowers on racemes, pink
or red, ¾–1¼″. Barrens, rocky ledges. The flowers are open for only
a few hours in full sun. Purslane family.

c. Petals 6
d. Flowers in heads (Bunch-flower family)
4. Swamp-pink *Helonias bullata*
Plants ≤3′, covered with short-bracts at base. **Leaves elongate-
spoon-shaped,** prostrate, expanding after flowering, ≤12″. **Flowers
in heads,** 1¼–4″ wide; flowers ⅜″; anthers blue. Swamps, bogs.

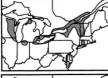

d. Flowers on umbels (several families)
*5. Flowering rush *Butomus umbellatus*
Plants 3–5′. **Leaves erect, floating, or submersed, linear,** ≤3″. **Flowers numerous on terminal umbels,** ¾–1″; pistils several, pink.
Shores, riverbanks. Flowering-rush family.

6. Meadow garlic *Allium canadense*
Plants erect, 8–24″; leaves on the lower third. **Leaves linear,** flat,
≤¼″. **Flowers in a small cluster,** pink or white, ⅜–½″, **usually replaced by bulblets.** Open woods, prairies. Onion family. See other
onions pp. 64, 396.

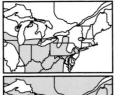

7. Nodding onion *Allium cernuum*
Plants 1–2′. **Leaves linear,** 4–10″. **Flowers 8–35, nodding** on an umbel, white or pink, ¼″. Woods, rocky banks, prairies. Onion family.

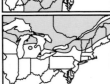

b. Petals irregular in size and shape
8. Calypso *Calypso bulbosa*
Plants 4–8″. **Leaves 1,** round or oval, 1¼–2″, **notched at base.**
Flowers solitary, ¾–1½″; sepals and lateral petals ⅜–¾″; lip ½–1″.
Moist conifer woods, often under cedar. Orchid family.

1. Purple pitcher-plant

2"

2. Thread-leaved sundew

3. Rockpink fame-flower

4. Swamp-pink

*5. Flowering rush

6. Meadow garlic

7. Nodding onion

8. Calypso

**LEAVES basal or
absent
LEAVES simple
PETALS various**

Miscellaneous species with basal or absent leaves
Composite, Poppy and Orchid families
a. Leaves basal
   b. Petals absent (Composite family)

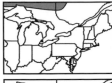

1. Rosy pussytoes *Antennaria microphylla*
Plants mat-forming, 2–16″. Leaves narrowly oval, broader toward
the tip, ¼–1¼″, gray-hairy. Flowers in heads; heads in small clusters, ⅛–¼″. Dry, open places. See other pussytoes p. 374.

   b. Petals irregular
      c. Flowers heart-shaped (Poppy family)

2. Eastern bleeding-heart *Dicentra eximia*
Plants 8–20″. Leaves highly dissected; lobes narrowly oval or oblong, ≤16″. Flowers on panicles, pink, sometimes white, ¾–1″. Dry
or moist mountain woods. See also p. 390.

      c. Flowers various shapes (Orchid family)

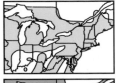

3. Grass-pink *Calopogon tuberosus*
Plants 3–36″. Leaves 1, linear, 8–12″. **Flowers** 3–15 on racemes,
1–2″; sepals and lateral petals pointed; lip ½–¾″, crested with
magenta and yellow-tipped hairs. Acid bogs, swamps.

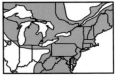

4. Pink lady-slipper *Cypripedium acaule*
Plants 6–16″. Leaves 2, narrow elliptic, 4–8″. **Flowers** solitary,
terminal, **2½–4″;** sepals and lateral petals yellowish green or green-
brown, narrowly oval, 1¼–2″; 2 lower sepals united; **lip pouch-like,**
pink (occasionally white) with red veins, 1¼–2½″. Acid swamps,
bogs, dry woods, dunes. See also pp. 86, 166, 298.

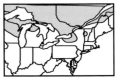

5. One-leaf orchis *Amerorchis rotundifolia*
Plants 4–10″. Leaves elliptic or oval, broader at either end, 1¼–4½″.
Flowers few on terminal racemes, ½–¾″; sepals and petals oblong,
¼–⅜″; outer petals and upper sepal fused; lateral sepals spreading;
lip purple spotted, ¼″; 3-lobed; terminal lobe triangular. Calcareous
conifer forests, thickets, fens.

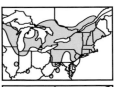

***6. Helleborine** *Epipactis helleborine*
Plants ≤32″. Leaves clasping, oval or narrowly oval, ≤4″, smaller
upward. Flowers on racemes, 4–12″ tall; **flowers ¾–1″, pink or
green;** sepals and petals narrowly oval, ⅜–½″. Roadsides, woods.

a. Leaves absent (cont. on next page)
7. Dragon's mouth *Arethusa bulbosa*
Plants 4–12″. Leaves basal, absent at flowering time, 4–12″. **Flowers solitary,** 1½–4″; sepals and petals narrowly oval, ¾–2″; lip
pink-white spotted with purple and yellow, ¾–1¼″. Sphagnum bogs,
swampy meadows.

8. Striped coral-root *Corallorhiza striata*
Plants 6–20″. Leaves absent. **Flowers 7–25 on terminal spikes,**
2–8″ tall; flowers ½–1½″; sepals and lateral petals ¼–¾″; oblong; lip
declined, ¼–½″, white and red-striped. Rich woods. See also p. 398.

1. Rosy pussytoes

2. Eastern bleeding-heart

3. Grass-pink

4. Pink lady-slipper

3"

5. One-leaf orchis

*6. Helleborine

7. Dragon's mouth

3"

8. Striped coral-root

LEAVES absent, float-
ing or submersed
LEAVES misc.
PETALS various

Miscellaneous aquatic and leafless species
Several families
a. Leaves absent (cont. from previous page)

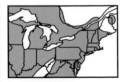

*1. Asparagus *Asparagus officinalis*
Plants ≤6′, highly branched; ultimate branches flattened and appearing to be fine leaves. **Leaves represented by very small scales.**
Flowers solitary or paired, greenish-white or maroon, ¼″. Escape
from cultivation, weedy places, salt-marshes. Asparagus family.

a. Leaves floating or emerging out of the water

2. Water-shield *Brasenia schreberi*
Plants ≤6′, submersed. **Leaves floating, elliptic**, 1½–4¾″, reddish-
brown and coated with "jelly" below; stalk attached at center of leaf.
Flowers solitary on stalks ≤6″; flowers 1–1¼″. Quiet water. Water-
shield family.

*3. Oriental sacred lotus *Nelumbo nucifera*
Plants emergent. **Leaves held above the water, round**, ≤24″,
stalked ≤7′. Flowers solitary, terminal, pink or white, ¾–10″. Escape
near gardens. See yellow-flowered American lotus lily p. 228. Indian
lotus family.

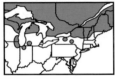

4. Mare's tail *Hippuris vulgaris*
Plants erect, 8–24″. **Leaves in whorls of 6–12**, submersed leaves
soon degenerating; **emergent leaves linear**, ⅜–1¼″. Flowers solitary in upper axils, ⅛″. Shallow water, mud. Plantain family.

a. Leaves submersed

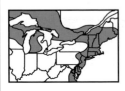

5. Purple bladderwort *Utricularia purpurea*
Plants submersed, ≤3′. Leaves in whorls of 5–7, divided into thread-
like segments, many of which have a bladder at the end. **Flowers
1–4 on racemes**, irregular, violet or red-violet with a yellow spot,
⅜″. Quiet water. The bladders capture small aquatic animals, which
are a source of nutrients for the plant. See other bladderworts p. 226.
Bladderwort family.

6. Pink bog-button *Sclerolepis uniflora*
Plants emergent 4–12″. Leaves in whorls of 4 or 6, linear, ¼–1″.
Flowers in solitary heads, ⅛–⅜″; rays absent. Still, shallow water,
wet soil. Submersed plants look different, often lacking leaves.
Composite family.

*7. Eurasian water-milfoil *Myriophyllum spicatum*
Plants submersed. Leaves whorled, pinnately compound, ⅜–2″, leaf-
lets filiform. Flowers on emergent spikes 1½–4″ tall, flowers <⅛″.
Lakes. See also p. 254. Water-milfoil family.

8. Farwell's water-milfoil *Myriophyllum farwellii*
Plants fully submersed. Leaves alternate or opposite, pinnately com-
pound, ⅜–1¼″; leaflets thread-like. Flowers solitary in axils, <⅛″.
Lakes. Water-milfoil family.

*1. Asparagus

2. Water-shield

*3. Oriental sacred lotus

4. Mare's tail

5 Purple bladderwort

6. Pink bog-button

*7. Eurasian water-milfoil

8. Farwell's water-milfoil

LEAVES alternate, opposite or basal
LEAVES simple
PETALS various

Miscellaneous orange-flowered species (1)
Several families

a. Leaves alternate or opposite
  b. Petals 5

1. Butterfly milkweed *Asclepias tuberosa*
Plants, 1–3′, often branched; **sap not milky. Leaves alternate, sometimes upper opposite,** linear or narrowly oval, broader toward either end, 2–5″. Flowers 1-many in terminal umbels, ½–¾″. Prairies, upland woods. See also pp. 110, 246, 330, 354. Dogbane family.

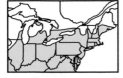

2. Few-flowered milkweed *Asclepias lanceolata*
Plants ≤3′, unbranched; **sap milky. Leaves opposite, linear or narrowly oval,** 2¾–8″. Flowers few on 1–4 umbels, red-orange with purple center, ¼–⅜″. Swamps, bogs, brackish marshes. Dogbane family.

***3. Pimpernel** *Anagallis arvensis*
Plants 4–12″. Leaves elliptic or oval, ⅜–¾″, stalkless. Flowers solitary in leaf axils, red-orange or white, ⅛–¼″. Roadsides, disturbed areas. Myrsine family.

  b. Petals irregular
4. Orange touch-me-not *Impatiens capensis*
Plants 2–5′, stems succulent. Leaves oval or elliptic, 1¼–4″. Flowers solitary, ¾–1¼″, drooping, **cornucopia-shaped; spur ¼–⅜″.** Moist woods, brook-sides, wet roadside ditches, springy places. The fruit explodes on touch. See yellow-flowered touch-me-not p. 168. Touch-me-not family.

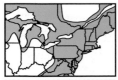

  b. Petals absent
5. Golden saxifrage *Chrysosplenium americanum*
Plants decumbent, 2–8″. Lower leaves opposite; upper alternate; leaves oval or round, ¼–½″. Flowers solitary at ends of branches, ¼″; **petals absent; anthers orange.** Springy or muddy soil. Saxifrage family.

a. Leaves basal
***6. Orange hawkweed** *Hieracium aurantiacum*
Plants 4–28″, long-hairy. Basal leaves oval, broader toward the tip or narrowly elliptic, 2–10″. **Flowers in 5–30 heads** on flat-topped panicles; heads ¾–1″; rays many. Fields, roadsides, meadows. See yellow hawkweeds pp. 162, 220. Composite family.

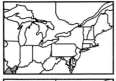

7. Copper iris *Iris fulva*
Plants 20–60″. Leaves sword-shaped, 20–40″. Flowers 1–2 per branch, 2¾–3¾″; **sepals 3, spreading.** Swamps. See other irises pp. 56, 214. Iris family.

8. Orange milkwort *Polygala lutea*
Plants 4–19″. Leaves narrowly oval, oval, or oblong, ½–1¾″. Flowers on dense, head-like racemes, ⅜–1¼″ tall; flowers, ¼″; petals irregular; wings pointed. Pine barrens; sandy, acid swamps; bogs. See other milkworts pp. 82, 282. Milkwort family.

1. Butterfly milkweed 2. Few-flowered milkweed *3. Pimpernel

4. Orange touch-me-not 5. Golden saxifrage

*6. Orange hawkweed 7. Copper iris 8. Orange milkwort

LEAVES whorled or basal
LEAVES simple
PETALS 6 or irregular

Lilies and Orchids
Several families
Lilies are pollinated by larger butterflies, such as Swallowtails. Critical characters are in the shape and coloring of the flower. See the yellow lily p. 210.

a. Petals 6

b. Leaves whorled (Lily family)

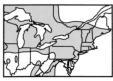

1. Wood lily *Lilium philadelphicum*
Plants 1–3′. Leaves whorled, narrowly oval or linear, 2–4″, pointed. Flowers erect, 1–5, 1½–3″; petals oval or narrowly oval, recurved, purple-spotted, hairless. Dry, open woods; meadows; shores.

2. Michigan lily *Lilium michiganense*
Plants ≤10′. Leaves whorled, narrowly oval, **rough along margin and midvein. Flowers nodding, 1–2, 2¼–3½″; petals strongly recurved, narrowly oval, purple spotted; green star at flower center not clearly defined.** Bogs, meadows, wet woods, wet prairies.

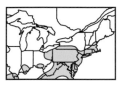

3. Turk's-cap lily *Lilium superbum*
Plants 3–8′. Leaves whorled, narrowly oval, **smooth or hairy but not rough along margin and midvein. Flowers nodding, 1–2, 2½–3½″; petals strongly recurved, narrowly oval, spotted with purple, green star at center of flower clearly defined.** Bogs, meadows, wet woods, wet prairies.

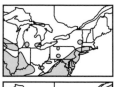

b. Leaves basal, narrow (Day-flower and Iris families)
*4. Blackberry-lily *Belamcanda chinensis*
Plants 1–4′. Leaves sword-shaped, 6–10″. **Flowers** on terminal cymes, 1¼–2″, with red or purple spots. Pastures, roadsides, thickets, hillsides. Iris family.

*5. Day-lily *Hemerocallis fulva*
Plants 1½–6′. Leaves linear, 2′. **Flowers** in terminal cluster of 3–15, **3–5″.** Roadsides, borders of fields, thickets. Day-lily family.

a. Petals irregular (Orchid family)
6. Yellow fringed orchid *Platanthera ciliaris*
Plants 1–3′. **Basal leaves 1–3,** narrowly oval, ≤6″. Flowers on spikes, 2–6″ tall; flowers ½″; sepals oval, broader toward either end, ¼″; lateral petals linear, appearing torn; **lip** oblong, ⅜–½″, **long-fringed; spur** ½–1¼″. Bogs, fields, woods. See also pp. 36, 248, 298.

7. Crested fringed orchid *Platanthera cristata*
Plants 10–30″. **Basal leaves 1–3,** narrowly oval or linear, ≤8″. Flowers on spikes ¾–5″ tall; flower ¼–½″; sepals oval, broader toward either end, ⅛–¼″; lateral petals oblong, apearing torn; **lip** oblong, **long-fringed,** ¼–⅜″; **spur** ¼–⅜″. Meadows, damp pine woods.

8. Yellow fringeless orchid *Platanthera integra*
Plants 10–24″. **Stem leaves 2,** narrowly oval, folded, recurved, ≤10″. Flowers on spikes, 1–3½″ tall; flowers ½″, divergent; **lip** oval, **toothed or entire,** ¼″; **spur** ¼″. Open, acid bogs; pine barrens.

1. Wood lily   2. Michigan lily   3. Turk's-cap lily

*4. Blackberry-lily   *5. Day-lily

6. Yellow fringed orchid   7. Crested fringed orchid   8. Yellow fringeless orchid

LEAVES alternate
LEAVES simple
PETALS 4

Evening-primroses amd Sundrops
Evening-primose family

Evening-primroses are open at night and are pollinated by moths; sundrops are open during the day and are pollinated by bees. The stigmas are characteristically 4-lobed, forming an **X**. Critical characters are fruit shape and size, sepal shape, and habit. See pink-flowered evening-primrose p. 112.

a. Fruit with rounded sides
  b. Fruit oval, broader toward base
1. Common evening-primrose *Oenothera biennis*
Plants erect, **1–6′, branched**. Leaves narrowly oval or oblong, 4–8″, stalkless or short-stalked. Flowers on terminal spikes, 1–2″. Fields, roadsides, prairies, weedy places.

2. Small-flowered evening-primrose *Oenothera parviflora*
Plants erect, **4–32″, unbranched**. Leaves narrowly oval, 4–8″. Flowers on terminal spikes, ¾–1½″. Gravelly shores, talus, sands, and dry openings. Similar to no. 1, but smaller, with tip of inflorescence nodding.

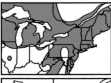

  b. Fruit cylindrical
3. Cleland's evening-primrose *Oenothera clelandii*
Plants erect, 16–40″. **Leaves** linear or narrowly oval, 1¼–3″, **unlobed.** Flowers numerous on a terminal spike, 4–12″ tall; flowers ½–1″. Fields, prairies in sandy soil.

4. Cut-leaved evening-primrose *Oenothera laciniata*
Plants prostrate to erect, 4–30″. **Leaves** oblong or narrowly oval, broader toward either end, ¾–4″, **lobed.** Flowers few, unstalked, in leaf axils, ½–1½″. Dry, usually sandy, open ground.

a. Fruit with pronounced wings
  b. Flowers 4–5″
5. Big-fruited evening-primrose *Oenothera macrocarpa*
Plants prostrate or erect, ≤20″, hairy. Leaves linear, 2–4″, long-stalked. Flowers stalkless in leaf axils, 4–5″. Rock barrens.

  b. Flowers ≤2½″
6. Sundrops *Oenothera fruticosa*
Plants 1–3′, **hairless or nearly so.** Leaves oval or linear, 1¼–4″, toothless. **Flowers** few at stem tip, **1–2″.** Fields, meadows, open woods.

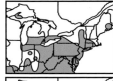

7. Midwestern sundrops *Oenothera pilosella*
Plants 1–3′, **long-hairy.** Leaves narrowly oval or elliptic, 1¼–5″, toothed. **Flowers** many in compact cluster, 1–2½″. Wet meadows, fields, open woods.

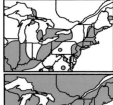

8. Little sundrops *Oenothera perennis*
Plants 4–24″, hairy or hairless. Leaves linear, elliptic or narrowly oval, broader toward tip, 1–2″. Flowers on a nodding or straight, leafy raceme; **flowers** ½–¾″. Fields, meadows, open woods.

1. Common evening-primrose

2. Small-flowered evening-primrose

3. Cleland's evening-primrose

4. Cut-leaved evening-primrose

5. Big-fruited evening-primrose

4½"

6. Sundrops

7. Midwestern sundrops

8. Little sundrops

LEAVES alternate and basal
LEAVES simple or compound
PETALS 4

Celandine and Mustards (1)
Poppy and Mustard families

Mustards have a distinctive flower with 4 sepals, 4 petals, and 6 stamens, 4 long and 2 short, and a distinctive fruit, which retains a papery membrane after the seeds are shed.

a. Sepals 2; plants with orange sap (Poppy family)

*1. Celandine *Chelidonium majus*
Plants 3–32″; **sap orange.** Leaves pinnately divided into 5–7 lobes; lobes ¾–3″. Flowers few on an umbel, ½–¾″. Rich, moist soil.

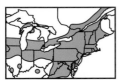

a. Sepals 4; plants with clear sap (Mustard family)
  b. Fruit not more than 3 times longer than wide

2. Common yellow-cress *Rorippa palustris*
Plants ½–4″. **Leaves** narrowly oval or oblong, 1¼–8″, **compound, lobed or toothed.** Flowers on racemes, ⅛–¼″. **Fruit cylindrical, spreading, ⅛–⅜″.** Shores, openings, weedy places. See also p. 300.

*3. Yellow alyssum *Alyssum alyssoides*
Plants 2–12″. **Leaves linear, spoon-shaped, or narrowly oval, broader toward tip,** ¼–½″, toothless, blunt. Flowers on racemes, pale yellow to nearly white, <⅛″. **Fruit round, divergent, ⅛″.** Grasslands, roadsides, weedy places.

*4. Woad *Isatis tinctoria*
Plants 1½–4′. **Basal leaves narrowly oval, broader toward tip,** long-stalked; **stem leaves narrowly oval or oblong,** ¾–4″, stalkless. Flowers on large, terminal panicles made-up of individual racemes, ¼″. **Fruit oblong, drooping, ¼–½″.** Weedy areas.

  b. Fruit more than 3 times longer than wide
    c. Fruit without beak (cont. on next page)

5. Drummond's rock-cress *Arabis drummondii*
Plants 1–3′, hairless. Leaves narrowly oval or narrowly oblong, ¾–3¾″, lobed at base, hairless. Flowers on racemes, ⅜–¾″. Fruit erect, 1½–2½″. Ledges, gravels, thickets. See also p. 268.

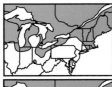

6. Tower rock-cress *Arabis glabra*
Plants 1–5′, hairy below. **Lower leaves spoon-shaped or narrowly oval, broader toward tip,** 2–4¾″, withering early; **stem leaves narrowly oval,** ≤4¾″, lobed at base. Flowers on racemes, ¼–⅜″. Fruit erect, 2–3½″. Ledges, cliffs, thickets, fields.

*7. Early winter-cress *Barbarea verna*
Plants 12–32″, hairless. Basal leaves oblong, ≤8″, **with 8–40 lobes;** stem leaves smaller. Flowers on racemes, ½″. Fruit erect to spreading, ½–1¼″. Damp soil, fields, roadsides.

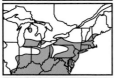

*8. Common winter-cress *Barbarea vulgaris*
Plants 8–32″, hairless. Basal leaves violin-shaped, with 2–8 lobes; terminal lobe large, oval or round; stem leaves reduced, clasping. Flowers crowded on racemes, ½″. Fruit ascending, 1½–2½″. Wet meadows, damp fields, roadsides, gardens.

*1. Celandine

2. Common yellow-cress

*3. Yellow alyssum

ʌ4. Woad

5. Drummond's rock-cress

6. Tower rock-cress

*7. Early winter-cress

*8. Common winter-cress

LEAVES alternate
LEAVES lobed
PETALS 4

Mustards (2)
Mustard family
c. Fruit without beak (cont. from previous page)

*1. Slim-leaf wall-rocket *Diplotaxis tenuifolia*
Plants **shrubby**, 1–2½′. Leaves fleshy with unpleasant aroma, narrowly oval, broader toward tip, or narrowly oblong, toothed or lobed. Flowers on racemes, ¼–½″. **Fruit linear, ascending, ¾–2″.** Weedy places, roadsides.

*2. Tumble mustard *Sisymbrium altissimum*
Plants 2–4′. Leaves lobed, lower with 5–8 pairs of linear to narrowly oval, toothed lobes; upper leaves smaller with few lobes, stalked. Flowers on loose racemes, ¼–⅜″. **Fruit slender, ascending or spreading, 2–4″.** Fields, roadsides, weedy places.

*3. Hedge mustard *Sisymbrium officinale*
Plants 1–3′. Lower leaves deeply lobed, stalked; lobes oblong or oval; terminal lobe round, toothed; upper leaves stalkless, lobeless. Flowers on stiff spike-like panicles, ⅛″. **Fruit narrowly cylindrical, pointed, appressed to stem, ¼–½″.** Waste places.

c. Fruit with distinct beak
d. Sepals erect

*4. Brown mustard *Brassica juncea*
Plants 1–4′, hairless. Lower leaves 2½–12″, violin-shaped, lobed or toothed; **upper leaves smaller, stalked, not clasping.** Flowers on a raceme, ½″. **Fruit cylindrical, ½–1½″.** Weed of cultivated fields. Note the fruit curves out from the stem.

*5. Field mustard *Brassica rapa*
Plants 24–32″, hairless, succulent. Lower leaves violin-shaped, 2½–12″, lobed, stalked; **upper leaves** oblong or narrowly oval, toothed, **stalkless and clasping.** Flowers on a raceme, ½″. Fruit cylindrical, 1¼–2″. Cultivated fields.

*6. Star mustard *Coincya monensis*
Plants prostrate to erect, ≤3′, hairy upward. Leaves pinnately lobed, 2–9″, blue-green; upper leaves less lobed. Flowers on racemes, yellow with brown or violet veins, ½″. **Fruit narrowly cylindrical, 1–3″.** Highways.

d. Sepals spreading

*7. White mustard *Sinapis alba*
Plants 1–2′, rough-hairy. Leaves stalked, lower ≤8″, violin-shaped, upper leaves smaller. Flowers on a raceme, ½″. **Fruit bristly, cylindrical, ¾–1¼″.** Fields, weedy places.

*8. Charlock mustard *Sinapis arvensis*
Plants 8–30″, rough-hairy to hairless. Lower leaves oval, broader toward tip, coarsely toothed or lobed; upper leaves smaller. Flowers on a raceme, ½″. **Fruit hairless, cylindrical, 1¼–1½″.** Fields, gardens, weedy places.

*1. Slim-leaf wall-rocket

*2. Tumble mustard

*3. Hedge mustard

*5. Field mustard

*6. Star mustard

*4. Brown mustard

*7. White mustard

*8. Charlock mustard

LEAVES alternate
LEAVES simple
PETALS 4–5

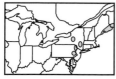

Miscellaneous alternate-leaved species with 4–5 petals
Several families
a. Petals 4 (Evening-primrose family)
1. Seedbox *Ludwigia alternifolia*
Plants erect, freely branching, 2–4′, hairless. Leaves narrowly oval or linear, 2–4″, stalked. **Flowers axillary,** stalked ⅛″, with 2 bracts near top, **flowers ½–¾″; petals 4,** same length as sepals. Swamps, shores. See other Ludwigias pp. 168, 244.

a. Petals 4–5 (Stonecrop family)
2. Rose-root stonecrop *Rhodiola rosea*
Plants erect, 1–16″. Leaves oval or narrowly oval, broader toward tip, ⅜–1¼″, stalkless. **Flowers unisexual, in compact clusters,** ⅜–2¼″ wide; **flowers ¼″,** yellow or reddish; **petals 4–5.** Cliffs, ledges.

a. Petals 5
  b. Flowers on racemes (several families)
3. Ditch-stonecrop *Penthorum sedoides*
Plants 8–28″. **Leaves** narrowly oval or narrowly elliptic, 2–4″, **sharply toothed.** Flowers on racemes, 1–3¼″ long; **flowers ¼″;** petals often absent. Marshes, muddy shores. Ditch-stonecrop family.

4. Hoary puccoon *Lithospermum canescens*
Plants 2–20″, densely **soft-hairy. Leaves** narrowly oval or narrowly oblong, ¾–2½″, toothless. Flowers on curved racemes; **flowers** ¼–½″. Prairies; dry, open woods. See also p. 280. Borage family.

5. Plains puccoon *Lithospermum caroliniense*
Plants 8–40″, **rough-hairy. Leaves** linear or narrowly oval, 1¼–2½″, **toothless.** Flowers on curved racemes, 4–12″ long, **flowers** ½–1″. Upland woods, shores, prairies, especially in sandy soil. Similar to no. 4, but the hairs are rough to the touch and the flowers are larger. Borage family.

***6. Moth mullein** *Verbascum blattaria*
Plants 1–5′, smooth below, sticky-hairy above. Leaves variable, narrowly triangular, oblong, or narrowly oval, **coarsely toothed or toothless, stalkless or lower stalked. Flowers** on elongate, loose racemes, ¾–1¼″, yellow or white. Roadsides, old fields. See also p. 284. Scroph family.

  b. Flowers on spikes (Scroph family)
***7. Clasping mullein** *Verbascum phlomoides*
Plants 1½–4′. **Leaves** oblong or narrowly oval, broader toward either end, toothless or nearly so, **stalkless, running down stem below leaf blade. Flowers** on spikes, 1–1½″. Weedy areas.

***8. Common mullein** *Verbascum thapsus*
Plants 2–8′, densely gray-woolly. **Lower leaves** flannel-like, oblong or narrowly oval, broader toward tip, 4–12″, entire or toothed, **stalked;** upper leaves smaller, stalkless. Flowers on dense spikes, 4–20″ tall; **flowers** ½–1″. Roadsides, fields, weedy places.

1. Seedbox

2. Rose-root stonecrop

3. Ditch-stonecrop

4. Hoary puccoon

5. Plains puccoon

*6. Moth mullein

*7. Clasping mullein

*8. Common mullein

LEAVES alternate, and sometimes opposite
LEAVES simple
PETALS 5

Miscellaneous alternate-leaved species with 5 petals
Several families
a. Flowers with dark markings at center (Nightshade family)
Ground-cherries The distinctive markings in the flower center guide visiting bees to nectar and pollen. Critical characters are flower size and the hairs. See white-flowered ground-cherry p. 280.

1. Clammy ground-cherry *Physalis heterophylla*
Plants 8–36″, sticky-hairy. Leaves oval or diamond-shaped, 1¼–4¼″, toothed or toothless. Flowers axillary, on stalks ≤⅜″; **flowers ½–1″**; sepals triangular or oval. Upland woods, prairies, clearings.

2. Long-leaved ground-cherry *Physalis longifolia*
Plants 12–32″, hairy or hairless. Leaves narrowly oval or linear, ⅜–4″, few-toothed. Flowers axillary, on stalks ⅜–¾″; **flowers ⅜–½″**; sepals triangular or oval. Fields, open woods, prairies.

3. Virginia ground-cherry *Physalis virginiana*
Plants 8–36″, long, soft-hairy. Leaves oval or narrowly oval, 1–2½″, toothed or toothless. Flowers axillary, on stalks ⅜–¾″; **flowers ½–¾″**; sepals narrowly oval. Fields, upland woods, prairies.

***4. Tomatillo** *Physalis philadelphica*
Plants 8–24″, nearly hairless. Leaves oval or diamond-shaped, ¾–2½″. Flowers axillary, on stalks ¼″; **flowers ¼–½″**; sepals triangular. Escape from gardens.

a. Flowers without central markings
(Evening-primrose and Flax families)
5. Creeping water-purslane *Ludwigia peploides*
Plants prostrate or floating, 8–24″ long. Leaves narrowly oval, broader toward either end, 1¼–3½″, stalked 1–1½″. Flowers axillary, stalked ≤¼″; flowers 1″; stamens 10. Ponds, swamps. See other Ludwigias pp. 168, 244. Evening-primrose family

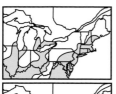

Flax Lower leaves are often opposite. See also p. 12. Flax family.
6. Common yellow flax *Linum medium*
Plants erect, 6–28″. Leaves narrowly oval, broader toward either end, ⅜–1″; **most leaves alternate, lowest opposite.** Flowers on a flat-topped panicle, stalked ⅛″; flowers ½″. Dry, upland woods; beaches.

7. Ridgestem yellow flax *Linum striatum*
Plants erect, 8–48″. Leaves elliptic or narrowly oval, broader toward tip, ½–1½″; **most leaves opposite, only uppermost alternate.** Flowers on elongate panicles, stalks ≤⅛″; flowers ¼–½″. Damp or wet woods, swamps, bogs.

8. Grooved yellow flax *Linum sulcatum*
Plants erect, 6–32″. Leaves linear, ⅜–1″. Flowers on panicles; stalks ≤⅛″; flowers ⅜–¾″. Dry, open prairies; woods. **Differs from nos. 6 and 7 in being branched and having linear leaves.**

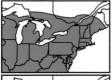

1. Clammy ground-cherry

2. Long-leaved ground-cherry

3. Virginia ground-cherry

*4. Tomatillo

5. Creeping water-purslane

6. Common yellow flax

7. Ridgestem yellow flax

8. Grooved yellow flax

LEAVES alternate or opposite
LEAVES simple
PETALS 5–6

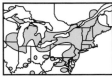

Miscellaneous alternate-leaved species with 5–6 petals
Several families
a. Stamens 10 or fewer (Purslane and Stonecrop families)
1. Common purslane *Portulaca oleracea*
Plants forming mats. **Leaves** alternate or occasionally opposite, fleshy, spoon-shaped or oval, broader toward tip, ⅜–1¼″, rounded. Flowers solitary or in terminal clusters, ¼–⅜″, open only in morning. Cultivated; weedy areas. Purslane family.

*2. Mossy stonecrop *Sedum acre*
Plants forming moss-like mats, flowering stems 1–4″. **Leaves** oval, ⅛–¼″, blunt. Flowers on 2 short-branched cymes, ⅜″. Rocks; walls; dry, open places. See also pp. 284, 354. Stonecrop family.

a. Stamens many (occasionally 10 in *Helianthemum***)**
b. Stamens fused around style (Mallow family)
*3. Velvet-leaf *Abutilon theophrasti*
Plants 2–6′, soft-hairy. **Leaves** heart-shaped, 4–6″, sharp-pointed, toothless or toothed, velvety-hairy. Flowers single in axils, or clustered, ½–1½″. Fields, weedy places.

*4. Prickly sida *Sida spinosa*
Plants 8–40″. **Leaves narrowly oval, oblong, or elliptic, ¼–1½″,** toothed, truncate or notched at base, stalked. Flowers clustered in leaf axils, ¼–½″. Fields, roadsides, weedy places. Note the spines at the base of each leaf.

b. Stamens free of style (Poppy and Rockrose families)
*5. Mexican prickly-poppy *Argemone mexicana*
Plants 10–32″. Leaves lobed, 2½–8″, clasping, usually blotched with paler green. **Flowers** terminating branches, 1½–2½″. Weedy areas, roadsides, fields. Poppy family.

Rockroses superficially look like roses. The first flowers of the season open only in sun for a single day, these are showy; later they produce closed flowers that are self-fertilizing (See inset in no. 6). Critical characters are flower number and habit. Rockrose family.

6. Frostweed *Helianthemum canadense*
Plants erect, 6–24″, later much branched, **branches strongly ascending.** Leaves of main stem narrowly oval, broader toward either end, 1–1½″, densely hairy. **Flowers solitary, terminal,** ¾–1″; calyx hairy. Dry sands, rocky woods, clearings, barrens.

7. Bushy rockrose *Helianthemum dumosum*
Plants <4″, later ≤12″ **with widely spreading branches.** Leaves similar to no. 6, but smaller, densely hairy above. **Flowers solitary or paired at tips,** ¾–1″; calyx hairy. Dry, sandy soil; barrens.

A. Bicknell's rockrose *Helianthemum bicknellii* (not illus.)
Differs from nos. 6 and 7 in having panicles of 2–10 flowers. Dry, usually sandy soil.

1. Common purslane

*2. Mossy stonecrop

*3. Velvet-leaf

*4. Prickly sida

*5. Mexican prickly-poppy

6. Frostweed

7. Bushy rockrose

LEAVES alternate or basal
LEAVES simple
PETALS 6

Fairy-bells, Bellworts, and Bog-asphodel
Colchicum and Bog-asphodel families

Bellworts characteristically have long, narrow, nodding flowers. They appear in spring, before the trees have grown leaves; once the leaves have matured, the bellworts die back. They are pollinated by bumblebees that enter the flower, grasp the stigma, and vibrate (buzz) the flower to release the pollen. As with many spring forest wildflowers, their seeds are collected by ants which take them back to their nest, removing a food body attached to the seed and dropping the seed outside their nest.

a. Leaves alternate (Colchicum family)
    b. Leaves not clasping the stem
1. Yellow fairy-bells *Disporum lanuginosum*
Plants 12–36″, downy. Leaves oval or narrowly oval, 2–4¾″, sharply pointed, stalkless, woolly below. **Flowers 1–3 in clusters at ends of branches**, ½–1″. See also p. 286. Rich woods.

2. Sessile bellwort *Uvularia sessilifolia*
Plants **solitary**, 4–13″, hairless. Leaves 0–2 below fork, narrowly oval, **gray-green below**, pointed at both ends. **Flowers nodding, solitary**, ½–1″ long. Woods, thickets, clearings.

3. Mountain bellwort *Uvularia puberula*
Plants **clumped**, 4–18″, slightly hairy or hairless. Leaves 0–2 below fork, narrowly oval, **bright-green**, pointed, rounded at base. **Flowers nodding, solitary**, ½–1¼″ long. Mountain woods. **Similar to no. 2, but leaves shiny and plants clumped.**

    b. Leaves clasping the stem
4. Large-flowered bellwort *Uvularia grandiflora*
Plants 6–26″. **Leaves** 1–2 below fork; 4–8 leaves on sterile branches, green, broadly oval or oblong, 1¾–5″, clasping, the leaves wrapping about the stem and appearing to be perforated by the stem, **white downy below**. Flowers nodding, solitary, 1–2″ long; petals twisted, hairless within. Rich woods, thickets.

5. Perfoliate bellwort *Uvularia perfoliata*
Plants 6–24″. **Leaves** 2–4 below fork; 1–4 leaves on sterile branches, gray-green, oval or narrowly oval, 1½–4¼″, clasping, the leaves wrapping about the stem and appearing to be perforated by the stem, **hairless**. Flowers nodding, solitary, ¾–1½″ long; petals sticky-hairy within. Moist, thin woods; thickets; clearings.

a. Leaves basal (Bog asphodel family)
6. Bog-asphodel *Narthecium americanum*
Plants 10–18″. Basal leaves linear, 4–8″; stem leaves few, bract-like. Flowers on crowded racemes, ¾–2½″ tall; flowers ½″. Pine barren bogs.

1. Yellow fairy-bells

2. Sessile bellwort

3. Mountain bellwort

4. Large-flowered bellwort

5. Perfoliate bellwort

6. Bog-asphodel

LEAVES alternate and basal
LEAVES simple
RAYS many

Goldenrods with heads in axils, racemes, or spikes (1)
Composite family
Goldenrods are easily recognized by their small, yellow, composite heads and alternate or basal, simple leaves. They are related to the asters. Critical characters are inflorescence shape, and leaf size and shape. Goldenrods are an important late-season nectar source for a wide variety of insects. See white-flowered goldenrods p. 290.

a. Heads in axillary clusters

1. Zigzag goldenrod *Solidago flexicaulis*
Plants 1–4', **tending to zigzag. Leaves oval to elliptic, 2¾–6″**, sharp-pointed, toothed, hairy below. Heads in short axillary clusters and a terminal cluster; heads ¼″; **rays 3–4**. Rich woods, thickets, cool slopes. One of the few goldenrods found in shaded places.

2. Blue-stemmed goldenrod *Solidago caesia*
Plants 1–3', often bluish. **Leaves narrowly oval or narrowly elliptic,** 1½–7″, sharp-pointed, narrowed to base, toothed, hairless or nearly so. Heads in axillary clusters; heads ¼″; **rays 3–8**. Rich woods. One of the few goldenrods found in shaded places.

A. New England goldenrod *Solidago cutleri* (not illus.)
Plants 2–14″. **Leaves mostly near base, spoon-shaped, elliptic, or narrowly oval, broader toward tip,** ¾–6″, blunt or pointed. Heads in axillary clusters and a terminal cluster; heads ⅛–¼″; **rays 10–13**. Rocky, alpine areas.

a. Heads on wand-like or club-shaped inflorescences (cont. on next page)
   b. Leaves broad, generally elliptic or oval
3. Downy goldenrod *Solidago puberula*
Plants 1–3', **hairy;** stems purplish. Leaves narrowly oval, elliptic, or oval, broader toward tip, 2–8″, pointed or blunt, sharp-toothed. Heads on elongate, narrow racemes; heads ¼″; rays 9–16. Open sands, rocks.

   b. Leaves narrow, generally narrowly oval or narrowly elliptic
4. Bog goldenrod *Solidago uliginosa*
Plants ½–5', **hairless. Leaves** narrowly oval, broader toward either end, or narrowly elliptic, 2½–18″, **pointed; upper leaves gradually reduced.** Heads on elongate, sometimes branched, racemes, 1–18″; branches straight or recurved; heads ¼″; rays 1–8. Bogs.

5. Wand goldenrod *Solidago stricta*
Plants 1–8', **hairless. Leaves** narrowly oval, broader at either end, or elliptic, 2½–12″, **blunt; upper leaves abruptly reduced to numerous bracts.** Heads on narrow, elongate, wand-like spikes, sometimes nodding at the tip; heads ¼″; rays 3–7. Sandy, moist places, especially among pines.

1. Zigzag goldenrod

2. Blue-stemmed goldenrod

3. Downy goldenrod

4. Bog goldenrod

5. Wand goldenrod

LEAVES alternate
LEAVES simple
RAYS many

Goldenrods with pyramidal or nodding inflorescences (2)
Composite family
a. Heads on pyramidal panicles or panicles nodding at tip (cont.)
b. Leaves 3-nerved

1. Canada goldenrod *Solidago canadensis*
Plants 1–6½', **hairless below, downy above the middle; stem green. Leaves mostly on stem, linear, narrowly oval, or elliptic,** 1¼–6", tapering at both ends. Heads on panicles with recurved one-sided branches, 2–16" tall; heads <⅛". Open places, thin woods.

2. Smooth goldenrod *Solidago gigantea*
Plants 1–8', **hairless; stem whitish. Leaves mostly on stem, linear, narrowly oval or narrowly elliptic,** 2½–6¾", sharp-pointed, sharp-toothed. Heads on pyramidal panicles with recurved, one-sided branches, 2–20" tall; heads ¼"; rays 9–17. Moist, open places.

3. Rough-leaved goldenrod *Solidago patula*
Plants 1½–7', **hairless below. Leaves mostly near base, elliptic, oval, broader toward either end,** 3–16", tapering abruptly to the stalk, toothed, rough above, smooth below. Heads on panicles; branches wide-spreading one-sided; heads ⅛". Swamps, meadows.

4. Gray goldenrod *Solidago nemoralis*
Plants ½–4½', **densely gray-hairy. Leaves mostly near base, narrowly oval, broader toward tip,** 2–10", tapering to base, toothed. Heads on narrow to broad panicles; branches recurved, one-sided; heads ¼–⅜". Dry woods, fields.

b. **Leaves not 3-nerved; plants hairy or rough**
5. Sweet goldenrod *Solidago odora*
Plants 2–5', rough hairy, **anise-like odor. Leaves linear,** 1½–4¾", stalkless, toothless, hairy below, dotted. Heads on panicles with recurved, one-sided branches; heads ⅜". Dry, open woods; sandy soil.

6. Wrinkle-leaved goldenrod *Solidago rugosa*
Plants 1–6½', spreading hairy or rough. **Leaves elliptic, narrowly oval, or oval,** 1½–5", **not clasping,** rough, **toothed,** hairy on ribs below. Heads on panicles with recurved, one-sided branches, 2–16" tall; heads ¼"; rays 6–11. Thickets, roadsides, banks, open places.

7. Pine-barren goldenrod *Solidago fistulosa*
Plants 2–6½', spreading hairy. **Leaves narrowly oval or elliptic,** 1½–4¾", **broad-based and clasping, toothless or nearly so,** hairy on midrib below. Heads on dense panicles with recurved, one-sided branches; heads ¼"; rays 7–12. Wet or dry sandy places; pinelands.

b. **Leaves not 3-nerved; plants hairless**
c. **Leaves mostly near the base** (cont. on next page)
8. Seaside goldenrod *Solidago sempervirens*
Plants 1–8', **succulent,** hairless. Leaves narrowly oval, broader toward tip, 4–16", toothless. Heads on dense panicles, 2–18" tall; heads ¼". **Salty places.**

1. Canada goldenrod

2. Smooth goldenrod

3. Rough-leaved goldenrod

4. Gray goldenrod

5. Sweet goldenrod

6. Wrinkle-leaved goldenrod

7. Pine-barren goldenrod

8. Seaside goldenrod

Goldenrods with pyramidal or flat-topped inflorescences (3)
Composite family
c. Leaves mostly near the base (cont. from previous page)

1. Early goldenrod *Solidago juncea*
Plants 1–4′, hairless. **Leaves** narrowly elliptic or narrowly oval, broader toward tip, 6–16″, **tapering to stalk**, sharp-toothed, hairless, small leaves in axils of leaves. Heads on dense panicles with recurved, one-sided branches; heads ¼″. Dry fields, open woods.

2. Elm-leaved goldenrod *Solidago ulmifolia*
Plants 1–5′, hairless below inflorescence. **Leaves** elliptic or oval, broader toward tip, 2½–4¾″, **abruptly contracted to stalk,** coarsely toothed, soft-hairy below. Heads on panicles with recurved one-sided branches, 1½–16″ tall; heads ⅛″. Dry woods.

c. Leaves mostly on the stem
3. Showy goldenrod *Solidago speciosa*
Plants 1–6′, reddish, hairless. Leaves mostly near the base, elliptic or oval, broader toward tip, 2¾–12″, pointed. Heads on narrow pyramidal panicles; heads ¼″; rays 5–9. **Woods, fields, prairies.**

4. Coastal swamp goldenrod *Solidago latissimifolia*
Plants 1½–10′, hairless. Leaves elliptic or narrowly oval, broader at either end, 2–6″, short-pointed, rounded at base, toothed or toothless, stalkless or nearly so. Heads on pyramidal panicles with short or elongate, slightly recurved, one-sided branches; heads ¼″; rays 6–12. **Fresh or brackish swamps.**

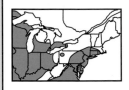

a. Heads on flat-topped panicles
5. Stiff goldenrod *Oligoneuron rigidum*
Plants 1–5′, downy to hairless. **Leaves elliptic, oblong, or oval,** 2¼–10″, rounded or pointed, toothless or nearly so, rough; upper leaves rigid and oval. Heads on dense compound flat-topped panicles, 2–10″ wide; heads ¼–½″; rays 7–14. See also p. 290. Prairies; dry, open places.

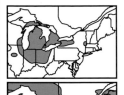

6. Ohio goldenrod *Oligoneuron ohioense*
Plants 1–2½′, hairless. **Leaves narrowly elliptic or narrowly oval,** 3–9″, **3-nerved,** toothless or nearly so. Heads many on dense compound flat-topped panicles; heads ¼″; rays 6–8. Swamps, beaches, moist places.

7. Common flat-topped goldenrod *Euthamia graminifolia*
Plants 1–5′, hairless or hairy. **Leaves narrowly oval,** 1-6″, **3–5-nerved**, toothless, **with resinous dots**. Heads on flat-topped panicles; heads ⅛″; rays 15–25. Open, usually moist ground.

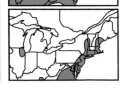

8. Coastal plain flat-topped goldenrod *Euthamia tenuifolia*
Plants 1–3′, hairy or hairless. **Leaves linear,** ¾–6″, **1-nerved,** toothless**, with resinous dots**. Heads on flat-topped panicles; heads ¼″; rays 6–15″. Open, sandy places.

1. Early goldenrod

2. Elm-leaved goldenrod

3. Showy goldenrod

5. Stiff goldenrod

6. Ohio goldenrod

4. Coastal swamp goldenrod

7. Common flat-topped goldenrod

8. Coastal plain flat-topped goldenrod

LEAVES alternate
LEAVES simple
RAYS many

Miscellaneous composites with disks and rays (1)
Composite family

a. Heads with dome-like disks
Sneezeweeds get their name from their previous use as snuff. Critical characters are head color, ray number, and leaf shape and size.

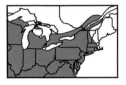

1. Common sneezeweed *Helenium autumnale*
Plants 2–6′, stem winged. Leaves linear, elliptic, or narrowly oval, 1½–6¼″, toothed. Heads several to many in hemispheric or spherical clusters; heads ¾–2″; **disk yellow**, ¼–1″; **rays** turning back, **13–21**, 3-lobed. Rich thickets, swamps, meadows, shores.

2. Narrow-leaved sneezeweed *Helenium amarum*
Plants 8–24″, hairless. Leaves linear, ½–3″. Heads on long stalks. ½–1″; **disk yellow**, ¼–½″; **rays 5–10″**. Prairies, open woods, fields, weedy places. Differs from nos. 1 and 3 in having leaves <¼″ wide.

3. Purple-headed sneezeweed *Helenium flexuosum*
Plants 8–40″, minutely hairy; stem winged. Leaves oblong, linear, or narrowly oval, broader toward either end, 1¼–4¾″. Heads, many in flat-topped clusters; heads ¾–1¼″; **disk red-brown or purple**, ¼–½″; rays 8–13, 3-lobed. Openings, meadows, weedy places, fields.

a. Heads with flat disks
  b. Flowers ≥ 1″

*4. Curly-top gum-weed *Grindelia squarrosa*
Plants 4–40″. **Leaves** linear to oval, ½–2¾″, **resin-dotted**. Heads several to many; **involucral bracts reflexed and gummy**, 1–2″; disk ⅜–¾″; rays 25–40. Weedy places, roadsides, fields.

*5. Elecampane *Inula helenium*
Plants ≤7′, spreading-hairy. **Leaves** elliptic or oval, ≤20″, **densely woolly below**. Heads few, 2¼–4″; disk 1¼–2″; rays numerous. Moist and wet, disturbed sites.

  b. Flowers ≤ 1″
6. Falcate golden-aster *Pityopsis falcata*
Plants 4–14″, white-woolly. **Leaves linear, ≤4″**, hairless, often folded. Heads several at ends of branches, ¾″; rays ⅜″. Dry sands.

7. Camphor-weed *Heterotheca subaxillaris*
Plants 8–36″, sticky-hairy. **Leaves oval or oblong**, ¾–3¾″, broad-based, wavy-edged, **stalkless, clasping**. Heads many on flat-topped panicles; heads ½–¾″, 15–30. Dry, sandy places; roadsides.

8. Shaggy golden-aster *Chrysopsis mariana*
Plants 8–32″, hairy. **Leaves oblong or narrowly oval, broader toward tip**, ≤ 2½″, blunt or rounded, toothless, **stalked**. Heads often congested, ¾–1″; disk ⅜–¾″; rays 13–21. Woods, sandy places.

1. Common sneezeweed

2. Narrow-leaved sneezeweed

3. Purple-headed sneezeweed

*4. Curly-top gum-weed

*5. Elecampane

6. Falcate golden-aster

7. Camphor-weed

8. Shaggy golden-aster

LEAVES alternate or basal
LEAVES simple or lobed
RAYS many

Miscellaneous composites with rays only
Composite family

a. Seeds with long hairs (pappus)
  b. Plants hairy but not sticky-hairy

1. Canada hawkweed *Hieracium kalmii*
Plants ½–5′, **white-hairy**. Leaves elliptic, oval, narrowly oval, or oblong, 1¼–4¾″, cleft or entire, toothed, stalkless; upper leaves with rounded or notched base. **Heads in a loose, flat-topped cluster**, sometimes solitary; heads 1″; rays 40–110. See also pp. 136, 220. Woods, beaches, fields.

2. Hairy hawkweed *Hieracium gronovii*
Plants 1–5′, **rough, white-hairy**. Leaves mostly below the middle of the stem, elliptic, oval, or narrowly oval, broader toward tip, 1½–8″, rounded or blunt. **Heads on elongate cylindrical panicles;** heads ½–1″; rays 20–40. Dry, open woods.

3. Long-haired hawkweed *Hieracium longipilum*
Plants 2–7′, **densely, long, rusty-hairy**. Leaves chiefly crowded at the base, narrowly elliptic or narrowly oval, broader toward tip, 3½–12″, pointed, long-hairy. **Heads on elongate cylindrical panicles**, hairy; heads ½–1″; rays 40–90. Dry prairies, open woods, fields.

4. False dandelion *Pyrrhopappus carolinianus*
Plants 4–40″, **white-hairy**. Leaves oblong or narrowly oval, broader toward tip, entire or lobed, ≤ 10″. **Heads 1–few at ends of branches;** heads ⅜–1″. Fields, dry woods, bottomlands, weedy places. **Note the seed has a long beak.**

  b. Plants sticky-hairy

*5. Stinkwort *Dittrichia graveolens*
Plants ≤3′, sticky-hairy, stinking. **Leaves** linear or narrowly elliptic, ≤4″, **toothless or nearly so.** Heads many on diffuse panicle; **heads** ¼″. Weedy areas. Note the long hairs on seeds of fruiting heads.

*6. Smooth hawksbeard *Crepis capillaris*
Plants 4–36″, sticky-hairy. **Leaves** narrowly oval, broader at either end, ≤ 12″, stalked, hairless or hairy, **toothed or lobed.** Heads several–many; heads ½–¾″. Meadows, pastures, lawns, weedy places.

a. Seeds without long hairs

7. Orange dwarf dandelion *Krigia biflora*
Plants 8–32″, hairless. **Leaves elliptic or narrowly oval, broader toward tip**, sometimes lobed, 1½–10″, toothed or toothless; upper leaves oval, stalkless and clasping. **Heads several, 1–1½″.** See also p. 224. Woods, prairies, roadsides, fields.

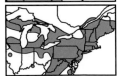

*8. Nipplewort *Lapsana communis*
Plants ½–5′, hairy or hairless. **Leaves oval or round,** 1–4″, blunt or rounded, toothed or lobed. **Heads** on loose panicles, ¼–½″. Woods, fields, weedy places.

1. Canada hawkweed

2. Hairy hawkweed

3. Long-haired hawkweed

4. False dandelion

*5. Stinkwort

*6. Smooth hawksbeard

7. Orange dwarf dandelion

*8. Nipplewort

**LEAVES alternate or opposite
LEAVES simple
RAYS many**

Large-flowered, alternate-leaved composites
Composite family
Coneflowers have a conical disk, see p. 178 for other coneflowers with lobed leaves. Sunflowers are similar but have a flat or button-shaped disk. Sunflowers generally have opposite leaves, but often the upper leaves are alternate. See pp. 202–4 for other sunflowers.

a. Disk flat

*1. Common sunflower *Helianthus annuus*
Plants 1–12′, rough-hairy. **Leaves** alternate except lowermost, **oval**, 4–16″, long-stalked, rough on both sides. Heads solitary; involucral bracts broadly oval; **heads 3–6″; disk reddish, 1–2″;** rays >12. Weedy areas; moist, low areas.

*2. Common blanket-flower *Gaillardia aristata*
Plants ¾–2′, hairy. **Leaves all alternate, linear or narrowly oval, broader at either end,** ≤6″, toothless or sometimes lobed. Heads 1–few, long-stalked; involucral bracts narrowly triangular, hairy; heads purple to yellow-orange, 1¼–4″; **disk purple or brown,** ½–1¼″; rays 6–16. Garden escape. See page 80 for another gaillardia.

a. Disk dome-shaped or cone-shaped
  b. Disk reddish, purple, or brown

3. Black-eyed Susan *Rudbeckia hirta*
Plants 1–3′, rough-hairy. Leaves elliptic, narrowly oval, or oval, broader at either end, 2–5½″, coarsely toothed. Heads solitary; **involucral bracts** narrowly oval or linear, **spreading, hairy; heads 2–4″;** disk ½–¾″; rays 8–15. Meadows, pastures, fields, roadsides, woods.

4. Eastern coneflower *Rudbeckia fulgida*
Plants 1–4′, hairy. Leaves elliptic or narrowly oval, ½–4″, few-toothed or toothless. Heads solitary; **involucral bracts** narrowly oval, **reflexed,** hairy; **heads 1–2″;** disk ⅜–½″; rays 8–21. Woods, moist places.

A. Piney-woods coneflower *Rudbeckia heliopsidis* (not illus.)
Plants 2–3′, hairy or not. Leaves oval or narrowly oval, 2½–4″, toothed or toothless. Heads several; **involucral bracts** narrowly oval, **spreading or reflexed,** hairy; **heads 1½–2½″;** disk ⅜–½″; rays 13. Pine and oak-hickory woods. Difficult to distinguish from no. 4, though generally with longer-stalked, broader leaves.

b. Disk yellow

5. Wingstem *Verbesina alternifolia*
Plants 3–10′, hairy; stem winged. Leaves narrowly elliptic, narrowly oval, or oval, 4–10″, pointed at both ends. Heads 20–100 on an open panicle; **involucral bracts** narrowly oval; **reflexed; heads 1–2″;** disk ⅜–½″; rays 2–10, **drooping.** Thickets, woods, bottomlands.

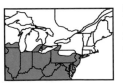

*1. Common sunflower

*2. Common blanket-flower

3. Black-eyed Susan

4. Eastern coneflower

5. Wingstem

LEAVES alternate
LEAVES simple or lobed
PETALS irregular

Miscellaneous irregular flowers
Several families
a. Lower petal represented by pouch-like lip (Orchid family)
Lady-slippers are distinguished by the pouch-like lip. See other lady-slippers pp. 86, 136, 298.

1. Yellow lady-slipper *Cypripedium parviflorum*
Plants 8–32″. Leaves oval or narrowly oval, 2½–8″. Flowers 1–2, terminal, **2½–6″;** sepals and lateral petals greenish-yellow to purplish-brown; sepals oval or narrowly oval, 1¼–3″; pointed, lateral petals narrowly oval, 1¼–3″, usually twisted; lip yellow with purple veins, ¾–2″. Woods, bogs, shady swamps.

Yellow lady-slipper comes in two forms, the Large yellow lady-slipper (*Cypripedium parviflorum* var. *pubescens*), 1a, has lips ¾–2″ and the sepals and petals are less suffused with purple. It grows throughout the range. The Small yellow lady-slipper (*Cypripedium parviflorum* var. *makasin*), 1b, has lips ½–1¼″ and the sepals and petals are shorter and deeply suffused with purple. It is found in the northern half of the range.

2. Ivory lady-slipper *Cypripedium kentuckiense*
Plants 24–30″. Leaves 3–6, oval, narrowly oval or elliptic, 5–10″. Flowers 1–2, terminal; **flowers 6½–7″;** upper sepal 2¾–3¼″; the lower 2 fused, 2–2¾″; lateral petals long-pendant, 3¼–3½″; lip ivory or dull yellow, 2–2¼″. Ravines, low woods.

a. Lower petal not pouch-like (cont. on next page)
  b. Plants ≤1½′ (Bean and Violet families)
3. Weedy rattlebox *Crotalaria sagittalis*
Plants erect, 4–16″, hairy. **Leaves linear or narrowly oval, 1¼–3″.** Flowers 2–4 on racemes attached opposite the leaves, ¼–½″; bean flower standard large. Dry, open areas; weedy areas. The dry seeds rattle in the pods. Bean family.

Yellow violets Mainly pollinated by small, greenish mason bees. These and other violets produce unopened flowers during the summer at the base of the plant, these flowers produce seed by self-fertilization. See other violets pp. 12, 58, 216, 296, and 378. Violet family.

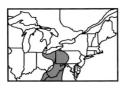

4. Spear-leaved yellow violet *Viola hastata*
Plants 2–12″, hairless or slightly hairy. Leaves 2–4 near top of stem, triangular, arrow-shaped, or heart-shaped, ¾–3″, sharp-pointed, dark green with silvery blotches. Flowers on axillary stalks, ¼–½″; sepals narrowly oval; petals yellow with brown-purple lines. Rich, deciduous woods; ravines.

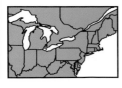

5. Downy yellow violet *Viola pubescens*
Plants 4–18″, softly hairy to hairless. Leaves 2–4 near the top, kidney-shaped, 2–4¾″; stem leaves round to broadly oval; stipules toothed. Flowers axillary, ⅜–¾″; petals clear yellow with brown-purple veins. Rich woods, meadows.

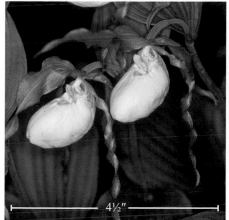

1a. Large yellow lady-slipper

1b. Small yellow lady-slipper

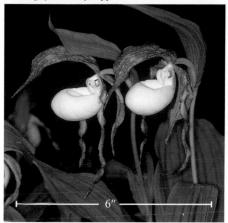

2. Ivory lady-slipper

3. Weedy rattlebox

4 Spear-leaved yellow violet

5. Downy yellow violet

LEAVES alternate
LEAVES simple or lobed
PETALS irregular

Miscellaneous irregular flowers
Several families

a. Lower petal not pouch-like (cont. from previous page)
b. Plants ≥1½' (several families)
1. Yellow touch-me-not *Impatiens pallida*
Plants erect, succulent, 2–6'. Leaves oval or elliptic, 1¼–4".
Flowers few in upper axils, 1–1½"; spur ⅛–¼". See also p. 136.
Wet woods, meadows. Touch-me-not family.

*2. Birthwort *Aristolochia clematitis*
Plants ascending or reclining vines, 1½–3'. Leaves heart-shaped, 2–4", blunt or broadly pointed, pale below. Flowers in axillary clusters, 1¼". Near gardens. See also p. 260. Dutchman's-pipe family.

3. Pipe-vine *Aristolochia tomentosa*
Plants woody vines. Leaves heart-shaped, 4–8", hairy below.
Flowers axillary, "pipe"-shaped, yellow-green with purple lobes, 1½". Floodplain woods. Caterpillars of Pipevine Swallowtail butterflies feed on pipe-vines. Dutchman's-pipe family.

*4. Straw foxglove *Digitalis lutea*
Plants erect, 2–3', hairless. Leaves oblong or narrowly oval, 4–6".
Flowers many on one-sided racemes; flowers ½–¾". See also p. 86.
Garden escape. Plantain family.

5. Northeastern paintbrush *Castilleja septentrionalis*
Plants 6–24", hairless except woolly in inflorescence. Leaves linear or narrowly oval, 1¼–4". **Flowers on compact spikes;** bracts pale yellow or white; flowers ½–1". Damp, rocky soil; gravelly soil. See also p. 84. Broom-rape family.

a. Flowers without petals
b. Leaves oval (Evening-primrose family)
6. Round-pod water-purslane *Ludwigia sphaerocarpa*
Plants profusely branched, 1–3', hairless or hairy. Leaves narrowly oval or linear, 2–4", tapering to base. Flowers axillary, ¼", **not surrounded by bracts.** Swamps and wet soils.

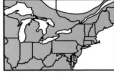

b. Leaves linear (Cat-tail family)
7. Common cat-tail *Typha latifolia*
Plants 3–10'. **Leaves** linear ⅜–1" wide. Flowers on dense spikes, the **male flowers** above, **not separated from the female part below.** Marshes.

8. Narrow-leaved cat-tail *Typha angustifolia*
Plants 3–5'. **Leaves** linear ¼–⅜" wide. Flowers on dense spikes, the **male flowers** above, **separated from the female part below.**
Marshes, sometimes saline areas.

1. Yellow touch-me-not

*2. Birthwort

3. Pipe-vine

*4. Straw foxglove

5. Northeastern paintbrush

6. Round-pod water-purslane

7. Common cat-tail

8. Narrow-leaved cat-tail

LEAVES alternate
LEAVES simple and
compound
PETALS 5

Alexanders, Parsnips, and Pimpernels
Umbel family

These species have an umbel inflorescence arranged like the spokes of an umbrella. Critical characters are in the fruit (not used here); less so, in leaf number and shape and umbel shape. See white umbels pp. 308–12.

a. Basal leaves heart-shaped

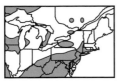

1. Heart-leaved golden alexanders *Zizia aptera*
Plants 1–2½'. Basal leaves heart-shaped, ¾–2½"; stem leaves compound; leaflets 3 or 5, narrowly oval or oval, broader toward tip; leaflets, ¾–2½", toothed. Flowers on umbels with several to many spokes, ½–2½" wide; flowers ⅛"; **central flower unstalked.** Moist meadows, thickets, open woods.

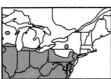

2. Smooth meadow-parsnip *Thaspium trifoliatum*
Plants ¾–3', hairless. Basal leaves heart-shaped, ¾–2½"; stem leaves compound, leaflets 3, oval or narrowly oval, 1½–3", toothed. Flowers on umbels with 6–10 spokes, ½–3" wide; flowers yellow or purple, ⅛"; **all flowers stalked.** Woods. Yellow flowers are more common in the West; purple flowers are more common in the East,

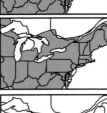

a. Basal leaves compound

3. Golden alexanders *Zizia aurea*
Plants 1–3', hairless. **Leaflets 9–27,** oval or narrowly oval, **toothed.** Flowers on umbels with 6–20 spokes, 1-2" wide; flowers ⅛"; **central flower stalkless.** Moist fields, meadows, shores, wet woods.

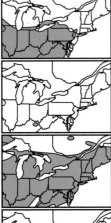

4. Bearded meadow-parsnip *Thaspium barbinode*
Plants 1–4', **hairy at joints. Leaflets 9–27,** oval or narrowly oval, ½–2½", **toothed.** Flowers on umbels with 8–16 spokes, 1¼–2½" wide; flowers ⅛"; **all flowers stalked.** Woods, thickets, talus.

5. Mountain meadow-parsnip *Thaspium pinnatifidum*
Plants ≤32", minutely hairy at joints. **Leaflets many,** narrow, ≤⅛" wide, **toothed.** Flowers on umbels ¾–2" wide; flowers ⅛"; **all flowers stalked.** Rich mountain woods.

*6. Parsnip *Pastinaca sativa*
Plants 2–5', grooved. **Leaflets 5–15,** oblong or oval, 2–4", **toothed.** Flowers on umbels with 15–25 spokes, 2–8"; flowers ⅛". Fields, roadsides, weedy places.

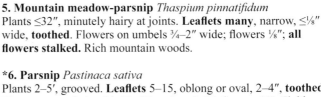

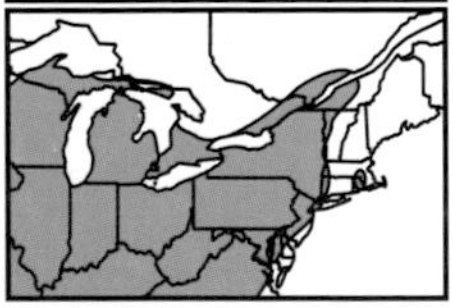

7. Yellow pimpernel *Taenidia integerrima*
Plants 1–3', hairless, **smelling of celery. Leaflets 9–27,** narrowly oval, oval, oblong, or elliptic, **toothless,** ⅜–1¼". Flowers on umbels with 7–16 spokes; flowers ⅛". Dry woods; open, rocky slopes.

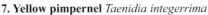

8. Shale-barrens pimpernel *Taenidia montana*
Plants 1–3', hairless, **smelling of anise. Leaflets 9–27,** narrowly oval, oval, oblong, or elliptic, **toothless,** ⅜–1¼". Flowers on umbels with 7–16 spokes; flowers ⅛". Dry woods; open, rocky slopes.

1. Heart-leaved golden alexanders

2. Smooth meadow-parsnip

3. Golden alexanders

4. Bearded meadow-parsnip

5. Mountain meadow-parsnip

*6. Parsnip

7. Yellow pimpernel

8. Shale-barrens pimpernel

LEAVES alternate or basal
LEAVES compound
PETALS 5

Cinquefoils
Rose family

Cinquefoils have numerous stamens, often 5 leaflets, and fruits lacking bristles or long hairs (compare next page). See a dwarf form with basal leaves only p. 222, white-flowered cinquefoils p. 294.

a. Leaves with 3 leaflets from a single point

1. Rough cinquefoil *Potentilla norvegica*
Plants 1–3′, mostly branched. **Leaflets 3**, elliptic or oval, broader toward tip, ≤3″, toothed. **Flowers many, ¼–½″**. Clearings, thickets, roadsides, weedy places.

a. Leaves with 5 leaflets from a single point

2. Common cinquefoil *Potentilla simplex*
Plants erect, eventually arching to ground and forming runners, ½–4′. Flowers and leaves on different stems. Leaflets 5, elliptic or narrowly oval, broader toward tip, ½–3″, **toothed for most of length. Flowers solitary** in axils, ⅜–½″. Dry woods, fields

3. Dwarf cinquefoil *Potentilla canadensis*
Plants erect, becoming prostrate with runners, 2–6″, densely silvery hairy. Leaflets 5, narrowly oval, broader toward tip, ½–2″, **without teeth below the middle. Flowers solitary** on long stalks from leaf axils, ¼–½″. Dry fields, woods.

*4. Silvery cinquefoil *Potentilla argentea*
Plants **freely branching,** depressed or ascending, 4–18″, **white-woolly**. Leaflets 5, linear or narrowly oval, broader toward tip, ½–1¼″, silvery beneath, deeply toothed, with inrolled margins. **Flowers in a cluster** at ends of white-woolly branches, ¼–½″. Dry fields, weedy places.

*5. Downy cinquefoil *Potentilla intermedia*
Plants **freely branching,** 1–6″ becoming 12–24″, **gray-hairy**. Leaflets 5 or 3, 1¼–2″, deeply and irregularly toothed. **Flowers on a diffuse panicle**, ¼–½″. Dry roadsides, weedy places.

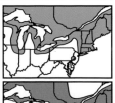

*6. Rough-fruited cinquefoil *Potentilla recta*
Plants erect, unbranched, 1–2′. Leaflets 5–6, narrowly oval, broader toward tip, 1–6″, deeply toothed. **Flowers on a flat-topped panicle**, ½–1″; petals notched, light yellow. Dry roadsides, weedy places.

a. Leaves with 5 or more leaflets arranged along a stalk

7. Silverweed *Argentina anserina*
Plants prostrate, 1–3′, with leaves and flowers on separate stalks. Basal leaves ≤12″; leaflets 7–25, narrowly oval, broader toward tip, ≤1½″, silvery-hairy beneath; stem leaves smaller. Flowers solitary, ¾–1″. Gravelly or sandy shores, banks, salt-marshes.

8. Shrubby cinquefoil *Potentilla fruticosa*
Plants shrubby, ≤3′. Leaflets 5–7, pinnately arranged, narrowly oval, broader at either end, ⅜–¾″, toothless, whitish below. Flowers solitary or few, ¾–1¼″. Wet meadows, bogs, shores.

1. Rough cinquefoil

2. Common cinquefoil

3. Dwarf cinquefoil

*4. Silvery cinquefoil

*5. Downy cinquefoil

*6. Rough-fruited cinquefoil

7. Silverweed

8. Shrubby cinquefoil

LEAVES alternate
LEAVES compound
PETALS 5

Agrimonies and Avens
Rose family

a. Flowers on spikes

Agrimonies have irregular leaflets and bristles on the outside of the fruits. Critical characters are leaflet number and hairs.

1. Common agrimony *Agrimonia gryposepala*
Plants 1–5', long-hairy. **Leaflets 5–9,** narrowly oval or oval, broader toward tip, sticky-hairy above, **hairless below. Flowers on sticky spikes,** ⅛–¼". Moist and dry woods, thickets, borders.

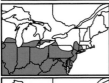

2. Southern agrimony *Agrimonia parviflora*
Plants ≤6', densely hairy. **Leaflets 11–23,** narrowly oval, sharp-pointed, **sticky. Flowers on sticky spikes,** ⅛". Damp woods, thickets, rocky slopes.

3. Downy agrimony *Agrimonia pubescens*
Plants 1–5', densely short-hairy. **Leaflets 5–13,** oblong or elliptic, blunt, hairless above, **velvety-hairy below. Flowers on hairy spikes,** ⅛". Dry woods, ledges.

4. Roadside agrimony *Agrimonia striata*
Plants 1–6', hairy below. **Leaflets 7–11,** narrowly oval, **sticky and slightly hairy below. Flowers on densely crowded, hairy spikes,** ¼". Dry and moist woods, thickets, borders.

a. Flowers in clusters

Avens have irregular leaves and long styles that form stick-tights. Critical characters are in the leaf shape and styles. See red avens p. 90; white-flowered avens p. 302

5. Yellow avens *Geum aleppicum*
Plants 1–4', hairy. Leaves highly variable; lobes and leaflets mostly oval, broader toward tip, tapering to base; lower leaves long-stalked; principal leaflets mixed with minute leaflets. **Flowers ½–1", petals equal to or longer than sepals.** Swamps, thickets, wet meadows.

6. White-Mountain avens *Geum peckii*
Plants 6–24", **hairless.** Basal leaves shallowly 5–7-lobed; terminal lobe 2–4" wide; other leaflets absent or up to 6, appearing torn; stem leaves similar to lateral leaflets of basal leaves. **Flowers 1–5, ½–1"; petals longer than sepals.** Damp slopes, gravels, cliffs.

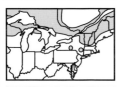

7. Big-leaved avens *Geum macrophyllum*
Plants 1–3', **bristly-hairy.** Basal leaves long-stalked; terminal leaflet 2–4½" wide, 3-lobed; lateral leaflets few, much smaller, interspersed with minute leaflets; upper leaves short-stalked or stalkless. **Flowers ¼–½"; petals much longer than sepals.** Rich woods, thickets, rocky ledges.

8. Spring avens *Geum vernum*
Plants 6–24", **slightly hairy.** Basal leaves long-stalked, unlobed or nearly so; stem leaves pinnately compound or with 3 leaflets. **Flowers yellow or white, ¼"; petals equalling sepals.** Woods.

1. Common agrimony

2. Southern agrimony

3. Downy agrimony

4. Roadside agrimony

5. Yellow avens

6. White-Mountain avens

7. Big-leaved avens

8. Spring avens

Miscellaneous alternate, compound-leaved species
Several families

a. Petals 5
  b. Leaflets 3 (Wood-sorrel family)
  1. Common yellow wood-sorrel *Oxalis stricta*
  Plants prostrate to erect, 6–20″, hairy. **Leaflets** 3, heart-shaped, ⅜–¾″ wide, hairless. Flowers 2–7 on umbels; **fruit stalks bent downward; flowers** ⅛–½″. Dry, open soil.

A. Creeping yellow wood-sorrel *Oxalis corniculata* (not illus.)
Plants prostrate, 6–20″, hairy. Leaflets 3, heart-shaped, ⅜–¾″ **wide,** hairless. Flowers 2–7 on cymes; **flower stalks erect or ascending; flowers** ⅛–½″. Weedy areas.

2. Big yellow wood-sorrel *Oxalis grandis*
Plants erect, 1–4′, hairy below. **Leaflets** 3, heart-shaped, ¾–2″ **wide,** purplish along margin. Flowers 1–few on panicles; **flowers** ½–1″. Rich woods. See pink-flowered wood-sorrels p. 90.

  b. Leaflets usually more than 3 (Rose and Umbel families)
  3. False goat's-beard *Astilbe biternata*
  Plants 3–6′, glandular-hairy above. Leaves 2–3 times compound; leaflets stalkless or stalked, oblong or oval, rounded or notched at base, sharp-pointed, sharp-toothed. Flowers on panicles, ⅛″. Moist, north-facing woods. Rose family.

4. Cluster sanicle *Sanicula odorata*
Plants 1–2½″. Leaves 3–5 lobed; lobes sharply toothed. Flowers 12–25, on head-like umbels, ¼″. Rich woods, thickets. See white-flowered sanicles p. 316. Umbel family.

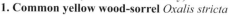

a. Petals 6 (Mignonette family)
*5. Yellow mignonette *Reseda lutea*
Plants erect or ascending, 8–32″, hairless. Leaves pinnately lobed; lobes few, narrow, toothless but cleft. Flowers on dense, terminal racemes, ¼″; sepals and petals 6, with 3 basal appendages. Weedy places. See white-flowered mignonette p. 316.

1. Common yellow wood-sorrel

2. Big yellow wood-sorrel

4. Cluster sanicle

3. False goat's-beard

5. Yellow mignonette

LEAVES alternate
LEAVES lobed or
compound
RAYS many

Coneflowers and Compass-plant
Composite family

Coneflowers are difficult to identify because the leaves are so variable. Those treated here generally have some leaves lobed, especially leaves near the base. Critical characters are in the disk shape and color and leaf lobing. See unlobed species p. 164.

a. Rays drooping

1. Cut-leaved coneflower *Rudbeckia laciniata*
Plants 1½–10′, **hairless**, gray-green. Leaves appearing cut, pinnately 3–7 lobed, hairless or rough-hairy; lower leaves stalked; upper leaves stalkless. Heads several to many, long-stalked; heads 2½–4″; **disk broadly cylindrical or nearly spherical, dull green-yellow,** ½–1″; rays 6–16, drooping. Swamps, moist thickets.

2. Gray-headed coneflower *Ratibida pinnata*
Plants 1–5′, **hairy**. Leaves pinnately 3–5 lobed; lobes narrowly oval, pointed, toothed. Heads solitary, 2–4″; **disk nearly spherical, gray or brown, anise-scented,** ⅜–¾″, ellipsoid; rays 6–13, drooping. Prairies, old fields, dry woods.

3. Prairie coneflower *Ratibida columnifera*
Plants 1–6′, **hairy**. Leaves pinnately lobed; lobes linear or narrowly oval, toothless. Heads several to many, yellow or purplish, 1¼–4″; **disk broadly cylindrical, purplish,** ½–1¾″; rays 4–11, drooping. Prairies.

a. Rays spreading

4. Sweet coneflower *Rudbeckia subtomentosa*
Plants 1–6′, **downy**. Leaves oval or narrow elliptic, 3-lobed, toothed. Heads several, 1¾–3″; **disk** nearly spherical or cone-shaped, **dark purple or brown,** ¼–½″; rays 12–21. Prairies, low ground. Leaves generally more deeply lobed than no. 5.

5. Three-lobed coneflower *Rudbeckia triloba*
Plants 1–5′, **hairy or hairless**. Leaves broadly oval or heart-shaped, usually some 3-lobed. Heads solitary at ends of branches, 1–2″; **disk** depressed, **black-purple,** ¼–½″; rays 6–13, shorter than no. 4. Woods, moist soil.

6. Compass-plant *Silphium lacinatum*
Plants 3–10′, **rough-hairy**. Leaves deeply pinnately lobed, ≤20″, stalked. Heads few, 2–4″; **disk yellow,** ¾–1¼″; rays 17–25. Prairies. Compass-plant gets its name from the belief that the basal leaves align in a north-south direction. See other silphiums p. 202–4.

1. Cut-leaved coneflower

2. Gray-headed coneflower

3. Prairie coneflower

4. Sweet coneflower

5. Three-lobed coneflower

6. Compass-plant

LEAVES alternate
LEAVES compound
RAYS many

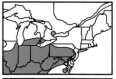

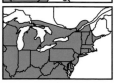

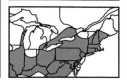

Lettuces, Sow-thistles, Thistles and Wormwood
Composite family

a. Plants with white or brown sap
  b. Heads ¼″, with 5–22 flowers
1. Tall lettuce *Lactuca canadensis*
Plants 1–10′, hairless or occasionally hairy. **Leaves** narrowly oval, variable, toothless, toothed, or lobed, 4–14″, **without prickles**, lobed or narrowed at base. **Heads** 50–100 on panicles, ¼″, **13–22-flowered**. See also p. 16. Fields, weedy places, woods.

*2. Willow-leaved lettuce *Lactuca saligna*
Plants 1–3′, hairless. **Leaves** linear with arrow-shaped base, toothless; or oblong and lobed, 2½–6″, **without prickles**. **Heads** 50–100 on long panicles, ¼″, **8–16-flowered**. Roadsides, weedy places.

*3. Prickly lettuce *Lactuca serriola*
Plants 1–4′, hairless, prickly. **Leaves** oblong or narrowly oval, lobed or not, 2½–6″, **prickly on margin and midrib below**. **Heads** 50–100 on diffuse panicles, yellow, blue when dried, ¼″, **5-flowered**. Fields, weedy places.

  b. Heads ≥½″, with 80–250 flowers
*4. Field sow-thistle *Sonchus arvensis*
Plants 16–80″, hairless. **Leaves** elliptic or narrowly oval, 2½–16″, toothless or usually lobed, **base with rounded or pointed lobes**; upper leaves heart-shaped, less lobed, clasping. **Heads** few to many on open, flat-topped panicles, sticky-hairy, 1¼–2″, **150–355-flowered**. Fields, roadsides, gravelly shores, weedy places.

*5. Common sow-thistle *Sonchus oleraceus*
Plants 4–80″. **Leaves** narrowly oval, broader toward either end, 2½–12″, lobed, margins weakly prickly, **base with rounded lobes, clasping**; upper leaves smaller. **Heads** several on flat-topped panicles, ½–1″, **80–250-flowered**. Weedy places, cultivated fields.

*6. Prickly sow-thistle *Sonchus asper*
Plants 4–80″, hairless. **Leaves** narrowly oval, broader toward either end, 2½–12″, lobed or not, spiny-margined, **base with pointed lobes. clasping stem. Heads** several on flat-topped panicles, ½–1″, **80–250-flowered**. Weedy places, roadsides.

a. Plants with clear sap
7. Yellow thistle *Cirsium horridulum*
Plants 8–60″, cobwebby when young. Basal leaves narrowly oval, broader toward tip; stem leaves lobed, spiny. **Heads** 1-several, yellow, white, lavender, or purple, 1½–3″. Open places, salt or freshwater marshes. See other thistles, pp. 24, 94, 318.

*8. Annual wormwood *Artemisia annua*
Plants 1–3′, sweet-smelling. Leaves ¾–4″, 2–3 times compound; leaflets linear or narrowly oval, toothed. **Heads** on open panicles, often nodding, <⅛″. See also pp. 182, 240. Fields, weedy areas.

1. Tall lettuce

*2. Willow-leaved lettuce

*3. Prickly lettuce

*4. Field sow-thistle

*5. Common sow-thistle

*6. Prickly sow-thistle

7. Yellow thistle

*8. Annual wormwood

2"

LEAVES alternate
LEAVES compound
RAYS many or absent

Miscellaneous composites with alternate, compound leaves
Composite family

a. Rays present

1. Yellowtop *Packera glabella*
Plants 6–40″, succulent, hairless or nearly so. Leaves deeply 1–2 times divided into a few triangular lobes, ≤8″; lobes rounded, toothed; upper leaves smaller. Heads 8–20, on flat-topped panicles, ¾–1¼″; disk ¼–⅜″; rays present. See also p. 218. Moist places, often weedy.

***2. Tansy-ragwort** *Senecio jacobaea*
Plants 8–40″, cobwebby at first, then hairless. Leaves deeply 2–3 times divided into numerous irregular lobes, ragged, 2–9″; lobes 1½–8″. Heads numerous on short, broad panicles, ½–1″; disk ¼–⅜″; rays present. Dry soils, weedy places. **Similar to no. 1, but leaves more divided.**

a. Rays absent or very small
    b. Heads ½–1″; rays present, minute

3. Eastern tansy *Tanacetum bipinnatum*
Plants 4–32″, soft-hairy or woolly. Leaves 2–3 times compound with finely dissected margins, 2–8″. **Heads** 1–15, on irregular flat-topped panicles, ½–1″; disk ⅜–¾″; rays numerous, minute. Beaches, riverbanks.

    b. Heads ¼–½″; rays absent

***4. Common groundsel** *Senecio vulgaris*
Plants 4–65″, cobwebby or hairless. Leaves oval or narrowly oval, broader toward tip, toothed to pinnately lobed, ¾–4″, stalked; upper leaves stalkless and clasping. Heads several to many; involucral bracts black-tipped; heads ⅜–½″; **disk cylindrical.** Disturbed soil, weedy places. **Note the distinctive black tips on the bracts surrounding the heads.**

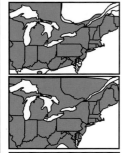

***5. Common tansy** *Tanacetum vulgare*
Plants 1–5′. Leaves deeply lobed with finely dissected margins, 4–8″, toothed, aromatic. Heads 20–100 on **flat-topped panicles,** ¼–⅜″; **disk button-shaped.** Roadsides, fields, weedy places.

***6. Pineapple-weed** *Matricaria discoidea*
Plants low-sprawling, 2–18″, with pineapple odor. Leaves deeply lobed, ⅜–2″; lobes short, linear. Heads several to many, **at ends of branches, ¼–⅜″; disk cone-shaped.** Roadsides, waste places.

***7. Lavender cotton** *Santolina chamaecyparissus*
Plants shrubby, 4–12″. Leaves 1¼–4″, deeply lobed; lobes ≤1¼″. **Heads solitary,** ¼–½″; **disk hemispherical.** Weedy places.

***8. Dusty miller** *Artemisia stelleriana*
Plants 12–28″, woolly. Leaves few-lobed, 1¼–4″; lobes rounded, slightly lobed. **Heads on narrow, dense panicles,** ¼″; disk rounded. See also pp. 180, 240. Sandy beaches.

1. Yellowtop

*2. Tansy-ragwort

3. Eastern tansy

*4. Common groundsel

*5. Common tansy

*6. Pineapple-weed

*7. Lavender cotton

*8. Dusty miller

LEAVES alternate
LEAVES compound
PETALS irregular

Yellow Clovers, Medick and Trefoils
Bean family

Clovers are distinctive in having three, toothed leaflets. Critical characters are inflorescence shape and leaflet shape and size. See red and pink clovers, p. 100; white clovers, p. 320.

a. Leaflets 3
  b. Flowers in heads

*1. Hop-clover *Trifolium aureum*
Plants erect, much branched, 6–16″, hairy. Leaf stalks ¼–½″; **leaflets 3, narrowly oval, broader toward tip, ½–¾″, stalkless**, truncate or notched at tip. Flowers on short, cylindrical heads, ½–¾″; stalks ½–1½″; flowers ¼″. Roadsides, fields, weedy places.

*2. Low hop-clover *Trifolium campestre*
Plants prostrate, much branched, 4–16″, hairy. Leaf stalks ¼–½″; **leaflets 3, oval, broader toward tip, ¼–½″, often notched; terminal leaflet stalked <⅛″**. Flowers on spherical or short, cylindrical heads, ¼–½″; flowers ⅛–¼″. Roadsides, overgrown fields, weedy places.

*3. Little hop-clover *Trifolium dubium*
Plants much branched, 2–14″, hairy. Leaf stalks ¼–½″; **leaflets oval, broader toward tip, ¼–½″, notched; terminal leaflet stalked <⅛″**. Flowers on spherical or short, cylindrical heads, ¼″; flowers ⅛″. Roadsides, overgrown fields, weedy places. **Very similar to no. 2, but flowers 5–15 per head (vs. 20–40).**

*4. Black medick *Medicago lupulina*
Plants prostrate or ascending, ≤3′, downy. Leaf stalks absent or ≤1¼″; leaflets 3, elliptic, round, oval, broader toward tip, ½–1″, **short spine-tipped; terminal leaflet stalked**. Flowers on globose or short, cylindrical heads, ≤½″; flowers ⅛″. Roadsides, weedy places. The fruit is a black, twisted pod. See also p. 36.

  b. Flowers on long racemes

*5. Yellow sweet-clover *Melilotus officinalis*
Plants ascending or erect, 2–8′, hairless. Leaflets 3, narrowly oval, broader toward tip, ½–1″, blunt, toothed. **Flowers on racemes 2–4″ tall;** flower ¼″. Roadsides, weedy places. See also p. 338.

a. Leaflets 5

*6. Bird's-foot trefoil *Lotus corniculatus*
Plants prostrate or erect, 6–24″, hairless. Leaflets 5, **elliptic or narrowly oval, broader toward tip, ¼–½″. Flowers 4–8,** on heads, ⅜–½″, sometimes becoming orange. Meadows, roadsides, disturbed sites.

*7. Narrow-leaved trefoil *Lotus tenuis*
Plants prostrate or erect, 6–24″, hairless. Leaflets 5, ¼–½″, **linear or narrowly oval. Flowers 1-4,** on heads, ⅜–½″. Roadsides, weedy places.

*1. Hop-clover

*2. Low hop-clover

*3. Little hop-clover

*4. Black medick

*5. Yellow sweet-clover

*6. Bird-s-foot trefoil

*7. Narrow-leaved trefoil

LEAVES alternate
LEAVES compound
FLOWERS irregular

Sensitive plants, Sennas, and related species
Bean family

The species on this page have flowers more regular than most legumes. They all have small glands near the base of the leaf (see inset in no. 2) that secrete sugars. These attract ants, ladybugs, and some other insects. One thought is that these sugar sources discourage ants from seeking other sources in the flower. Pollination is predominantly by bumblebees seeking pollen. Note that the pistil is usually twisted to one side and the stamens to the other, preventing self-fertilization. Nos. 1 and 2 have sensitive leaves that close after being touched.

a. All stamens fertile

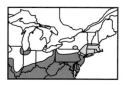

1. Wild sensitive plant *Chamaecrista nictitans*
Plants 6–15″, hairless or hairy. **Leaflets 14–40, oblong,** ¼–½″, sharply pointed, sensitive to touch. Flowers 1–3, in axillary clusters, ¼″; petals very unequal; **stamens 5.** Dry woods, dunes, weedy sites. **Note stalked gland at base of leaf stalk.**

2. Partridge-pea *Chamaecrista fasciculata*
Plants 1–2′, hairy. **Leaflets 10–32, linear or oblong,** ½–1″, pointed or blunt, with a bristle tip, sometimes sensitive to touch. Flowers 1–6, on axillary racemes, 1–1½″; petals about equal in size, some with red marks at the base; **stamens 10, dark, drooping, 4 with yellow anthers, 6 with purple anthers.** Sandy fields, weedy areas. **Note low gland at base of leaf stalk.**

a. Upper 3 stamens sterile

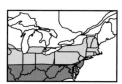

3. Northern wild senna *Senna hebecarpa*
Plants ¼–6′, hairless or hairy above. **Leaflets 10–20, linear or oblong,** 1–2″, pointed or blunt. Flowers many, on axillary racemes, ¾″; petals slightly dissimilar; **stamens 10 with dark brown anthers; upper 3 sterile.** Moist, open woods; roadsides; stream banks; thickets. **Note the long, hairy ovary.**

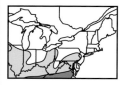

4. Southern wild senna *Senna marilandica*
Plants 3–6′, hairless or slightly hairy above. **Leaflets 8–16, oblong or elliptic,** 1–2″, pointed or blunt, with a short bristle tip. Flowers few, on axillary racemes, ¾–1¼″; petals slightly dissimilar; **stamens 10 with dark brown anthers; upper 3 sterile.** Moist, open woods; stream banks; roadsides. **Note the appressed hairs on the ovary.**

*A. Sickle-pod *Senna obtusifolia* (not illus.)
Plants ≤3′, bad smelling. **Leaflets 4–6, elliptic or oval, broader toward tip,** 1¼–2¾″, narrowly pointed. Flowers 1–2 in upper leaf axils, ¾–1¼″; petals slightly dissimilar; **stamens 10 with dark brown anthers; upper 3 sterile.** Moist woods.

1. Wild sensitive plant

gland

2. Partridge-pea

3. Northern wild senna

4. Southern wild senna

Wild Indigo, Buckbean, Vetches, and Corydalis
Bean and Poppy families

a. Leaflets 3 (Bean family)

1. Yellow wild indigo *Baptisia tinctoria*
Plants 1–3′, blue-gray. **Leaflets** 3, oval, broader toward tip, ⅜–1¼″, rounded to slightly notched, tapering to base. Flowers on racemes terminating most branches, 1¼–4″ tall; flowers ¼–½″. Dry, sterile soil; sandy soil. See also p. 188, 320.

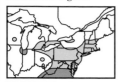

*2. Piedmont-buckbean *Thermopsis villosa*
Plants 2–6′. **Leaflets** 3, narrowly oval, **1¼–4″,** hairy below. Flowers on dense, hairy racemes, 4–12″ tall, flowers ¾–1¼″. Dry woods, ridges, clearings.

a. Leaflets 7–58
  b. Leaves pinnately compound (Bean family)
*3. Big-flowered vetch *Vicia grandiflora*
Plants ascending, ≤24″. Leaflets 7–14, linear or oblong, ⅜–¾″, **terminating in tendrils. Flowers 1–2 in upper axils,** 1–1¼″; calyx hairy. Fields, open woods. See also p. 34, 320.

4. Canada milk-vetch *Astragalus canadensis*
Plants erect, 1–5′, hairless or nearly so. **Leaflets 15–35,** oblong or elliptic, ⅜–1½″, hairy below. **Flowers on dense racemes, ¼– ½″ tall;** flowers white or yellow, ½–¾″; calyx tube minutely hairy. Open woods, riverbanks, shores.

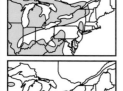

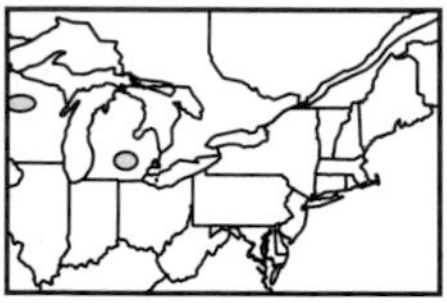

*5. Chick-pea milk-vetch *Astragalus cicer*
Plants prostrate or ascending, 10–28″. **Leaflets 34–58,** narrowly oval or oblong, ¼–1½″, hairy beneath. **Flowers on racemes, 1½–4¼″ tall;** flowers ⅜–½″. Disturbed moist areas.

  b. Leaves 2 times dissected (Poppy family)
Corydalis has asymmetrical flowers: the upper two petals are spurred at the base and dilated at the tip; the lower two petals are smaller and not spurred. Critical characters are flower size and spur length. See also p. 92.

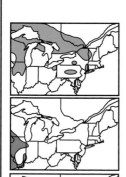

6. Golden corydalis *Corydalis aurea*
Plants prostrate to ascending, 4–24″. Leaves 2× dissected. Flowers on dense racemes, ⅜–1¼″ tall; **flowers ½″; spur ⅜ the flower length;** sepals broadly oval, <⅛″. **Rocky banks, sandy soils.**

A. Slender corydalis *Corydalis micrantha* (not illus.)
Plants erect or ascending, 4–12″. Leaves 2× dissected. Flowers on racemes; **flowers ⅜–½″; spur ⅜ the flower length. Moist, sandy soils.**

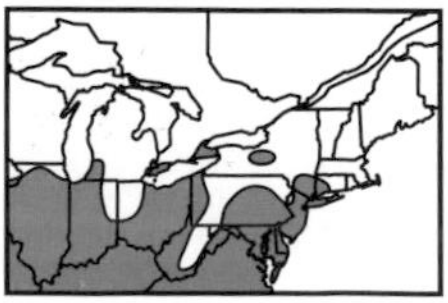

7. Short-spurred corydalis *Corydalis flavula*
Plants erect, becoming prostrate, 4–20″. Leaves 2× dissected. Flowers on racemes; **flowers ¼″; spur less than ¼ the flower length.** Rocky slopes, sandy slopes, shores, open woods.

1. Yellow wild indigo

*2. Piedmont-buckbean

*3. Big-flowered vetch

4. Canada milk-vetch

*5. Chick-pea milk-vetch

6. Golden corydalis

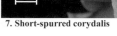

7. Short-spurred corydalis

LEAVES opposite
LEAVES simple
PETALS 4–5

St. John's-worts (1)
St. John's-wort family

St. John's-wort refers to a superstition that dew that fell on the plant the night before June 24 (St. John's day) was efficacious for preserving the eyes from disease. The flowers do not produce nectar, so visiting insects are seeking pollen. The leaves have translucent dots, visible when held to the light. Critical characters for St. John's-worts are habit, style number, and fruit characteristics.

a. Shrubs (cont. on next page)
   b. Petals 4

1. St. Peter's-wort *Hypericum crux-andreae*
Plants 12–32″. **Leaves elliptic**, ½–1½″, rounded, slightly clasping at base. Flowers solitary, slightly stalked, 1–1½″; sepals round; petals oval, broader toward tip; **styles 3**. Sandy areas.

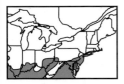

2. St. Andrew's-cross *Hypericum hypericoides*
Plants ½–4′, sometimes forming mounds 1′ wide. **Leaves linear, narrowly oblong, or narrowly oval, broader toward tip**, ⅜–1¼″, blunt or rounded, narrowed at base. Flowers solitary, ½–1″; sepals rounded, the inner 2 much smaller; petals narrowly oblong; **styles 2**. Sandy areas, rocky areas.

   b. Petals 5
3. Kalm's St. John's-wort *Hypericum kalmianum*
Plants ≤3′. Leaves linear or narrowly oblong, ¾–1½″, margins often curled under. **Flowers 3–7**, on small racemes at ends of branches; flowers ¾–1½″; **styles 5**. Rocky shores, sandy shores.

4. Bushy St. John's-wort *Hypericum densiflorum*
Plants ≤7′. Leaves linear or narrowly oval, broader toward tip, ¾–1½″. **Flowers 7-many** on panicles; flowers ⅜–½″; **styles 3**. Swamps, wet meadows. Note the small flowers.

5. Cedarglade St. John's-wort *Hypericum frondosum*
Plants 1½–3′. Leaves narrow oblong, elliptic, or narrowly oval, 1¼–2½″. **Flowers 1–3 (usually 1)** at ends of branches, ¾–1½″; **styles 3**. Bluffs, cliffs, rocky hills.

6. Shrubby St. John's-wort *Hypericum prolificum*
Plants ≤7′. Leaves linear or oblong, 1¼–2½″, abruptly narrowed to stalk. **Flowers 3–7** on racemes, ½–1″; **styles 3**. Swamps, cliffs, woods, other habitats.

1. St. Peter's-wort

2. St. Andrew's-cross

3. Kalm's St. John's-wort

4. Bushy St. John's-wort

5. Cedarglade St. John's-wort

6. Shrubby St. John's-wort

LEAVES opposite
LEAVES simple
PETALS 5

St. John's-worts (2)
St. John's-wort family

a. Herbs (cont. from previous page)
   b. Flowers > 1″

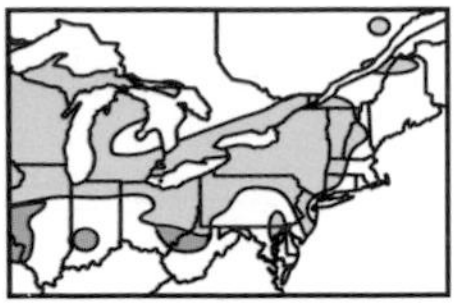

1. Great St. John's-wort *Hypericum ascyron*
Plants 2–6′, branched, 2–4-angled. Leaves narrowly oval or elliptic, 2–4″, blunt or pointed, stalkless, clasping. **Flowers** solitary at ends of branches, 1½–2″; **styles 5**. Moist thickets, meadows.

   b. Flowers ⅜–1″

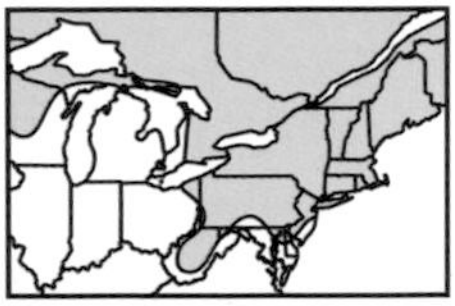

2. Pale St. John's-wort *Hypericum ellipticum*
Plants ½–1¾′, unbranched, obscurely 4-angled. **Leaves elliptic,** ½–1″, blunt or rounded at both ends, slightly clasping. Flowers few to many; **bracts linear or narrowly oval,** ½–¾″; sepals oblong or narrowly oval; **styles 1.** Sandy shores, gravelly shores.

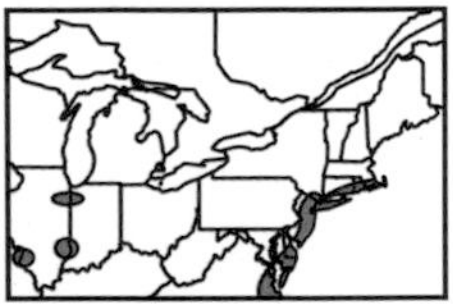

3. Creeping St. John's-wort *Hypericum adpressum*
Plants 1–2½′, unbranched, obscurely 4-angled. **Leaves linear or narrowly elliptic,** 1–2½″, broadly pointed, tapering at base, margins curled under. Flowers many; **bracts minute,** ½–¾″; sepals narrowly oval or oval; **styles 1.** Marshes, shores, wet meadows.

***4. Common St. John's-wort** *Hypericum perforatum*
Plants 1–2½′, with many branches. **Leaves** linear-oblong, 1–1½″, half this size on lateral branches, stalkless. Flowers many, ¾–1″; sepals narrowly oval, without black glands; **petals with black dots along margins; styles 3.** Roadsides, fields.

5. Spotted St. John's-wort *Hypericum puctatum*
Plants 1–3′, with few branches. **Leaves** elliptic or narrowly oblong, 1½–3″, blunt or notched; base narrow, with conspicuous black dots. Flowers crowded, ⅜–⅝″; sepals oblong, heavily black-dotted; **petals with numerous black dots; styles 3.** Thickets, fields, open woods.

6. Coppery St. John's-wort *Hypericum denticulatum*
Plants 10–30″, branched or unbranched, 4-angled. **Leaves** elliptic to oval, ⅜–¾″, blunt, stalkless. Flowers few to many, coppery-colored; bracts narrowly oval or narrowly triangular, ½–1″; sepals narrowly oval, broader toward either end; **styles 3.** Bogs, marshes, wet woods.

   b. Flowers ⅛–¼″

7. Dwarf St. John's-wort *Hypericum mutilum*
Plants 12–36″, much-branched. **Leaves** narrowly to broadly oval, ½–1½″, **3–5-nerved, rounded at the barely clasping base.** Flowers many, ⅛–¼″; bracts scale-like; sepals linear; styles 3. Wet areas.

8. Canada St. John's-wort *Hypericum canadense*
Plants 4–25″, branched. **Leaves** linear or narrowly oval, broader toward tip, ½–1½″, **1–3-nerved, tapering to base, stalkless.** Flowers many, ⅛–¼″; bracts very narrowly oval; sepals narrowly oval; styles 3. Wet shores, meadows.

1. Great St. John's-wort

2. Pale St. John's-wort

3. Creeping St. John's-wort

*4. Common St. John's-wort

5. Spotted St. John's-wort

6. Coppery St. John's-wort

7. Dwarf St. John's-wort

8. Canada St. John's-wort

LEAVES opposite or whorled
LEAVES simple
PETALS 5

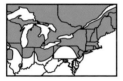

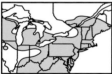

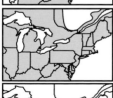

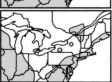

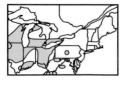

Loosestrifes
Myrsine family

Loosestrife flowers are often marked with red or dark spots. Critical characters are flower size and color and leaf shape and position.

a. Plants creeping

*1. Moneywort *Lysimachia nummularia*
Plants creeping, ½–2′. **Leaves opposite**, round, ½–1″, short-stalked, dotted. Flowers solitary in axils, ¾–1¼″. Moist, often weedy, places such as lawns, roadsides, shores.

a. Plants erect
   b. Flowers on terminal racemes
2. Bulbil loosestrife *Lysimachia terrestris*
Plants erect, 1–3′. **Leaves opposite**, narrowly oval, 1½–4″, dotted, stalkless. **Flowers on terminal racemes, 2–12″ tall**; flowers ½–¾″. Open swamps, shores. **Note the axillary, reddish bulblets found in the leaf axils in the summer (see insert).**

   b. Flowers in axillary clusters
*3. Spotted loosestrife *Lysimachia punctata*
Plants erect, ≤3′. **Leaves 3–4-whorled**, narrowly oval, 2–4″, dotted, short-stalked. Flowers in axillary clusters, 1–1½″. Roadsides, weedy places.

4. Swamp loosestrife *Lysimachia thyrsiflora*
Plants erect, 1–2½′. **Leaves opposite**, narrowly oval or linear, 2–6″, dotted, stalkless. Flowers in dense, axillary clusters, ½″; petals marked with black. Cold swamps, springy marshes, bogs. **Note the stamens are twice the length of the petals.**

   b. Flowers solitary, axillary
5. Whorled loosestrife *Lysimachia quadrifolia*
Plants erect, 1–3′. **Leaves 4-whorled, narrowly oval**, 1¼–3½″, dotted, stalkless. Flowers axillary, usually in whorls of 4, ½–¾″. Moist or dry upland woods, thickets, shores.

6. Fringed loosestrife *Lysimachia ciliata*
Plants erect, 1–4′. **Leaves opposite, oval or narrowly oval**, 1¼–6″, undotted; **leaf-stalks fringed**. Flowers solitary in axils, ½–1″, nodding; petals irregularly toothed. Thickets, swamps, shores.

7. Mississippi-valley loosestrife *Lysimachia hybrida*
Plants erect, 1–5′. **Leaves opposite, narrowly oval**, 1½–4″, undotted; **leaf-stalks fringed below the middle**. Flowers solitary, axillary, ½–1″, nodding; petals toothed. Wet woods, prairies, swamps. Similar to no. 6, but leaves narrower and leaf-stalks not as fringed.

8. Smooth loosestrife *Lysimachia quadriflora*
Plants erect, 1–3′. **Leaves stiff, opposite, linear**, 1¼–3½″, **stalkless**. Flowers solitary in upper axils, ½″, petals slightly toothed. Moist or wet prairies, bogs, swales, shores. Similar to nos. 6–7, but with narrower, stalkless leaves

*1. Moneywort

2. Bulbil loosestrife

*3. Spotted loosestrife

4. Swamp loosestrife

5. Whorled loosestrife

6. Fringed loosestrife

7. Mississippi-valley loosestrife

8. Smooth loosestrife

LEAVES opposite
LEAVES simple. lobed
or compound
PETALS 5

Miscellaneous yellow-flowered species
Several families
a. Leaves unlobed (Lopseed and Plantain families)

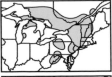

1. Musky monkey-flower *Mimulus moschatus*
Plants creeping at base, tips ascending, 4–8″, sticky and hairy, musky smelling. Leaves oval or narrowly oval, 1¼–2½″, toothless. **Flowers** solitary in axils, ½–1″; **stamens 4**. Shallow water, springy places. See also p. 54. Lopseed family.

2. Yellow hedge-hyssop *Gratiola aurea*
Plants ascending or creeping, 1½–16″, hairless or sticky-hairy. Leaves narrowly oval or oval, 3-nerved, ¼–1½″, sticky-hairy, barely toothed or toothless, **somewhat clasping**. **Flowers** solitary in axils, ½″; **stamens 2**. Muddy or sandy shores. See also p. 348. Plantain family.

***3. Common yellow monkey-flower** *Mimulus guttatus*
Plants erect, 12–24″, hairless or sticky-hairy above. Leaves round or oval, broader toward tip, ¼–½″, toothed. **Flowers** solitary in axils, 1–1¾″; **stamens 4**. Wet places. Lopseed family.

a. Leaves deeply lobed or compound (Broom-rape family)
  b. Stems hairless
The false foxgloves are pollinated by bees. Critical characters are in the hairs and flower stalks.

4. Smooth false foxglove *Aureolaria flava*
Plants 2–6′, **hairless**, often purplish. Lower leaves deeply lobed, elliptic or narrowly oval; upper leaves less lobed, hairy or hairless. **Flowers** solitary, **stalked ⅛–⅜″**; flower 1½–2″. Dry, upland, deciduous woods.

A. Appalachian false foxglove *Aureolaria laevigata* (not illus.)
Similar to no. 4, but the lower leaves are rarely lobed, and the flower stalk ≤⅛″. Upland woods.

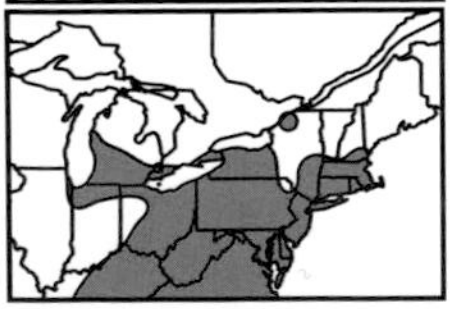

  b. Stems hairy
5. Fern-leaf false foxglove *Aureolaria pedicularia*
Plants 1–4′, **hairy and sticky**, bushy. Leaves 2× pinnately lobed, 1¼–2½″, stalkless or nearly so; lobes 5–8, finely hairy. **Flowers** solitary, **stalked ⅜–1″**; flowers 1–1½″. Dry, upland, deciduous woods; clearings.

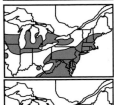

6. Downy false foxglove *Aureolaria virginica*
Plants 1–6′, **downy**. Lower leaves narrowly oval, unlobed or with 1–2 blunt lobes, pointed, downy, stalked; upper leaves smaller. Flowers solitary, **stalked <⅛″**; flowers 1¼–1½″. Dry, oak woods.

B. Western false foxglove *Aureolaria grandiflora* (not illus.)
Similar to no. 6, but the **flower stalk** is ⅛–½″, and **the flower is 1½–2″**. Upland woods.

1. Musky monkey-flower

2. Yellow hedge-hyssop

*3. Common yellow monkey-flower

4. Smooth false foxglove

5. Fern-leaf false foxglove

6. Downy false foxglove

Miscellaneous yellow-flowered shrubs and vines
Several families

a. Vines (Dogbane and Logania families)

***1. Pale swallow-wort** Cynanchum rossicum*
Plants erect, or scrambling vines. Leaves oval, 3½–4¾″. **Flowers on terminal and axillary panicles ,** ¼″, yellowish or greenish-white. Thickets, weedy places. See other swallow-worts p. 256. Dogbane family.

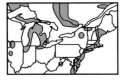

2. Yellow jessamine *Gelsemium sempervirens*
Plants climbing to 15′. Leaves narrowly oval, 1½–2½″. Flowers axillary, ¾–1¼″. Woods, Thickets. Logania family.

a. Shrubs 4–8″ (Rock-rose family)
Hudsonias are small bushes in sandy, open areas; the flowers stay open for only a day, and then only in sunlight.

3. False heather *Hudsonia tomentosa*
Plants shrubby, 4–8″, **covered with a white down.** Leaves narrowly oval or triangular, **scale-like,** ⅛″, closely overlapping, densely white-hairy. Flowers solitary, axillary, **stalked ≤⅛″; flowers** ¼″. Coastal dunes; beaches; inland, sandy areas.

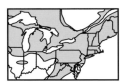

4. Golden heather *Hudsonia ericoides*
Plants bushy, 4–8″, **green. Leaves linear,** ⅛–¼″, loosely erect or spreading, green. Flowers solitary, axillary, **stalked ¼–½″; flowers** ¼″. Dunes, pine barrens, rocks.

a. Shrubs to 4′ (Bush-honeysuckle and Honeysuckle families)
5. Bush-honeysuckle *Diervilla lonicera*
Shrubs, ≤4′. **Leaves** narrowly oval or oval, **3–6″,** sharp-pointed. **Flowers 3–7 on terminal racemes; flowers** ½–¾″. Dry, rocky soil. Pollinated by bees. After pollination the flower changes to a deeper yellow-orange. Bush-honeysuckle family.

6. Mountain fly-honeysuckle *Lonicera villosa*
Shrubs, ≤3′. **Leaves** oval or oblong, ¾–3″, blunt or rounded at tip. **Flowers paired on axillary stalks,** ⅜–½″. Swamps, wet woods. Many other honeysuckles occur in the region, but all are larger shrubs. Honeysuckle family.

*1. Pale swallow-wort

2. Yellow jessamine

3. False heather

4. Golden heather

5. Bush-honeysuckle

6. Mountain fly-honeysuckle

LEAVES opposite
LEAVES simple
Rays absent or 5–10

Beggar-ticks, Golden-star, Bur-marigolds, and Tickseeds
Composite family

Bur-marigolds, Beggar-ticks, and Tickseeds are similar plants with seeds that have burs that allow them to stick to passing animals. Critical characters are leaf shape, head shape and size, and seeds. See others with compound leaves, p. 208; pink-flowered, p. 116.

a. Rays absent or nearly so

1. Purplestem beggar-ticks *Bidens connata*
Plants 4–40″, hairless, green or purple. Leaves narrowly oval or elliptic, 1¼–8″, sharp-pointed, toothed or often deeply cleft, stalked. Heads erect, ¼–¾″; rays absent or ≤¼″. Wet, weedy places.

2. Southern estuarine beggar-ticks *Bidens bidentoides*
Plants 4–36″, hairless, purple or green. Leaves narrowly oval, 1½–6¼″, sharp-pointed, toothed. Heads erect, ¼–½″; rays absent. Estuaries. **Similar to no. 1, but with generally smaller heads with 7–30 flowers (vs. 30–150).**

a. Rays 5

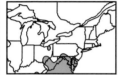

3. Golden-star *Chrysogonum virginianum*
Plants erect, 3–24″; often flowering while still developing, hairy and sticky. Leaves oval or round, 1–4″, hairy, toothed; upper leaves notched at base. **Heads 1–few, terminal and axillary, 1–1½″;** disk ¼–⅜″; rays 5. Rich woods.

a. Rays more than 5

4. Showy bur-marigold *Bidens laevis*
Plants 4–40″, hairless. **Leaves linear or narrowly oval, 1½–8″,** toothed, stalkless. **Heads** slightly nodding, **1½–2½″;** disk ½–1″; rays 8. Low, wet places.

5. Bur-marigold *Bidens cernua*
Plants ½–7′, hairless or hairy. **Leaves linear or narrowly oval,** 1½–8″, stalkless. **Heads** solitary, nodding, **½–1¼″;** disk ½–1″; rays 8. Low, wet places.

6. Lobed tickseed *Coreopsis auriculata*
Plants 4–24″, hairy, becoming hairless. **Leaves oval, elliptic or round, ≤3″, often lobed,** stalked. Heads solitary or few, 1¼–2½″; disk ¼–½″; rays 8. Rich hardwoods, openings.

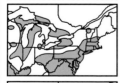

7. Long-stalked tickseed *Coreopsis lanceolata*
Plants 8–32″, hairless to spreading-hairy. **Leaves mostly near the base, spoon-shaped, linear, or narrowly oval,** 2–8″, unlobed or with small lateral lobes; lower leaves long-stalked. Heads 1–few, 1½–2½″; disk ⅜–¾″; rays 6–10. Dry, often sandy, fields; roadsides.

8. Finger tickseed *Coreopsis palmata*
Plants 1½–3½′, hairy at nodes. **Leaves trilobed, lobes oblong or linear,** 1¼–2¾″, with hairy margins, stalkless. Heads 1–few, 1¼–3″; disk ¼–½″; rays 8. Prairies, open woods.

1. Purplestem beggar-ticks

2. Southern estuarine beggar-ticks

3. Golden-star

4. Showy bur-marigold

5. Bur-marigold

6. Lobed tickseed

7. Long-stalked tickseed

8. Finger tickseed

LEAVES alternate and opposite
LEAVES simple
RAYS many

Sunflowers and similar species (1)
Composite family

a. Leaves opposite (lower) and alternate (upper)
  b. Disk red-purple

*1. Common sunflower *Helianthus annuus*
Plants 1–12′, rough-hairy. **Leaves oval**, 4–16″, long-stalked, rough on both sides. Heads solitary; involucral bracts oval; **heads 3–6″;** disk reddish, 1–2″; rays >12. Weedy areas; moist, low areas.

2. Narrow-leaved sunflower *Helianthus angustifolius*
Plants 1½–6′, hairy. **Leaves linear**, 2–8″. Heads few on flat-topped panicles; involucral bracts reddish, narrowly oval; **heads 2–3″;** disk reddish, ⅜–¾″; rays 10–15. Acid bogs, thickets, pinelands.

  b. Disk yellow

3. Ashy sunflower *Helianthus mollis*
Plants 1½–3′, **densely hairy. Leaves heart-shaped**, 2½–6″, nearly toothless, **stalkless**. Heads 1 or few; involucral bracts narrowly oval, densely white-hairy, heads 2–4″, disk ¾–1¼″, rays 16–35. Prairies, dry places.

4. Tall sunflower *Helianthus giganteus*
Plants 1½–10′, **spreading hairy or nearly hairless. Leaves oval or narrowly oval**, 2–8″, toothed or toothless, **stalked ⅛–½″**. Heads several to many; involucral bracts narrowly oval, often hairy; heads 1–3½″; disk ½–1″; rays 10–20. Swamps, moist places.

5. Sawtooth sunflower *Helianthus grosseserratus*
Plants 3–16′, **hairless** and blue-green. **Leaves narrowly oval**, 4–8″, sharp toothed; **stalk ⅜–1½″**. Heads numerous in open clusters; involucral bracts narrowly oval, slightly hairy; heads 2–4½″; disk ½–1″; rays 10–20. Bottomlands, damp prairies, moist places.

6. Jersusalem-artichoke *Helianthus tuberosus*
Plants 3–10′, **hairy, rough. Leaves narrowly oval or oval**, 4–10″, coarsely toothed, rough, **stalked ¾–3″**. Heads many on flat-topped panicles; involucral bracts narrowly oval, slightly hairy; heads 2–3½″; disk ½–1″; rays 10–20. Moist soil, weedy places.

a. Leaves all opposite
  b. Leaves stalkless (cont. on next page)
7. Woodland sunflower *Helianthus divaricatus*
Plants 2–6½′, hairless. Leaves oval or narrowly oval, 2–7″, shallow-toothed, rough on upper surface, stalkless. Heads 1–several at tips of branches; involucral **bracts narrowly oval**, fringed; heads 1¼–3″; **disk ⅜–½″; rays 8–15**. Dry woods, thickets, openings.

8. Cup-plant *Silphium perfoliatum*
Plants 4–8′, hairless. Leaves triangular or oval, 6–14″, rough, coarsely toothed; **bases fused around stem, forming cup**. Heads on open panicles, 2–3″; disk ½–1″; rays 16–35. Rich woods, thickets, low grounds.

*1. Common sunflower

2. Narrow-leaved sunflower

3. Ashy sunflower

4. Tall sunflower

5. Sawtooth sunflower

6. Jerusalem-artichoke

7. Woodland sunflower

8. Cup-plant

LEAVES opposite, whorled, or basal
LEAVES simple
RAYS many

Sunflowers and similar species (2)
Composite family

a. Leaves all opposite
    b. Leaves stalkless (cont. from previous page)

1. Prairie rosin-weed *Silphium integrifolium*
Plants 1½–6′, hairless. Leaves oval, narrowly oval, or elliptic, 2¾–6″, toothed or toothless, rough, stalkless, **often clasping**. Heads on short panicles; **involucral bracts oval or elliptic; heads 2–5″; disk ½–1″; rays 16–35**. Prairies; roadsides; sometimes, open woods. See also pp. 178, 202, 220.

2. Whorled rosin-weed *Silphium trifoliatum*
Plants 2½–10′, hairless. Leaves narrowly oval, oval, or triangular, 5–10″; **middle leaves whorled**, toothed or toothless, stalked. Heads several to many, on open panicles; involucral bracts broadly oval; heads 1½–2″; disk ⅜–½″; **rays 8–13**. Woodlands, prairies, pastures.

    b. Leaves stalked (at least the lower leaves)

3. Naked-stem sunflower *Helianthus occidentalis*
Plants 1½–5′, hairy at base, hairless above. Leaves **mostly toward base,** oval, narrowly oval or elliptic, 2½–6″, toothed or toothless; stem leaves smaller. Heads few; involucral bracts slender-tipped; heads 1¼–3″; disk ⅜–½″; rays 10–15. Dry, often sandy, soil.

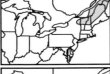

4. New England arnica *Arnica lanceolata*
Plants 2–30″, hairy and sticky. Leaves elliptic or narrowly oval, broader toward tip, ≤8″, toothed. Heads 1–10; involucral bracts pointed, ½–2″; disk ½–¾″; rays 10–15. Stream banks, moist areas.

5. Heart-leaved arnica *Arnica cordifolia*
Plants 1½–5′, hairy and sticky. Leaves heart-shaped, ≤4½″, toothed; basal and lower stem leaves stalked; upper leaves smaller and stalkless. Heads 1–3; involucral bracts 2–3½″; disk yellow, ½–1″; rays 10-15. Dry woods.

6. Rough-leaved sunflower *Helianthus strumosus*
Plants 3–8′, hairless. **Leaves** narrowly oval or oval, 3–10″, **shallow-toothed or toothless, rough;** stalk ¼–1¼″. Heads few to many; involucral bracts narrowly oval, very loose; heads 1¼–4″; disk yellow, ⅜–1″; rays 8–15, **stigmas of ray flowers unlobed**. Woods, openings.

7. Thin-leaved sunflower *Helianthus decapetalus*
Plants 2–5′, hairless. **Leaves** narrowly oval or oval, 3–8″, **sharp toothed, smooth**, stalk ½–2½″. Heads 1; involucral bracts narrowly oval, moderately loose; heads 1¼–3¾″; disk yellow, ⅜–¾″; rays 8–15; **stigmas of ray flowers unlobed**. Woods, banks, thickets.

8. Ox-eye *Heliopsis helianthoides*
Plants 1–5′, hairless. **Leaves** oval or narrowly oval, 2–6″, **sharply toothed, hairless or rough above**, long-stalked. Heads 1–several; involucral bracts oblong, spreading at tips, 1½–2½″; disk ⅜–1″; rays 8–16, persistent; **stigmas of ray flowers bilobed**. Woods, prairies, weedy places.

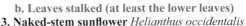
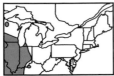

1. Prairie rosin-weed

2. Whorled rosin-weed

3. Naked-stem sunflower

4. New England arnica

5. Heart-leaved arnica

6. Rough-leaved sunflower

7. Thin-leaved sunflower

8. Ox-eye

LEAVES opposite and basal
LEAVES simple and sometimes lobed
PETALS irregular

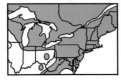

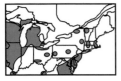

Miscellaneous species with irregular flowers
Several families

a. Basal leaves absent, stem leaves not lobed
  b. Flowers axillary

1. Cow-wheat *Melampyrum lineare*
Plants 2–20″. **Leaves** narrowly oval, broader toward tip, or spoon-shaped, **soon falling;** those below branches of inflorescences **linear or narrowly oval, ¾–2½″, toothless.** Flowers in upper axils, ¼–½″, white and yellow. Dry woods, bogs. Broom-rape family.

2. Yellow rattle *Rhinanthus minor*
Plants 4–40″, hairy. **Leaves narrowly triangular or oblong, ¾–2½″, toothed.** Flowers in upper axils, ⅜–½″. Fields, thickets, openings. The lowland populations are introduced; alpine populations are native. The seeds rattle in the fruit pod. Broom-rape family.

3. Lesser horse-gentian *Triosteum angustifolium*
Plants 12–32″, sparsely hairy and sticky-hairy. **Leaves narrowly oval or oval, broader at either end, 3–8″, toothless.** Flowers solitary in axils, ¼–½″, greenish-yellow, hairy. Moist woods, low ground. See red horse-gentians p. 116. Honeysuckle family.

4. Dotted monarda *Monarda punctata*
Plants 1–3′. **Leaves narrowly oval or narrowly oblong, ¾–3″. Flowers** in axillary clusters of 2–5, **subtended by whitish or lilac bracts;** flowers ½–1″. Dry fields, roadsides. See also pp. 118, 338. Mint family.

  b. Flowers on spikes or panicles
5. Northern horse-balm *Collinsonia canadensis*
Plants 2–5′, stem square. **Leaves oval or oblong,** the lower leaves 4–8″, sharp-pointed, hairless, toothed; upper leaves smaller. Flowers on pyramidal panicles 4–12″ tall; **flowers ½″,** paired, lemon-scented. Rich woods. Mint family.

6. Yellow giant-hyssop *Agastache nepetoides*
Plants 2–5′, stem square. **Leaves oval or narrowly oval, 2–6″,** smaller upward, toothed, rounded or notched at base; finely hairy beneath. Flowers on cylindrical spikes, 1¼–8″ tall; **flowers ⅜″.** Open woods, rich thickets. See also p. 48. Mint family.

*7. Butter-and-eggs *Linaria vulgaris*
Plants 1–4′, stem round. **Leaves linear, ½–2″,** narrowed at base. Flowers on terminal, dense racemes; flowers ¾–1¼″, long-spurred. Fields, roadsides, waste places. Plantain family.

a. Basal leaves present, pinnately lobed, stem leaves smaller
8. Forest-lousewort *Pedicularis canadensis*
Plants 6–24″. **Basal leaves** narrowly oblong or narrowly oval, broader toward tip, **pinnately lobed,** stem leaves smaller upward. Flowers on spikes 1¼–2″ tall; flowers ½–1″, yellow, purple, or deep red. Upland woods, prairies. Broom-rape family.

1. Cow-wheat

2. Yellow rattle

3. Lesser horse-gentian

4. Dotted monarda

5. Northern horse-balm

6. Yellow giant-hyssop

*7. Butter-and-eggs

8. Forest-lousewort

LEAVES opposite
LEAVES compound
RAYS absent or 8

Beggar-ticks and Tickseeds
Composite family

Beggar-ticks and Tickseeds are similar plants with seeds that have burs that allow them to stick to passing animals. Critical characters are leaf size, leaflet number, number of involucral bracts below head, and fruit characters. See others with simple leaves, p. 200; pink flowers, p. 116.

a. Rays absent

1. Few-bracted beggar-ticks *Bidens discoidea*
Plants 4–70″, hairless, stem red. **Leaflets 3**, narrowly oval, ≤4″, toothed. Heads ⅛–⅜″; disk ⅛–⅜″; rays absent; **outer involucral bracts 3–5, not fringed**. Wet places.

2. Devil's beggar-ticks *Bidens frondosa*
Plants 8–50″, hairless. **Leaflets 3–5**, narrowly oval, ≤4″, toothed. Heads ≤⅜″; disk ≤⅜″; rays absent; **outer involucral bracts 5–10, fringed**. Damp ground, fields, waste places.

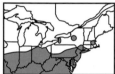

3. Spanish needles *Bidens bipinnata*
Plants 1¼–6¾″, hairless or minutely hairy. **Leaves 2–3 times compound, 1½–8″; leaflets numerous**. Heads ½″; disk ¼″; rays absent; **outer involucral bracts 7–10, fringed**. Roadsides, waste places.

4. Tall beggar-ticks *Bidens vulgata*
Plants 8–60″, hairless to densely hairy; stem reddish. **Leaflets 3–5**, narrowly oval, ≤4″, toothed. Heads ⅜″; disk ≤⅜″; rays absent; **outer involucral bracts 10–16, fringed**. Waste places.

a. Rays present
  b. Disk yellow

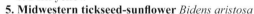

5. Midwestern tickseed-sunflower *Bidens aristosa*
Plants 1–5′, hairless or slightly hairy. Leaves 1–2 times compound, 2–6″; leaflets narrowly oval or linear, sharp-pointed, toothed. Heads 1–2″; disk yellow, ¼–½″; rays 8. Wet, usually shady, places. **Very similar to no. 6; differs in having a brittle border to the fruit.**

6. Northern tickseed-sunflower *Bidens coronata*
Plants 1–5′, hairless. Leaves pinnately compound, ≤6″, leaflets 3–7, narrowly oval or linear, toothed, pointed. Heads 1–2″; disk yellow, ¼–½″; rays 8. Wet places.

  b. Disk dark red or purple

7. Thread-leaved tickseed *Coreopsis verticillata*
Plants 6–44″, **hairy at nodes only**. **Leaflets 3**, linear, 1¼–2½″, central leaflet often lobed, stalkless. Heads 1¼–2½″; disk dark reddish-purple, ¼–⅜″; rays 8. Dry, open woods; clearings.

*8. Plains tickseed *Coreopsis tinctoria*
Plants 16–48″, **hairless**. **Leaves 1–2 times compound**, 2–4″; leaflets linear to narrowly oval. Heads numerous, 1–2″; disk red-purple, ¼–½″; rays 8. Moist, low places; disturbed sites.

1. Few-bracted beggar-ticks

2. Devil's beggar-ticks

3. Spanish needles

4. Tall beggar-ticks

5. Midwestern tickseed-sunflower

6. Northern tickseed-sunflower

7. Thread-leaved tickseed

8. Plains tickseed

LEAVES whorled
LEAVES simple
PETALS various

Miscellaneous species with whorled leaves
Several families

a. Petals 3 (Bunch-flower family)
1. Yellow trillium *Trillium luteum*
Plants 4–24″. **Leaves in whorls of 3, oval, mottled,** 2¾–5½″, stalkless. Flowers solitary, terminal, stalkless, 3–4½″, lemon-scented; sepals spreading, 1¼–1½″; petals 1½–2½″. Rich woods. See also pp. 64, 126, 258, 350.

a. Petals 4 (Madder family)
***2. Yellow bedstraw** *Galium verum*
Plants 8–30″, finely hairy. **Leaves in whorls of 6–8, linear,** ½–1½″, pointed, densely hairy. Flowers on dense, showy panicles; flowers <⅛″. Roadsides, fields. See also pp. 126, 244, 352.

A. Piedmont bedstraw *Cruciata pedemontana* (not illus.)
Plants 4–14″, hairy and prickly. **Leaves in whorls of 4, oval or elliptic,** 1¼–4¼″, pointed, slightly hairy. Flowers <⅛″. Disturbed areas.

a. Petals 5 (Smartweed family)
3. Shale-barren buckwheat *Eriogonum allenii*
Plants erect, 12–20″. **Leaves in whorls of 3–5, oval or oblong,** 2–6″, hairy, long-stalked. Flowers on broad, flat-topped panicles, ¼″. Shale-barrens.

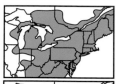

a. Petals 6 (Lily and Ruscus families)
4. Indian cucumber-root *Medeola virginiana*
Plants erect, 8–36″, woolly when young. **Leaves in whorls of 5–11, narrowly oval, broader toward tip,** 2½–4¾″, sharp-pointed. **Flowers** 3–9, nodding, on umbels, ½″. Rich woods. Ruscus family.

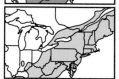

5. Wild yellow lily *Lilium canadense*
Plants erect, 1½–7′, hairless. **Leaves in whorls of 4–12, narrowly oval or linear,** 3–7″. **Flowers** 1–5, nodding, **2–3″;** petals slightly recurved. Moist or wet meadows. See orange lilies p. 138. Lily family.

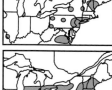

a. Petals irregular (Orchid family)
6. Small whorled pogonia *Isotria medeoloides*
Plants 3–6″, gray-green. Leaves in whorls of 5, drooping, oval, broader toward tip, 1¼–2½″. **Flowers** solitary, terminal, **1–2″,** light green or yellow-green; **sepals arching,** ½–1″; petals ½–⅝″; lip ⅜–½″. Open woods.

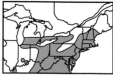

7. Large whorled pogonia *Isotria verticillata*
Plants 8–16″. Leaves in whorls of 5, oval, broader toward tip, 1¼–2″. **Flowers** solitary, terminal, **3–5″; sepals projecting forward to spreading, linear,** 1½–2½″, dark purple; lateral petals ½–1″; lip ¾–1″. Acid woods.

1. Yellow trillium

*2. Yellow bedstraw

3. Shale-barren buckwheat

4. Indian cucumber-root

5. Wild yellow lily

6. Small whorled pogonia

7. Large whorled pogonia

3½"

2½"

Yellow-eyed grasses
Yellow-eyed grass family

Yellow-eyed grasses have grass-like leaves and yellow flowers with three petals. They are difficult to identify without relying on technical characters of the sepals. The sepals are either inserted (not prolonged beyond the subtending bracts) or exserted (prolonged beyond the bracts), and they are folded with the central crease (keel) often fringed or ragged (see inserts).

a. Leaf sheath brown or straw-colored

1. Fringed yellow-eyed grass *Xyris fimbriata*
Plants 24–60″, rough. Leaves ascending, linear, strap-shaped, 2–28″, flat or twisted, **straw-colored or reddish at base.** Flowers many on spikes ⅜–1″ tall; **flowers ½″**; lateral sepals exserted; **keel long-fringed** above middle; opening in morning. Swamps, pond margins.

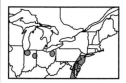

A. Carolina yellow-eyed grass *Xyris caroliniana* (not illus.)
Plants 12–44″, smooth, wiry. Leaves linear, 8–20″, twisted and flexuous, **dark brown at base.** Flowers few-many on spikes ¼–1¼″ tall; **flowers ¾″**, yellow or white; lateral sepals exserted; **keel fringed** below, long-fringed or jagged above; usually opening in the afternoon. Moist or well-drained, sandy or peaty soil.

B. Richard's yellow-eyed grass *Xyris jupicai* (not illus.)
Plants 8–32″. Leaves linear, 4–24″, yellow-green, flat, **pale at base.** Flowers on spikes ¼–½″; **flowers ¼″**; lateral sepals included; **keel jagged** more than half the length; opening in morning. Wet, sunny areas, often weedy.

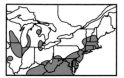

a. Leaf sheath red

2. Slender yellow-eyed grass *Xyris torta*
Plants 6–32″. **Leaves** linear, 8–20″, **twisted**, dark green. Flowers many on spikes ¼–1″; **flowers ¼″**; lateral sepals included; **keel minutely fringed with terminal tuft of hair;** opening in the morning. Wet, acid soil.

3. Small's yellow-eyed grass *Xyris smalliana*
Plants 12–60″. **Leaves** linear, 6–24″, deep green, **flat**, reddish at base. Flowers on spikes ⅜–1¼″; **flowers ½″**; lateral sepals exserted; **keel jagged;** opening in afternoon. Wet, low ground.

4. Bog yellow-eyed grass *Xyris difformis*
Plants 8–32″. **Leaves** linear, 2–24″, dark-green, **flat.** Flowers on spikes ¼–¾″; **flowers ¼″**; lateral sepals included; **keel jagged** in upper ½; opening in morning. Wet sands and peats, often in river-swamps.

C. Northern yellow-eyed grass *Xyris montana* (not illus.)
Plants 2–12″. **Leaves** linear, 1½–6″, dark green, **flat**, reddish at base. Flowers few on spikes <⅜″; **flowers ¼″**; lateral sepals included; **keel only slightly ragged toward tip;** opening in morning. Bogs and tamarack-swamps.

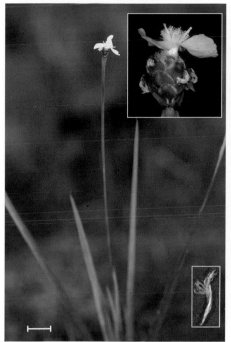

1. Fringed yellow-eyed grass

2. Slender yellow-eyed grass

3. Small's yellow-eyed grass

4. Bog yellow-eyed grass

LEAVES basal
LEAVES simple or
lobed
PETALS 3-6

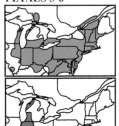

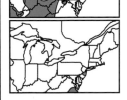

Miscellaneous basal-leaved species
Several families

a. Petals 3 (Iris family)
*****1. Yellow iris** *Iris pseudacorus*
Plants 1½–3′. **Leaves** stiff and erect, **sword-shaped**, 1½–3′. Flowers 4″, yellow or cream-colored; sepals recurved; petals erect. Swamps, streams, ponds. See blue iris p. 56; orange iris p. 136.

a. Petals 4 (Poppy family)
2. Celandine-poppy *Stylophorum diphyllum*
Plants 10–16″. Leaves long-stalked, oblong, or oval; **lobes 5–7,** oblong or oval, blunt; stem leaves 2, smaller than basal. Flowers few on terminal umbels, 2″; sepals hairy. Rich, moist woods.

a. Petals 5 (Pitcher-plant family)
3. Yellow pitcher-plant *Sarracenia flava*
Plants ≤3′, **Leaves forming pitchers**, erect, 12–36″, covered by a hood. Flowers 3½–4″. Sandy bogs, wet pinelands. See also p. 130.

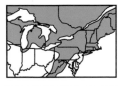

a. Petals 6
  b. Flowers hairless or nearly so (Lily and Star-grass families)
4. Corn-lily *Clintonia borealis*
Plants 6–16″, hairy above when young. **Leaves** dark green, **oblong, elliptic, or oval, broader toward tip**, 4–12″, sharp-pointed, fringed. **Flowers 2–8 on umbels,** nodding, ½–¾″, downy outside. Rich, moist woods; thickets; wooded bogs. Lily family. See also p. 368.

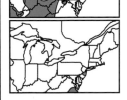

5. Yellow trout-lily *Erythronium americanum*
Plants 4–10″. **Leaves** mottled, **elliptic or narrowly oval**, ¾–2″, dying back after flowering. **Flowers solitary**, ½–2½″; petals recurved, often spotted within or with dark color on outside. Rich woods, bottomlands, meadows. The leaf speckling resembles the speckles on a trout. Lily family. See also p. 372.

6. Yellow star-grass *Hypoxis hirsuta*
Plants 2½–16″. **Leaves linear,** 4–24″, hairy, 5–9 nerved. **Flowers 2–7 on umbels,** ¼–1″. Dry, open woods; meadows. Star-grass family.

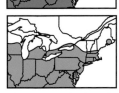

  b. Flowers densely woolly (Bog-asphodel and
  Bloodwort families)
7. Gold-crest *Lophiola aurea*
Plants 1–2′, hairy. Leaves erect, linear, ≤12″, upper leaves bractlike. Flowers in woolly clusters, 2–4″ wide; flowers ¼–⅜″; woolly; **stamens 6**. Acid swamps, pine-barren bogs. Bog-asphodel family.

8. Red-root *Lachnanthes caroliniana*
Plants 8–32″. Basal leaves erect, linear, ≤16″ wide; stem leaves smaller. Flowers in dense, flat or rounded, woolly heads, 1–3″ wide. Flowers ⅜″, woolly; **stamens 3**. Swamps, pine-barren bogs. Bloodwort family.

*1. Yellow iris

2. Celandine-poppy

3. Yellow pitcher-plant

4. Corn-lily

5. Yellow trout-lily

6. Yellow star-grass

7. Gold-crest

8. Red-root

LEAVES basal and alternate
LEAVES simple or lobed
PETALS 1–12 or irregular

Spearworts, Crowfoots, Marsh-marigold, and yellow Violet
Crowfoot and Violet families
a. Petals 1–12, regular (Crowfoot family)
  b. Leaves oval, narrowly oval, or linear
1. Creeping spearwort *Ranunculus flammula*
Plants prostrate or creeping, ½–1½'; flowering stems 4–18". **Leaves usually thread-like or linear,** ¼–2½"; sometimes oval or narrowly oval, broader at either end, 1–2", toothless or slightly toothed. **Flowers** solitary, ¼–½"; **petals 5–6**. Sandy, muddy, or gravelly areas.

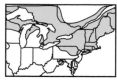

2. Low spearwort *Ranunculus pusillus*
Plants weakly ascending, 4–18". **Basal leaves oval or narrowly oval,** ½–1½"; base blunt or rounded, stalked; stem leaves narrower, toothless or finely toothed. **Flowers 1½"; petals 1–5**. Ditches, swamps, mud, shallow water.

  b. Leaves heart-shaped, kidney-shaped or round
    c. Basal and stem leaves similar in shape
*3. Cursed crowfoot *Ranunculus sceleratus*
Plants fleshy, erect, 8–24". **Leaves kidney-shaped, deeply 3-lobed,** ⅜–2", long stalked, succulent. **Flowers many,** ¼–⅜"; **petals 3–5**. Ponds, ditches, riverbanks.

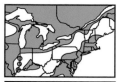

4. Hooked crowfoot *Ranunculus recurvatus*
Plants erect, ½–2', hairy. **Leaves kidney-shaped or round,** ⅜–4", stalked, deeply 3–5-lobed; stem leaves few. **Flowers** 2–16, ¼"; **petals 6**. Woods, stream sides. **Note the short, recurved sepals.**

*5. Lesser celandine *Ranunculus ficaria*
Plants prostrate, 4–14". **Leaves heart-shaped,** ¾–1½", blunt, toothless or bluntly toothed, long stalked; lower leaves apparently opposite. **Flowers 1"; petals 8–12**. Damp shade, weedy places, open woods, often carpeting large areas. **Note the 3 sepals.**

6. Marsh-marigold *Caltha palustris*
Plants erect, 8–24"; stem hollow. **Basal leaves round,** ¾–5", deeply notched at base. **Flowers** 1–7, ½–1½"; **petals 5–9**. Wet woods, meadows, swamps, bogs.

    c. Basal and stem leaves different in shape
7. Kidney-leaved buttercup *Ranunculus abortivus*
Plants ½–2', smooth. **Basal leaves kidney-shaped, round, or heart-shaped,** ½–4"; stem leaves deeply 3–5 lobed, stalkless. **Flowers** few-many, ¼"; **petals 5**. Woods, meadows, fields, clearings. See also pp. 222, 228, 394.

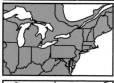

a. Petals 5, irregular (Violet family)
8. Round-leaved yellow violet *Viola rotundifolia*
Plants ¾–1½". **Leaves heart-shaped,** ¾–1¼", enlarging in summer to 2–4¾". **Flowers** solitary on short stalks, ½–¾"; **petals 5;** lateral petals hairy; lower petals with brown veins. Deep, rich woods. See also pp. 12, 58, 166, 296, 378.

1. Creeping spearwort

2. Low spearwort

*3. Cursed crowfoot

4. Hooked crowfoot

*5. Lesser celandine

6. Marsh-marigold

7. Kidney-leaved buttercup

8. Round-leaved yellow violet

LEAVES basal and alternate
LEAVES simple or lobed
RAYS many

Groundsels
Composite family

See other groundsels p. 182.

a. Basal leaves narrowly oval or heart-shaped

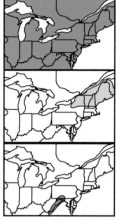

1. Heart-leaved groundsel *Packera aurea*
Plants 8–48″, hairy when young, then hairless. **Basal leaves heart-shaped,** ≤4¼″, blunt, toothed, long-stalked; stem leaves smaller, lobed. Heads ½–¾″; disk ¼–½″. Moist woods, swampy places.

2. New England groundsel *Packera schweinitziana*
Plants 12–40″, hairy when young, then hairless. **Basal leaves narrowly oval,** 1¼–4″, pointed, shallowly toothed, truncate or notched, long-stalked; stem leaves smaller, lobed. Heads ½–¾″; disk ¼–½″. Moist meadows, swampy woods. Heads paler than no. 1.

a. Basal leaves oblong or narrowly oval, broader toward the tip
  b. Plants persistently hairy throughout

3. Shale-barren groundsel *Packera antennariifolia*
Plants 4–16″, hairy. **Basal leaves** elliptic or oval, broader toward tip, ⅜–1½″, shallowly toothed, hairy below; **stem leaves** smaller, **toothless or lobed.** Heads 3–12, ½–1¼″; disk ¼–⅜″. Shale barrens.

A. Southern woolly groundsel *Packera tomentosa* (not illus.)
Plants 8–28″, woolly. **Basal leaves** narrowly oval, elliptic, or oval, ≤8″, toothed or toothless; **stem leaves** smaller, **toothless.** Heads several to many, ¼–⅜″; disk ⅜–½″. Dry, open places; pine woods.

B. Platte groundsel *Packera plattensis* (not illus.)
Plants 8–28″, woolly. **Leaves** elliptic, oval, or round, ≤4″, toothed or sometimes lobed; **stem leaves** smaller, **lobed.** Heads several to many, ¼–⅜″; disk ¼–⅜″. Prairies; dry, open places.

  b. Plants hairy when young, only persistently so at base

4. Appalachian groundsel *Packera anonyma*
Plants 8–32″, densely woolly only at base. **Leaves spoon-shaped or narrowly oval, broader toward tip**, 1¼–4″, toothed, stalked; stem leaves deeply lobed. Heads 20–100, ½–1″, disk ¼–⅜″. Meadows, pastures, roadsides, dry woods.

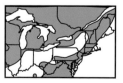

5. Northern meadow groundsel *Packera paupercula*
Plants 2–24″, hairy, becoming hairless. **Leaves elliptic or narrowly oval, broader toward either end,** ⅜–4¾″, stalked, tapering to base, toothed, stem leaves lobed. Heads 2–40 on flat-topped panicles, ⅜–1¼″; disk ¼–½″. Meadows, moist prairies, stream banks, beaches, cliffs.

6. Running groundsel *Packera obovata*
Plants 6–28″, hairy, becoming hairless. **Leaves narrowly oval, broader toward tip, or round**, ¾–8″, tapering or contracted to base, toothed; stem leaves smaller and lobed. Heads several to many on flat-topped panicles, ½–1¼″; disk ¼–½″. Rich woods, rocky outcrops. Northern Metalmark caterpillars feed on this species.

1. Heart-leaved groundsel   2. New England groundsel   3. Shale-barren groundsel

4. Appalachian groundsel   5. Northern meadow groundsel   6. Running groundsel

LEAVES basal
LEAVES simple
RAYS many

Miscellaneous composite species
Composite family

a. Disk present, sap clear
1. Lakeside daisy *Tetraneuris herbacea*
Plants 2–10″. Leaves tufted, narrowly oval, broader toward tip, ⅜–3″, densely hairy when young, dotted. **Heads solitary,** 1¼–1½″; disk ¼–¾″. Prairies, rocky areas.

2. Prairie rosin-weed *Silphium terebinthinaceum*
Plants 2–10′, hairless. Leaves mostly basal, oval, 12–24″, toothed, notched at base, long-stalked. **Heads numerous on panicles,** 1½–3″; disk ½–1½″. Prairies. See also pp. 178, 202, 204.

a. Disk absent; sap milky
   b. Leaves elliptic, narrowly oval, or paddle-shaped.
3. Rattlesnake weed *Hieracium venosum*
Plants erect, 8–32″. **Leaves** elliptic, oval, broader toward either end, 1¼–6¼″, toothed or toothless, **green- or purple-veined, and mottled,** hairy; stem leaves sometimes present. **Heads on open, flat-topped panicles,** ½–¾″. Dry, open woods; clearings. See also pp. 136, 162.

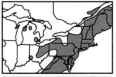

***4. Mouse-ear hawkweed** *Hieracium pilosella*
Plants carpet-forming, 1¼–10″. Leaves narrowly oval, broader toward tip, ½–4¾″, hairy on both surfaces. **Heads solitary, 1–1¼″.** Pastures, fields.

***5. Glaucous king-devil** *Hieracium piloselloides*
Plants 8–40″, **gray-green,** hairy or hairless. Leaves spoon-shaped or narrowly oval, broader toward tip, 1¼–7″, hairy or hairless. **Heads in compact or fairly open clusters,** ½–¾″. Fields, meadows, pastures, roadsides.

***6. Yellow king-devil** *Hieracium caespitosum*
Plants 6–40″, **green.** Leaves narrowly elliptic or narrowly oval, broader toward tip, 2–10″, hairy on both sides. **Heads 5–30 in compact clusters,** ½–¾″. Fields, pastures, along roadsides.

   b. Leaves linear
***7. Showy goat's beard** *Tragopogon pratensis*
Plants ½–2½′ with milky juice. Leaves grass-like, ≤ 12″, smaller above. Heads solitary, long-stalked; **stalk not swollen below head,** 1–2½″; **involucral bracts equal to the rays,** closed by noon. Road-sides, fields, waste places.

***8. Fistulous goat's beard** *Tragopogon dubius*
Plants 1–3′ with milky juice. Leaves grass-like, ≤ 12″. Heads soli-tary, long-stalked; **stalk expanded below head,** 1–2½″; **involucral bracts longer than the rays**. Roadsides; open, relatively dry, places.

1. Lakeside daisy

2. Prairie rosin-weed

3. Rattlesnake weed

*4. Mouse-ear hawkweed

*5. Glaucous king-devil

*6. Yellow king-devil

*7. Showy goat's beard

*8. Fistulous goat's beard

Page 222
LEAVES basal and alternate
LEAVES lobed or compound
PETALS 5–9

Buttercups, Globe-flower, Strawberries, and Cinquefoil
Crowfoot and Rose families

Buttercups have numerous pistils and stamens; the basal and stem leaves are often different in shape or size. Critical characters are in leaf shape and size and in the fruit. See also pp. 216, 228, 394.

a. Leaves irregularly lobed or compound; flowers ≥½″ (Crowfoot family)

1. Hispid buttercup *Ranunculus hispidus*
Plants erect or often reclining, 1–3′, hairy. Basal leaves round or with 3–7 leaflets, 1½–4″; **leaflets** wedge-shaped, **stalked**. Flowers 1–10, ½–1½″; **sepals spreading or reflexed, hairy;** petals 5–8. Swamps, marshes, dry woods, grasslands, roadbanks.

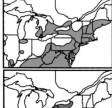

***2. Common buttercup** *Ranunculus acris*
Plants erect, 1–4½′, hairy. Basal leaves kidney-shaped, deeply 3–5-lobed; **lobes stalkless;** stem leaves much smaller. Flowers ½–1½″; **sepals spreading,** hairy; petals 5, notched. Fields, meadows, clearings. **Differs from no. 1 in that the leaf lobes are all unstalked.**

***3. Creeping buttercup** *Ranunculus repens*
Plants creeping and often forming patches; stems 6–18″ tall. Leaves dark green with pale blotches; leaflets 3; **terminal leaflet stalked;** largest leaflets 1–1½″. Flowers ½–1″; **sepals spreading or reflexed,** hairy; petals 5–9. Fields, lawns, roadsides.

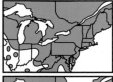

***4. Bulbous buttercup** *Ranunculus bulbosus*
Plants erect, 6–18″, silky-hairy, **bulbous at base.** Basal leaves 3-lobed; **terminal leaflets stalked;** stem leaves few. Flowers long-stalked, 1″; **sepals reflexed,** hairy; petals 5. Fields, meadows, lawns, disturbed areas.

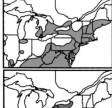

5. Globe-flower *Trollius laxus*
Plants erect, 4–20″. Leaves round, deeply 3–5-lobed, 2–12″; lobes further lobed or coarsely toothed; basal leaves long-stalked; stem leaves stalkless or nearly so. Flowers solitary, 1¼–1½″; **sepals absent; petals 5**–7. Swamps, wet woods, wet meadows.

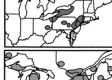

a. Leaves with 3 toothed leaflets; flowers ≤¾″ (Rose family)
6. Barren strawberry *Waldsteinia fragarioides*
Plants 3–8″. Leaves basal; leaflets 3, oval, broader toward tip, toothed and shallow-lobed. **Flowers on racemes,** ⅛–⅜″. Moist or dry woods.

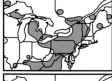

***7. Indian strawberry** *Duchesnea indica*
Plants 1–4″ Leaves basal; leaflets 3, oval or elliptic, ¾–1½″, toothed, sparsely hairy below. **Flowers solitary,** ½–¾″. Lawns, weedy places. Fruit strawberry-like but dry and inedible.

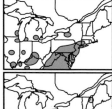

8. Dwarf mountain cinquefoil *Potentilla robbinsiana*
Plants tufted, 2–4″. Leaves only near base; leaflets 3, oval, broader toward tip, ¼–½″, with 2–4 deep teeth. **Flowers solitary,** ¼″. Rocky, alpine areas. See also pp. 172, 302.

1. Hispid buttercup

*2. Common buttercup

*3. Creeping buttercup

*4. Bulbous buttercup

5. Globe-flower

6. Barren strawberry

*7 Indian strawberry

8. Dwarf mountain cinquefoil

LEAVES basal and alternate
LEAVES lobed
RAYS many

Dandelions and related plants
Composite family
Dandelions and relatives have variously lobed, basal leaves and 1–many heads. Critical characters are in the fruit.

a. Heads solitary; plants unbranched

*1. Common dandelion *Taraxacum officinale*
Plants 2–20″, hollow. Leaves narrowly oval, broader toward tip, pinnately lobed, 2½–8″. **Heads solitary,** long-stalked, ¾–2″; **involucral bracts reflexed.** Lawns, weedy sites. Dandelion blooms every month of the year.

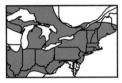

*A. Red-seeded dandelion *Taraxacum laevigatum* (not illus.)
Plants 2–20″, hollow. Leaves narrowly oval, broader toward tip, very deeply pinnately lobed, 2½–8″. **Heads solitary,** long-stalked, ⅜–¾″; **involucral bracts reflexed.** Fields, pastures, lawns. Similar to no. 1, but the flowers are smaller, the involucral bracts have a small hood-like part at the tip, and the seeds are red.

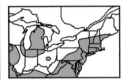

2. Virginia dwarf dandelion *Krigia virginica*
Plants 1¼–16″. First leaves round, toothless; later leaves linear, narrowly oval, or oval, broader toward tip, pinnately lobed, ½–4¾″. **Heads solitary,** long-stalked, ¼–½″; **involucral bracts reflexed later.** Dry, sandy places. See also p. 162.

*3. Big hawkbit *Leontodon hispidus*
Plants 4–24″, hairy or hairless. Leaves narrowly oval, broader toward tip, pinnately lobed or unlobed, ¾–10″. **Heads solitary,** ¾–1½″; **involucral bracts erect.** Fields, meadows, weedy places.

a. Heads several to many; plants often branching
*4. Fall dandelion *Leontodon autumnalis*
Plants 4–32″, **branching. Basal leaves** narrowly oval, broader toward tip, pinnately lobed, 1½–14″, **hairless or slightly hairy. Heads several, 1″.** Roadsides, pastures, fields, meadows, weedy places.

*5. Skeleton-weed *Chondrilla juncea*
Plants 1–5′, **branching.** Basal leaves pinnately lobed, 2–5″, stem leaves linear, ¾–4″. **Heads several to many, sparsely arranged along wiry stem,** ⅜–½″. Roadsides, fields, waste places.

*6. Spotted cat's ear *Hypochaeris radicata*
Plants 6–24″, **branching.** Basal **leaves** narrowly oval, broader toward tip, 1¼–5″, **very hairy. Heads several, 1–1½″.** Lawns, roadsides, pastures, fields, weedy places. Similar to no. 5, but leaves are very hairy.

*7. Asiatic hawk's-beard *Youngia japonica*
Plants 4–36″, **unbranched.** Leaves lobed or lobeless in small plants, ≤8″. **Heads many on panicles,** ≤¼″. Weedy areas.

*1. Common dandelion

2. Virginia dwarf dandelion

^3. Big hawkbit

*4. Fall dandelion

*5 Skeleton-weed

*6. Spotted cat's ear

*7. Asiatic hawk's-beard

LEAVES floating, sub-mersed, or absent
LEAVES lobed or compound
PETALS irregular

Bladderworts
Bladderwort family

Bladderworts have small, bladder-shaped traps that have sensitive hairs around the mouth, which when triggered by passing aquatic invertebrates cause the bladder to vacuum the invertebrate into the bladder where they are decomposed for nutrients. See purple-flowered bladderwort, p. 134.

a. Leaves minute and linear or absent; plants terrestrial

1. Horned bladderwort *Utricularia cornuta*
Plants erect; bladders mingled with roots. Leaves small, under-ground, linear or grass-like; bladders subterranean. Flowers 1–6 on racemes; **flowers ½–1″**; lower lip ¼–½″. Wet shores, peats, sands, mud.

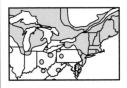

2. Slender bladderwort *Utricularia subulata*
Plants erect. Leaves absent or linear, ≤⅜″; bladders subterranean. Flowers 1–12 on erect racemes; **flowers ⅛–½″**; lower lip ⅛–¼″, 3-lobed. Wet peat, sand, shores, often among sphagnum.

a. Leaves alternate, not inflated; plants submerged or floating

3. Humped bladderwort *Utricularia gibba*
Plants submersed, forming mats. Leaves **dissected into thread-like segments**, ⅛–½″; bladders scattered among leaves. **Flowers 1–7** on erect racemes; **flowers ¼″;** lower lip ⅜″. Usually stranded when flowering. Shallow water.

4. Common bladderwort *Utricularia macrorhiza*
Plants submersed. Leaves numerous, elliptic or oval, ⅜–2″, **divided into thread-like segments**; bladders scattered among leaves, old bulblets on stem very black. **Flowers 6–20** on erect racemes; **flowers ½–1″**; lower lip ¼″, 3-lobed. Deep or shallow, quiet waters.

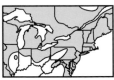

A. Northern bladderwort *Utricularia intermedia* (not illus.)
Plants submersed or stranded on surface of peat in bog depressions. Leaves numerous, 3-parted and **dissected into flat segments**, ¼–¾″; **bladders on branches separate from leaves.** Flowers 2–4 on lax racemes; **flowers ½–¾″**; lower lip ¼–½″. Shallow water.

a. Leaves in whorls of 3–10, inflated; plants floating

5. Inflated bladderwort *Utricularia inflata*
Plants submersed. Floating **leaves** inflated in whorls of 4–10, 1¼–3½″, **widest at or beyond the middle**, with finely dissected branches at tips; submersed leaves bearing bladders on thread-like segments. **Flowers 3–14** on erect racemes; **flowers ⅝–1″;** lower lip ⅜–½″; spur notched at tip, unlobed. Ponds, ditches.

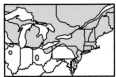

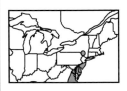

6. Floating bladderwort *Utricularia radiata*
Plants submersed. Floating **leaves** inflated in whorls of 3–8, **with all sides parallel, a little wider beyond the middle,** with finely dissected branches at tips, submersed leaves bearing bladders on thread-like segments. **Flowers 1–5** on erect racemes; **flowers ½″,** lower lip ¼–½″, 3-lobed. Spur not notched at tip. Ponds.

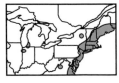

1. Horned bladderwort

2. Slender bladderwort

3. Humped bladderwort

5. Inflated bladderwort

4. Common bladderwort

6. Floating bladderwort

LEAVES emergent, floating, or submersed
LEAVES various
PETALS various

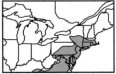

Miscellaneous plants with emergent, submersed, or floating leaves
Several families
a. Petals absent (Arum family)
1. Golden club *Orontium aquaticum*
Plants 8–16″. Basal leaves floating or emergent, elliptic, 2½–8″; floating leaves elongate, pointed. Flowers on dense spikes, ¾–2″ tall; **flowers minute, <⅛″**. Swamps, shallow water.

a. Petals 5–6 (several families)
*2. Yellow floating-heart *Nymphoides peltata*
Plants floating. Leaves floating, round, 2–6″. Flowers on umbels, ¾–1¼″; sepals ½–1″; petals 5. Quiet waters. See also p. 392. Buckbean family.

3. Yellow water-buttercup *Ranunculus flabellaris*
Plants submersed or floating, with a 6–24″ stem. Leaves ½–6″, usually all submersed, dissected into thread-like lobes ⅛″ wide; aerial leaves variable, often kidney-shaped, 3-parted; each leaflet 3-cleft. Flowers 1–7 on racemes, ½–1″; sepals 5, ⅛–¼″; petals 5–6. Quite water, muddy shores. See also pp. 216, 222, 394. Crowfoot family.

4. Water star-grass *Heteranthera dubia*
Plants usually submersed. Leaves linear, ≤6″, blunt. Flowers solitary, ⅜–¾″; tube ½–4¼″ long; outer petals linear; inner petals linear or narrowly oval. Quiet water, mud-flats. See also p. 64. Pickerel-weed family.

a. Petals more than 6 (several families)
5. American lotus *Nelumbo lutea*
Plants submersed or emergent. **Leaves held 1′ above the water,** bowl-shaped, 12–28″ wide; stalk attached at center. **Flowers** solitary, held above the water on stalks, **4–10″**. Ponds, streams, freshwater tidal river margins. See also p. 134. Lotus-lily family.

6. Spatterdock *Nuphar lutea* ssp. *advena*
Plants emergent. **Leaves held above the water,** oval or round, 5–16″, with deep notch. **Flowers** solitary at, or just above, the water surface, ½–1½″; sepals 6; petals numerous. Ponds, streams, tidal streams. Water-lily family.

7. Small pond-lily *Nuphar lutea* ssp. *pumila*
Plants floating. **Leaves mostly floating, sometimes submersed**, elliptic or oval, 1½–4″, with V-shaped notch. **Flowers** solitary at, or just above, the water surface, ⅜–¾″; sepals 5; petals numerous. Ponds, lakes, streams, etc. **Note the red stigma; nos. 6 and 8 have yellow stigmas.** Water-lily family.

8. Large pond-lily *Nuphar lutea* ssp. *variegata*
Plants floating or submersed as juveniles. **Leaves mostly floating, sometimes submersed**, oval or oblong, 2½–14″, with deep notch. **Flowers** solitary at, or just above, the water surface, **1–2″**; sepals 6; petals numerous. Ponds, lakes, streams, ditches. Water-lily family.

1. Golden club

*2. Yellow floating-heart

3. Yellow water-buttercup

4. Water star-grass

5. American lotus

6. Spatterdock

7. Small pond-lily

8. Large pond-lily

**LEAVES absent or
scale-like
PETALS various**

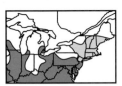

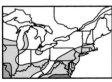

Miscellaneous leafless species
Various families

a. Petals 4
1. Pinesap *Monotropa hypopithys*
Plants 4–16″, hairy. Leaves represented by small scales. **Flowers**
3–10 on racemes, ¼–½″, **nodding;** petals 4. Rich woods. Pinesap is
a saprophyte often found under pines or oaks. See also the white-
flowered Indian-pipe p. 396. Heath family.

2. Bartonia *Bartonia virginica*
Plants erect, 1½–16″. Leaves represented by scales, ≤⅛″. **Flowers**
on racemes or panicles, ⅛″, **erect;** sepals narrowly triangular; petals
4. Sphagnum bogs, wet meadows. See white-flowered bartonia
p. 396. Gentian family.

a. Petals 5
3. Orange-grass *Hypericum gentianoides*
Plants erect, wiry, 4–20″, much branched. Leaves appressed to stem,
scale-like, ≤⅛″. **Flowers solitary in axils,** ⅛–¼″; sepals linear, <⅛″;
petals 5. Dry, open, sandy areas; rocky areas. See other St. John's-
worts p. 190–92. St. John's-wort family.

a. "Petals" many
*4. Coltsfoot** *Tussilago farfara*
Plants 2–20″. Leaves developing after flowering and enlarging
through the season, heart-shaped or round, 2–8″, white-woolly
beneath. **Heads** solitary, 1–1¼″. Disturbed areas, weedy areas.
Composite family.

5. Eastern prickly-pear *Opuntia humifusa*
Plants prostrate or spreading, forming mats; joints of stem succulent,
flattened, oblong or round, 2½–4¾″; **areoles** (areas with spines)
⅜–1″ apart, **spineless or with 1–2 spines;** spines ½–1¼″. **Flow-**
ers solitary, 1½–3″; petals many. Rocks, shores, and sand. Cactus
family.

6. Plains prickly-pear *Opuntia macrorhiza*
Plants prostrate and spreading, forming mats 5′ wide; joints of stem
succulent, flattened, round or oval, broader toward tip, 2½–4″; **areoles**
(areas with spines) ½–1¼″ apart, **with 3–6 spines;** spines 1½–2¼″.
Flowers solitary, 2–2¾″; petals many. Prairies, plains. Cactus family.

1. Pinesap

2. Bartonia

3. Orange grass

*4. Coltsfoot

5. Eastern prickly-pear

6. Plains prickly-pear

LEAVES alternate or basal
LEAVES simple or lobed
PETALS 3–4

Docks, Sorrel, Copperleaves, and Lady's mantle
Several families
The dock fruit is surrounded by the three sepals (called valves); often one or more of these valves has a small grain-like appendage on its side. Critical characters are in leaf size and shape and in the fruit. See also p. 66.

a. Fruit surrounded by 3 valves (Smartweed family)
1. Water dock *Rumex verticillatus*
Plants 1½–4½'. Leaves narrowly oval or linear, ≤12", narrowed at base, flat. Flowers on spike-like racemes, ⅛". **Fruit ¼"; valves broadly triangular, blunt, toothless; grains 3 (1 per valve).** Swamps; pools; streams; low, wet woods.

2. Bitter dock *Rumex obtusifolius*
Plants 1–4'. **Leaves** oval, oblong, or narrow, 6–10", blunt, rounded or lobed at base, **usually red-veined.** Flowers on spike-like racemes, ⅛". **Fruit ¼"; valves triangular, with 2–4 spiny teeth on margin; 1 valve with a grain.** Moist, weedy areas.

***3. Curly dock** *Rumex crispus*
Plants 1–4'. **Leaves** narrowly oval, 2½–5", pointed, rounded or notched at base; **margin very wavy.** Flowers on spike-like racemes, ⅛". **Fruit <¼"; valves heart-shaped, toothless; grains 3 (1 per valve).** Weedy areas. Bronze Copper caterpillars feed on nos. 2 and 3.

4. Wild sorrel *Rumex hastatulus*
Plants ½–4'. **Basal leaves numerous;** stem leaves narrowly oval, broader at either end, or linear, ¾–2", often lobed at base. Flowers on panicles 4–12" tall; flowers <⅛". **Fruit <⅛"; valves winged; grain absent.** Sandy soil.

a. Fruit surrounded by lobed bracts (Spurge family)
5. Rhombic copperleaf *Acalypha rhomboidea*
Plants 8–24", hairy or hairless. Leaves narrowly oval or oval, 1¼–3½". Flowers axillary, <¼"; male flowers at tip with 4 sepals; **female flowers in axils; sepals 3, surrounded by lobed bracts; bracts with 5–9 lobes.** Open woods, roadsides, gardens, weedy areas.

6. Virginia copperleaf *Acalypha virginica*
Plants 8–24", hairy. Leaves narrowly oval, 1¼–3". Flowers axillary, <¼", male flowers at tip with 4 sepals; **female flowers in axils; sepals 3, surrounded by lobed bracts; bracts with 10–15 lobes.** Dry or moist, open woods, fields, roadsides.

a. Fruit not surrounded by valves or bracts (Rose family)
***7. Lady's mantle** *Alchemilla monticola*
Plants 8–24". Leaves kidney-shaped, 1¼–4" wide, 5–9-lobed. Flowers in small clusters on panicles; flowers ⅛"; sepals 4. Fields, lawns.

1. Water dock

2. Bitter dock

Male fls

*3. Curly dock

Male fls

4. Wild sorrel

5. Rhombic copperleaf

6. Virginia copperleaf

*7. Lady's mantle

Lamb's quarters and related species
Amaranth family

Species on this page are part of a large group of plants with very small, green flowers. Critical characters are leaf size and shape and fruit characters.

a. Plants, especially the flowers, white-mealy
*1. Lamb's quarters *Chenopodium album*
Plants erect, ¼–3′. Leaves oval or narrowly oval, 1¼–4″, white-mealy, tapered to base. Flowers in small clusters, the clusters on panicles; flowers ⅛″, white-mealy. **Fruit not surrounded by bracts.** Fields, roadsides, waste places.

*2. Halberd-leaved orache *Atriplex patula*
Plants erect, 1–3′. **Leaves narrowly oval,** 1½–4¾″. Flowers numerous on spikes; flowers ⅛″. **Fruit enclosed in 2 oval or triangular bracts, ⅛–¼″, thin, not spongy-thickened.** Weedy areas.

3. Grass-leaved orache *Atriplex littoralis*
Plants erect, ≤5′. **Leaves linear or narrowly oval,** ¾–4¾″. Flowers on spikes; flowers ⅛″. **Fruit enclosed in 2 broadly triangular or oval bracts, ⅛–¼″, spongy-thickened.** Sea beaches, saline habitats.

*4. Triangle orache *Atriplex prostrata*
Plants erect 1–3′. **Leaves triangular,** ¾–4″, often with basal lobes. Flowers on spikes; flowers ⅛″. **Fruit enclosed in 2 triangular bracts, ⅛–⅜″, thin or spongy-thickened.** Sea beaches, salt marshes, saline habitats.

a. Plants with yellow, sticky hairs
*5. Clammy goosefoot *Chenopodium pumilio*
Plants spreading or prostrate, 8–16″. Leaves oblong or narrowly oval, ⅜–1¼″, coarsely toothed, densely yellow-hairy. Flowers in small clusters; clusters on axillary spikes; flowers ⅛″. Weedy places.

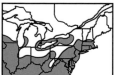

6. Mexican-tea *Chenopodium ambrosioides*
Plants erect, 1–3′. Leaves narrowly oval or oval, ¾–4¾″, pointed, pinnately lobed or toothed, sticky-hairy. Flowers in small clusters; clusters on long spikes; flowers ¼–⅜″. Roadsides, weedy places.

a Plants hairy or hairless
7. Winged pigweed *Cycloloma atriplicifolium*
Plants erect or spreading, 4–32″. **Leaves narrowly oval,** ¾–3″, **coarsely toothed,** falling early. Flowers on broad panicles or spikes; flowers ⅛–¼″, **winged in fruit (see inset).** Dry or sandy areas, weedy areas.

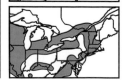

*8. Summer-cypress *Kochia scoparia*
Plants erect, 1–5′. **Leaves** linear or narrowly oval, ¼–2¾″, **toothless,** somewhat hairy. **Flowers** on hairy spikes; flowers ⅛″, **star-shaped in fruit.** Fields, roadsides, weedy places.

*1. Lamb's quarters

*2. Halberd-leaved orache

3. Grass-leaved orache

*4. Triangle orache

*5. Clammy goosefoot

6. Mexican-tea

7. Winged pigweed

*8. Summer-cypress

LEAVES alternate
LEAVES simple
PETALS 5

Pigweeds and related species
Amaranth family

a. Leaves fleshy or spiny
1. Common saltwort *Salsola kali*
Plants 2–20″, branches prostrate and ascending. Leaves linear,
≤1¼″. Flowers axillary, ¼″; sepals stiff, spiny-tipped. Sea-beaches.
The Russian thistle (*Salsola tragus*) is similar and found throughout
the range. It differs in having broader, winged fruit.

*2. White sea-blite *Suaeda maritima*
Plants prostrate to erect, 4–20″. Leaves linear, ⅜–2″, pointed.
Flowers in small clusters, mostly axillary; flowers ⅛″; sepals equal,
rounded. Salt-marshes.

3. Horned sea-blite *Suaeda calceoliformis*
Plants prostrate to erect, 2–32″. Leaves linear, ⅜–1½″, pointed.
Flowers on spikes ⅜–2½″ tall; flowers ≤⅛″; sepals unequal, pointed,
one larger than the other 4. Salt-marshes, roadsides.

a. Leaves not fleshy or spiny
  b. Male and female flowers on different plants
4. Salt-marsh water-hemp *Amaranthus cannabinus*
Plants erect, 3–8′. Leaves narrowly oval, ≤8″, blunt. Flowers on
terminal, narrow spikes or panicles; flowers ¼″; male flowers with
5 equal sepals; **female flowers without sepals**. Salt and brackish
marshes.

5. Rough-fruited water-hemp *Amaranthus tuberculatus*
Plants erect, 1½–7′. Leaves oblong or oval, ½–6″. Flowers on
elongate spikes or panicles; flowers ¼″; male flowers with 5 unequal
sepals; **female flowers with 1–2 sepals**. Sandy fields, weedy places.

*6. Sandhill amaranth *Amaranthus arenicola*
Plants erect, ½–7′. Leaves oval or narrowly oval, 1¼–4″, long-
stalked. Flowers on terminal panicles 4–16″ tall; flowers ¼″; male
flowers with 5 unequal sepals; female flowers with 5 recurved
sepals. Dry, sandy areas.

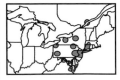

  b. Male and female flowers on the same plant
*7. Rough pigweed *Amaranthus retroflexus*
Plants erect, ≤7′. **Leaves** oval, ≤4″, **pointed**. Flowers on terminal
panicles and smaller axillary panicles, ⅛″; **sepals narrowly oval,
broader toward tip, slightly out-curved, rounded to spiny tip**.
Weedy areas.

*A. Purple amaranth *Amaranthus blitum*
Plants prostrate or erect, ≤2′. **Leaves** diamond-shaped or oval,
broader at either end, ⅜–1½″, **notched at tip**. Flowers mostly in
axillary clusters but also in short, terminal panicles; flowers ⅛″;
sepals oblong, erect. Weedy areas.

1. Common saltwort

*2. White sea-blite

3. Horned sea-blite

Male plant

Female plant

4. Salt-marsh water-hemp

5. Rough-fruited water-hemp

*6. Sandhill amaranth

*7. Rough pigweed

LEAVES alternate
LEAVES simple or
compound
PETALS 3–6

Miscellaneous alternate-leaved species (1)
Several families

a. Petals 3 (Meadow-foam family)
1. False mermaid *Floerkea proserpinacoides*
Plants prostrate or erect, 2–12″. **Leaves compound;** leaflets 3–7, linear, elliptic or narrowly oval, broader toward tip, ¼–¾″. Flowers axillary, ¼″; sepals narrowly oval; petals narrowly oval. Damp woods.

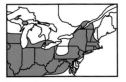

a. Petals 4 (Nettle family)
2. Pellitory *Parietaria pensylvanica*
Plants erect or ascending, 4–16″. Leaves narrowly oval, 1¼–3″, pointed, toothless. Flowers axillary, <⅛″. Dry woods, moist areas, sometimes weedy areas.

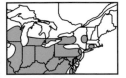

a. Petals 5 (Hemp and Violet families)
3. Green violet *Hybanthus concolor*
Plants erect, ≤3′. **Leaves broadly elliptic or oval,** 2¾–6¼″. Flowers 1–several in axils, ⅛″; sepals linear. Rich woods, ravines. Violet family.

4. Common hops *Humulus lupulus*
Plants vines, ≤40′. **Leaves 3-lobed,** 1¼–6″. Female flowers on spikes ⅜″ long, becoming 1¼–2¼″ tall in fruit; flowers ¼″. Moist sites. See another photo of this speices p. 346. Hemp family.

a. Petals 6 (cont. on next page) (Bunch-flower and Ruscus families)
Solomon's seal refers to the round "seal"-shaped scars left on the rhizomes where the earlier stems grew. Solomon's seal is pollinated by bumblebees.

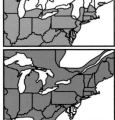

5. Smooth Solomon's seal *Polygonatum biflorum*
Plants arching or erect, 1–4′, **hairless.** Leaves narrowly oval or oval, 2–6″, often clasping. Flowers 1–15 in axillary clusters, ½–1″ long. Moist woods. Ruscus family.

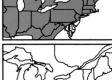

A. Hairy Solomon's seal *Polygonatum pubescens* (not illus.)
Similar to no. 5, but hairy. Moist woods and thickets. Ruscus family.

6. Appalachian bunch-flower *Melanthium parviflorum*
Plants 2–5′. **Leaves flat,** narrowly elliptic to oval, broader toward tip; upper leaves linear. Flowers on panicles 1–2′ tall; flowers ¼–½″. Moist, wooded slopes. See also p. 368. Bunch-flower family.

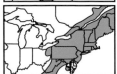

7. False hellebore *Veratrum viride*
Plants erect, ≤7′. **Leaves pleated,** oval or elliptic, ≤12″, often clasping. Flowers on panicles 8–20″ tall; flowers ½–1″. Swamps, wet woods. Among the first plants to send up leaves in the spring. Bunch-flower family.

1. False mermaid

2. Pellitory

3. Green violet

4. Common hops

5. Smooth Solomon's seal

6. Appalachian bunch-flower

7. False hellebore

LEAVES alternate
LEAVES simple
PETALS 6 or absent

Miscellaneous alternate-leaved species (2)
Several families

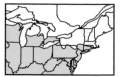

a. Petals 6 (Catbrier and Yam families)
1. Wild yam *Dioscorea villosa*
Plants twining, ≤20′. **Leaves alternate, opposite, or whorled,** oval, 2–5″, pointed. **Flowers on axillary spikes** 2–4″ tall; flowers ⅛–¼″. Open woods, thickets, roadsides. Yam family.

2. Carrion-flower *Smilax herbacea*
Plants herbaceous vines, ≤8′. **Leaves** oval or round, 2¾–4¾″, blunt or pointed; base notched, **green below,** most with tendrils. **Flowers 20–100 on axillary umbels,** ¼. Open woods, roadsides, thickets. Catbrier family.

3. Sawbrier *Smilax glauca*
Plants woody vines, ≤15′. **Leaves** oval, 2–3¾″, pointed, base notched, **whitened beneath. Flowers 5–12 on axillary umbels,** ¼–½″. Woods, roadsides, thickets, hedgerows. Catbrier family.

4. Catbrier *Smilax rotundifolia*
Plants woody vines, ≤20′. **Leaves** oval, round, or rounded-triangular, 2–5″, pointed or blunt, green below, 5–7-nerved. **Flowers 5–12 on numerous axillary umbels,** ¼″. Open woods, thickets, roadsides. Catbrier family.

a. Petals irregular (Orchid family)
5. White adder's-tongue *Malaxis brachypoda*
Plants 4–10″. **Leaf solitary, oval or elliptic,** 1¼–2½″, clasping. Flowers on very slender racemes; flowers ⅛″; sepals oval or narrowly oval; lateral petals narrowly oval, broader toward tip; lip greenish-white, ⅛″, heart-shaped. Damp woods, bogs.

a. Petals absent (Composite family)
6. Field sagewort *Artemisia campestris*
Plants 4–40″ tall, not strongly scented. **Basal leaves highly divided,** ¾–4″, hairy or hairless; stem leaves smaller and less divided. Heads on small spikes or diffuse panicles, ⅛″. Open places, sandy areas. See also pp. 180–82.

***7. Mugwort** *Artemisia vulgaris*
Plants 1½–6′, very aromatic. **Leaves** oval, broader toward either end, 2–4″, **deeply lobed,** densely white-hairy below. Heads in panicles, ¼″. Fields, roadsides, weedy places.

8. Common cocklebur *Xanthium strumarium*
Plants 8–60″. Leaves oval, round or kidney-shaped, ≤6″, often 3–5-lobed. **Heads** in axillary clusters, cylindrical or round, ⅜–1½″, covered with **prickles in fruit.** Fields, weedy places, beaches.

1. Wild yam

2. Carrion-flower

Male flowers

Female flowers

3. Sawbrier

4. Catbrier

5. White adder's-tongue

6. Field sagewort

*7. Mugwort

8. Common cocklebur

LEAVES opposite and alternate
LEAVES simple
"PETALS" various

Spurges
Spurge family

Spurges have unusual flowers; what appears to be the flower is an inflorescence; what appears to be the ovary is a female flower; and what appears to be a stamen is a male flower. Bracts, sometimes petal-like, are attached at a the top of a cup from which these highly modified flowers arise. Critical characters are leaf shape and size, inflorescence type, and "petal" number. See other spurges pp. 284, 344.

a. Plants erect
   b. Leaves alternate; opposite immediately below inflorescence

*1. Cypress spurge *Euphorbia cyparissias*
Plants erect, 6–12″. **Leaves** alternate, very numerous, crowded, linear, needle-like, ½–3″. Flowers on umbels with ≥10; flowers ⅛″; "petals" 2, moon-shaped. Roadsides, weedy areas, old cemeteries.

*2. Leafy spurge *Euphorbia esula*
Plants erect, 8–36″, hairless. **Leaves** alternate, linear or narrowly oval, broader toward tip, ⅜–1¼″, blunt or short-bristle-tipped. Flowers on umbels with 7–15 spokes; flowers ⅛″; "petals" 2-horned. Dry soil, roadsides, sandy banks, fields.

   b. Leaves mainly opposite
*3. Caper spurge *Euphorbia lathyris*
Plants unbranched, ≤6′, hairless. **Leaves opposite,** linear or narrowly oblong, 2½–6″, truncate or clasping at base, stalkless. **Flowers on umbels;** flowers ⅛″; "petals" 2-horned. Roadsides, weedy places.

4. Purple spurge *Euphorbia purpurea*
Plants branching toward top, 2–3′, hairless. **Leaves opposite,** elliptic or narrowly oval, broader at either end, 2–4″, blunt, entire. **Flowers on umbels with 5–8 spokes;** flowers ⅛″; "petals" kidney-shaped. Dry or moist woods; rich, swampy woods; thickets.

5. Tinted spurge *Euphorbia commutata*
Plants branched from base, 6–16″, hairless; young stems and leaves red-tinted. **Lower leaves opposite**, numerous, narrowly oval, broader toward tip, ≤½″; **upper leaves alternate**, oval, blunt or notched, stalkless; those subtending the inflorescences are broadly elliptic or oval. **Flowers on umbels with 3 spokes;** flowers ⅛″; "petals" whitish with slender horns. Moist woods, shaded hillsides, streams, rocks.

6. Ipecac spurge *Euphorbia ipecacuanhae*
Plants repeatedly branched, ≤1′, hairy or hairless. Leaves opposite, variable, linear to oval, ≤ 2″. **Flowers solitary on long stalks**, ¼″; "petals" 5, oblong. Sand, pinelands, barrens.

   a. Plants prostrate
7. Seaside spurge *Chamaesyce polygonifolia*
Plants prostrate, forming mats; stems red, hairless. Leaves opposite, linear or oblong, ¼–½″, entire. Flowers solitary in leaf axils, ⅛″; "petals" very small, broadly oval. Sand dunes, beaches.

*1. Cypress spurge

*2. Leafy spurge

*3. Caper spurge

4. Purple spurge

5. Tinted spurge

6. Ipecac spurge

7. Seaside spurge

LEAVES opposite or whorled
LEAVES simple
PETALS 3-5 or irregular

Miscellaneous opposite or whorled-leaved species
Several families

a. Petals 3
1. American Christmas-mistletoe *Phoradendron leucarpum*
Plants 8–16″, brittle. Leaves oblong or oval, broader toward the tip, ¾–2½″, blunt or rounded. Flowers on spikes ⅜–2″ tall from axils; male flowers in 3–6 segments, each with 15–60 flowers; female flowers in 3–5 segments, each segment with 4–11 flowers; flowers ¼″; parasitic on many species of trees. Sandalwood family.

a. Petals 4
2. Common water-purslane *Ludwigia palustris*
Plants prostrate or floating, reddish. **Leaves opposite,** oval or narrowly oval, ½–1″, abruptly narrowed to base, red-veined. Flowers axillary, stalkless, ¼″. Mud, shallow water. See also pp. 146–48, 168. Evening-primrose family.

3. Forest bedstraw *Galium circaezans*
Plants erect or ascending, 8–24″, hairy. **Leaves in whorls of 4,** oval or elliptic, ¾–2″, blunt. Flowers terminal and from upper axils; flowers green-purple, ⅛″, hairy. Dry woods, thickets. See also pp. 126, 210, 352. Madder family.

4. Stinging nettle *Urtica dioica*
Plants erect, 1–10′, stinging-hairy or hairless. **Leaves opposite,** oval or narrowly oval, 2–7″, pointed, toothed, stinging-hairy at least below, rounded or notched at base; stalks ½–2¼″. **Flowers many on racemes or panicles** in leaf axils; flowers ⅛″. Thickets, roadsides, shores, weedy places. Nettle family.

5. False nettle *Boehmeria cylindrica*
Plants erect, 12–40″. **Leaves opposite,** oval or narrowly oval, pointed, toothed, long-stalked. **Flowers on spikes from upper axils;** flowers ⅛″. Moist or wet, shady places. Nettle family.

a. Petals 5
6. Northern stitchwort *Stellaria borealis*
Plants weak, 1–18″, hairless or rough. **Leaves narrowly oval or oblong,** ½–1½″. Flowers 5–50 in terminal cymes, ¼″; petals shorter than sepals or usually lacking. Moist, usually shaded, places. See also pp. 326–28. Pink family.

*7. Knawel *Scleranthus annuus*
Plants low, spreading ≤6″. **Leaves linear,** ¼–1″. Flowers stalkless, ⅛″. Fields, roadsides, weedy places. Pink family.

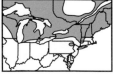

8. Naked mitrewort *Mitella nuda*
Plants erect, 2–8″. **Basal leaves round to kidney-shaped,** toothed; stem single or absent, oval. Flowers on racemes ¾–4″ tall; flowers ¼″. Bogs, wet woods. See also p. 330. Saxifrage family.

1. American Christmas-mistletoe

2. Common water-purslane

3. Forest bedstraw

4. Stinging nettle

5. False nettle

6. Northern stichwort

*7. Knawel

8. Naked mitrewort

LEAVES opposite or alternate
LEAVES simple
PETALS 5 or irregular

Miscellaneous opposite-leaved species
Several families

a. Petals regular (equal size and shape)
1. Green milkweed *Asclepias viridiflora*
Plants 12–32″. Leaves narrowly oval, linear, broadly oblong, or elliptic, 1½–4½″. **Flowers** on lateral umbels, ½″. Dry, upland woods; prairies; barrens. See also pp. 110, 136, 330, 354. Dogbane family.

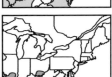

2. Ozark milkweed *Asclepias viridis*
Plants prostrate or ascending, 8–24″. Leaves oblong, narrowly oval or elliptic, 2½–4¾″, rounded or blunt, narrowed to base, stalked ⅛–⅜″. **Flowers** on umbels, ¾–1½″, hoods purple. Prairies, barrens, upland woods. Dogbane family. Note the upright petals, unique among the milkweeds.

a. Petals absent
3. Common ragweed *Ambrosia artemisiifolia*
Plants 12–40″, hairy or hairless. Leaves opposite below, alternate above, **pinnately lobed**, oval or elliptic, 1¼–4″. Heads ¼″. Weedy places. Composite family. This species and the next are the source of a fine pollen causing hay fever.

4. Giant ragweed *Ambrosia trifida*
Plants ≤80″, hairy above, hairless below. **Leaves** opposite, **3–5-lobed**, oval or round, ≤8″. Heads ¼″. Weedy places. Composite family.

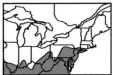

a. Petals irregular
5. Stone-mint *Cunila origanoides*
Plants woody at base, 8–16″, hairless. **Leaves** oval or triangular, ¾–1½″, pointed, hairless. Flowers 3–9 in axillary clusters, ¼″. Dry or rocky woods. Mint family.

6. Heart-leaved twayblade *Listera cordata*
Plants 4–10″. **Leaves** heart-shaped, ⅜–1¼″. **Flowers up to 25 on terminal racemes**, ¼″; sepals and lateral petals oval, reflexed; lip ⅛″. Deep, wet woods; sphagnum bogs. See also p. 258. Orchid family.

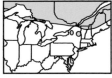

A. Auricled twayblade *Listera auriculata* (not illus.)
Plants 4–10″. **Leaves** oval or round, ¾–2″. **Flowers up to 20 on terminal racemes**, ¼″; sepals and lateral petals narrowly oval or oblong; **lip pale green**, ¼–⅜″, **with two lobes in the middle**. Wet woods. Orchid family.

*7. European twayblade *Listera ovata*
Plants 8–24″. **Leaves** oval or elliptic, **2–6″**. **Flowers 25–60 on terminal racemes**, ¼″; sepals oval; lateral petals linear; lip yellow-green, ¼–⅜″. Wet woods. Orchid family.

1. Green milkweed

2. Ozark milkweed

3. Common ragweed

4. Giant ragweed

5. Stone-mint

6. Heart-leaved twayblade

*7. European twayblade

LEAVES alternate or basal
LEAVES simple
PETALS irregular

Orchids
Orchid family
Bog orchids and fringed orchids have lips prolonged near the base into a spur. Critical characters are the shape and size of the lip and the spur. See other platantheras pp. 36, 138, 298.

a. Leaves on stem
  b. Lip deeply 3-lobed or fringed
1. Ragged fringed orchid *Platanthera lacera*

Plants 12–32″. Leaves narrowly oval or oval, ≤6″; upper leaves much smaller. Flowers on compact spikes, 2–6″ tall; flowers ⅜–½″; sepals oval, ¼–⅜″, reflexed behind the lip; lateral petals linear, ¼–⅜″; **lip deeply 3-lobed and fringed**, ⅜–½″; **spur ½–¾″**. Moist woods.

  b. Lip shallowly lobed or toothed
2. Green woodland orchid *Platanthera clavellata*

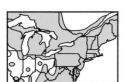

Plants 4–16″. Main leaves solitary, linear, 1¼–8″; upper leaves reduced. Flowers on open racemes, ¾–2¼″ tall; flowers ¼–½″, divergent, **twisted to one side;** sepals arching forward; lateral petals oval, broader at either end; **lip oblong, ⅛–¼″, shallowly 3-lobed; spur ¼–⅜″.** Wet woods, thickets, acid bogs.

3. Tubercled orchid *Platanthera flava*

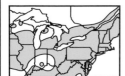

Plants 12–28″. Leaves narrowly oval or elliptic, ≤8″; the upper leaves much reduced. Flowers on loose or compact spikes, 2–8″; flowers ¼″; sepals erect or reflexed; lateral petals oblong or oval; **lip** deflexed, ¼″; **margin irregular with a fin-like protrusion on the upper side; spur ⅛–½″.** Bogs, swamps, flood plains.

4. Bracted orchid *Coeloglossum viride*

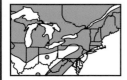

Plants 8–20″. Leaves narrowly oval or oval, broader toward tip, 2–4¾″, reduced upward. Flowers on terminal spikes, 2–8″ tall, with conspicuous bracts; flowers ¼–½″; sepals oval, ⅛–¼″, forming a hood; lateral petals narrowly oval, ⅛″; **lip ¼–⅜″, terminating in 3 lobes; spur ⅛″.** Moist woods.

  b. Lip unlobed, toothless
5. Tall northern bog-orchid *Platanthera aquilonis*

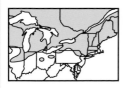

Plants 12–40″. Leaves narrowly oval, broader toward either end, ≤10″. Flowers on compact spikes 2½–8″ tall; flowers ¼–½″, erect or appressed; sepals spreading or reflexed; lateral petals narrowly oval, directed forward; **lip narrowly oval, ⅛–¼″, toothless; spur ≤¼″.** Bogs, wet woods.

a. Leaves basal
6. Hooker's orchid *Platanthera hookeri*

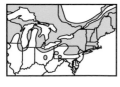

Plants 2¾–16″. **Leaves 2, prostrate**, broadly elliptic or round, 2½–6¼″. Flowers on terminal spikes 2–10″ tall; flowers ¾″, fragrant, ascending; sepals widely reflexed behind flower; lateral petals narrowly oval, incurved, directed downward; lip narrow triangular, directed outward, up-curved, ¼–⅜″; **spur ½–1″.** Rich, dry to moist woods.

1. Ragged fringed orchid

2. Green woodland orchid

3. Tubercled orchid

4. Bracted orchid

5. Tall northern bog-orchid

6. Hooker's orchid

LEAVES basal
LEAVES simple or
compound
PETALS absent or 6

Sweet-flag, Aroids, and Blue Cohosh
Several families

Arums have minute flowers on a thickened spike, called a spadix, usually this is surrounded by a leafy cup-like bract called a spathe. Acorus was once put in the Arum family, but DNA shows it is very distinctive.

a. Leaves simple

1. Sweet-flag *Acorus americanus* & *Acorus calamus*
Plants ≤7′. **Leaves linear,** sword-shaped, ≤7′. Flowers on dense spikes ⅜–3″ tall; flowers minute. Wet, open areas; marshes; swales; edges of quiet water. Plants in the southern part of the range are sterile introductions from Europe. They have a prominent vein on the leaf and are called *Acorus calamus*. Plants in the northern part of the range (from mid-Pennsylvania to mid-Illinois northward) are native and fertile, with obscure veins. They are called *Acorus americanus*. Sweet-flag family.

a. Leaves lobed

2. Arrow-arum *Peltandra virginica*
Plants 8–16″ Leaves arrow-shaped; blade narrow to broad, 4–12″; stalks 8–16″. Flowers on dense, white or orange spikes, 4–8″ tall, surrounded by a green, **ruffled spathe with a pale margin,** 4–8″ tall; flowers minute. Swamps, shallow water. Pollinated by flies. After pollination the spadix curls downward and the spathe helps it auger its way into the mud, where the seeds mature and germinate. Arum family.

a. Leaves compound

3. Jack-in-the-pulpit *Arisaema triphyllum*
Separate male and female plants; female plants larger, 8–24″. Leaves usually 2; leaflets 3, the middle one elliptic to diamond-shaped, the others asymmetric; stalks 8–24″. Flowers on dense, yellow spikes 1¼–8″ tall, surrounded and surpassed by a green or purple spathe, convoluted or often fluted below, 1¼–3½″ tall, **terminal part overarching opening;** flowers minute. Rich or wet woods, thickets, swamps, peat bogs. Pollinated by flies. Arum family.

4. Green dragon *Arisaema dracontium*
Plants 8–20″. Leaves usually 1; leaflets 7–15, elliptic or narrowly oval, broader toward tip; middle ones 4–8″; stalks 8–20″. Flowers on dense, green spikes 3–6″ tall, surrounded at base by a **much shorter, convoluted spathe,** 1¼–2½″ tall; flowers minute. Rich or alluvial woods, thickets, swales. Arum family.

5. Blue cohosh *Caulophyllum thalictroides*
Plants erect, 12–32″. Leaflets oval, broader toward tip, 2–5-lobed, 2–3″. Flowers on panicles or racemes, 1¼–2½″ tall; flowers ¼″; **sepals are petal-like; petals are small, yellow structures.** Rich, moist woods. Recently the plants with purple, ½″ flowers, blooming earlier have been recognized as a separate species, *Caulophyllum giganteum*. It has about the same range. Barberry family.

1. Sweet-flag

2. Arrow-arum

3. Jack-in-the-pulpit

4. Green dragon

5. Blue cohosh

**LEAVES submersed
and floating
LEAVES simple
PETALS 3**

Pondweeds
Pondweed family
Pondweeds are a difficult group of plants to identify because the characters are difficult to see and the differences subtle. Pondweed pollen is buoyant and floats to nearby flowers resulting in pollination. Underwater flowers are pollinated by pollen in bubbles floating to the stigma.

a. Floating leaves absent
*1. Curly pondweed *Potamogeton crispus*
Plants 12–32″. **Submersed** leaves linear, 1¼–3″, rounded, **margins wavy;** floating leaves absent. Flowers on dense spikes ⅜–¾″ long; flowers ⅛″. Waters high in nutrients.

2. Sago pondweed *Stuckenia pectinatus*
Plants 12–32″. **Submersed** leaves linear, 1¼–4″, 1-nerved; **margins not wavy,** pointed; floating leaves absent. Flowers submersed, ⅛″. Calcareous, shallow water. A major food for ducks.

a. Floating leaves present
3. Floating pondweed *Potamogeton natans*
Plants 3–7′. **Submersed** leaves linear, 4–16″, **3–5-nerved**, pointed; floating leaves oval or elliptic, 2–4″. Flowers on dense, cylindric spikes, ⅛″. Ponds and slow streams. **Similar to no. 4, but unbranched or nearly so.**

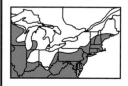

4. Common snailseed pondweed *Potamogeton diversifolius*
Plants ≤5′. **Submersed linear, or thread-like,** ⅜–4″, blunt to sharp-pointed, **1-nerved;** floating leaves narrowly oval to round, 2–16″. Flowers on spikes of 2 types, submersed, axillary and nearly spherical or floating, axillary and cylindric; flowers ⅛″. Shallow water. Note the submersed leaves are generally narrower than in other species.

5. Ribbon-leaved pondweed *Potamogeton epihydrus*
Plants ≤7′. **Submersed leaves linear ≤8″, 5–13-nerved;** floating leaves linear, paddle-shaped or narrowly oval, broader toward tip, 1¼–2½″. Flowers numerous on cylindrical spikes; flowers ⅛″. Ponds, slow water.

6. Variable pondweed *Potamogeton gramineus*
Plants 12–28″, branched. **Submersed leaves linear or narrowly oval, broader toward either end,** 1¼–3″, pointed, **3–7-nerved;** floating leaves narrowly to broadly elliptic, ¾–2″. Flowers on dense cylindric spikes, ⅛″. Lakes, ponds, streams.

*1. Curly pondweed

2. Sago pondweed

3. Floating pondweed

4. Common snailseed pondweed

5. Ribbon-leaved pondweed

6. Variable pondweed

**LEAVES emergent,
submersed or scale-like
LEAVES various
PETALS various**

Miscellaneous aquatic or leafless species
Several families

a. Leaves emergent or submersed (Water-milfoil family)
1. Water-milfoil *Myriophyllum heterophyllum*
Plants ≤4'. Leaves whorled, ¾–1½", pinnately divided into thread-like leaflets. Flowers on emergent spikes, 2–6" tall; flowers ⅛".
Ponds, ditches, streams. See also p. 134.

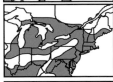

2. Common mermaid-weed *Proserpinaca palustris*
Plants prostrate, flowering branches erect, 4–16". Submersed leaves (if present), oval or broadly oblong, ¾–1½", deeply pinnately lobed into narrow leaflets; emergent leaves linear, ¾–2¼", toothed. Flowers on 1–3 in axils, ⅛". Swamps, wet shores, shallow water.

3. Coastal plain mermaid-weed *Proserpinaca pectinata*
Plants prostrate. Submersed leaves (if present) oval or broadly oblong, ¾–1½", deeply pinnately lobed into narrow leaflets; emergent leaves oval, ½–1¼", deeply pinnately lobed. Flowers 1–3 in axils, ⅛". Sandy bogs.

a. Leaves scale-like (Amaranth family)
Glassworts have succulent stems with opposite, scale-like leaves; the flowers and fruit are sunken into the flesh of the spikes.

4. Dwarf glasswort *Salicornia bigelovii*
Plants erect, 4–16". **Leaves** scale-like, **sharp-pointed**. Flowers on spikes ¾–4" tall; joints ⅛". Salt-marshes.

5. Samphire *Salicornia maritima*
Plants erect or lower branches prostrate, 4–20". **Leaves** scale-like, **rounded**. Flowers on spikes ¾–2½" tall; joints ⅛". Salt-marshes.

6. Perennial glasswort *Salicornia virginica*
Plants woody at base, prostrate and forming mats; flowering stems erect 4–20". **Leaves** scale-like, **blunt**. Flowers on spikes ⅜–1½" tall; joints ⅛". Salt-marshes.

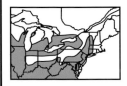

a. Leaves absent (Orchid family)
7. Putty-root *Aplectrum hyemale*
Plants 12–24". Leaves produced late in summer and lasting through winter but gone by flowering-time, elliptic, 4–8", dark green with white veins. Flowers 7–15 on racemes; flowers ¾–1¼"; sepals and lateral petals green tinted with brown, ⅜–⅝"; lip white with purple, oval, broader toward tip, ⅜–½". Rich woods. The flowers do not produce nectar and therefore do not attract many insects. It is believed that most of the seed produced is through self-fertilization.

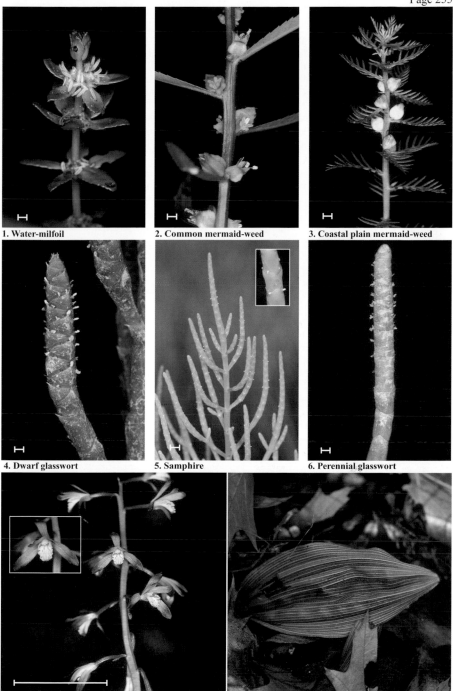

1. Water-milfoil

2. Common mermaid-weed

3. Coastal plain mermaid-weed

4. Dwarf glasswort

5. Samphire

6. Perennial glasswort

7. Putty-root

LEAVES alternate, opposite or whorled
LEAVES simple or compound
PETALS various

Miscellaneous brown-flowered species
Several families

a. Leaves alternate
  b. Leaves simple (Heath and Rock-rose families)
See information on pinweeds on p. 84.

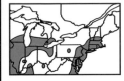

1. Hairy pinweed *Lechea mucronata*
Plants 8–32″, spreading-hairy. Leaves on basal shoots sometimes whorled, oval, ≤½″; **stem leaves elliptic or narrowly oval, broader toward either end**, ⅜–1¼″. Flowers aggregated on short, lateral branches; flower ⅛″; **outer sepals the same length as inner sepals;** inner sepals mostly hairless, keeled. Fields, open woods. Rock-rose family.

2. Narrow-leaved pinweed *Lechea tenuifolia*
Plants 4–12″, with few hairs. Leaves on basal shoots crowded, linear, ⅛–¼″; **stem leaves linear, ≤¾″**. Flowers on panicles, ⅛″; **outer sepals longer than inner sepals;** inner sepals hairy, not keeled. Dry, upland woods. Rock-rose family.

3. Broom crowberry *Corema conradii*
Plants shrubby, ≤20″. Leaves linear, ⅛–¼″. Flowers few in dense clusters; flowers ¼″; petals absent. Sandy areas, exposed rock areas. Heath family.

  b. Leaves compound (Bean family)
4. Small-flowered woolly bean *Strophostyles leiosperma*
Plants hairy. **Leaflets 3**, narrowly to broadly oblong or narrowly oval, ¾–2″, hairy on both sides. Flowers in small clusters on long stalks, ⅛–¼″. Dry or moist, sandy soil; upland woods; dunes; shores. See also p. 100.

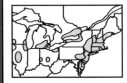

a. Leaves opposite
  b. Petals regular (equal size and shape) (Dogbane family)
5. Black swallow-wort *Cynanchum louiseae*
Plants climbing, 3–6′. Leaves oblong or oval, 2–4″, pointed. Flowers 6–10 on umbels, ≤¼″, purple-black or dull reddish-brown; flower shape and color may vary. Woods; moist, sunny places. See also p. 256.

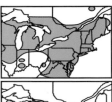

  b. Petals irregular (Scroph family)
6. American figwort *Scrophularia lanceolata*
Plants erect, ≤7′. Leaves oval or narrowly oval, 3–8″, toothed. Flowers on **cylindrical panicles;** flowers ¼–⅜″; **sterile filament (see arrow) yellow-green**. Open woods, roadsides.

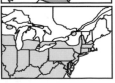

7. Eastern figwort *Scrophularia marilandica*
Plants erect, ≤10′. Leaves oval or narrowly oval, ≤10″, pointed, toothed. Flowers on **pyramidal panicles;** flowers ¼–⅜″; **sterile filament (see arrow) dark purple or brown.**. Open woods.

1. Hairy pinweed

2. Narrow-leaved pinweed

3. Broom crowberry

4. Small-flowered woolly bean

5. Black swallow-wort

6. American figwort

7. Eastern figwort

Page 258
LEAVES opposite or whorled
LEAVES simple
PETALS 3 or irregular

Pogonias, Twayblades and Trilliums
Bunch-flower and Orchid families

a. Leaves alternate (Orchid family)

1. Spreading pogonia *Cleistes divaricata*

Plants 8–24″. **Leaf solitary**, linear, 2–4″. Flowers solitary, 2½–4″; sepals linear, 1¼–2½″; lateral petals pink to purple, narrowly oval, broader toward tip, 1¼–2″, recurved at tip; lip greenish, purple-veined. Swamps, wet woods. The spreading pogonia is pollinated by bumblebees. It does not produce nectar but does produce sticky pollen.

a. Leaves opposite (Orchid family)

2. Large twayblade *Liparis liliifolia*

Plants 4–10′. **Leaves** oval to elliptic, **5–6″**. Flowers 5–30, on loose racemes, ¾″; sepals green-white, ⅜″; lateral petals green to pale purple, ⅜″; lip pale purple, ⅜″, oval, broader toward tip, nearly flat. Rich woods.

3. Southern twayblade *Listera australis*

Plants 6–12″. **Leaves** oval, ⅜–1½″, abruptly pointed. Flowers ≤25 on racemes, ¼″; sepals and lateral petals oval, reflexed, <⅛″; lip ¼–⅜″, narrowly oblong. Shaded bogs, wet woods. See also p. 246. The flowers have a fetid odor; they are believed to be pollinated by fungus gnats. The lip has 4 hairs which are triggered by the visiting insect. Once triggered a dab of glue is squirted onto the insect body, followed by the pollen-mass.

a. Leaves whorled (Bunch-flower family)
See other trilliums on pp. 64, 126, 210, 358.

b. Leaves stalked

4. Prairie trillium *Trillium recurvatum*

Plants 6–18″. Leaves elliptic, oval, or round, usually mottled, 2½–7″, pointed, stalked. **Flowers** solitary, 1½–2½″, erect, stalkless; petals erect, ¾–1¼″, narrowly oval or oval and usually spirally twisted. Moist woods.

b. Leaves stalkless

5. Purple trillium *Trillium cuneatum*

Plants 8–16″. Leaves broadly oval, mottled, 2¾–5½″, pointed, base rounded, stalkless. **Flowers** solitary, **3–5½″**, erect, stalkless, musk-scented; petals erect, 1½–2¾″, elliptic or narrowly oval, broader toward tip. Rich woods. Pollinated by large, green fleshflies.

6. Toad trillium *Trillium sessile*

Plants 4–12″. Leaves oval or round, often mottled, 1½–4″, rounded at both ends, stalkless. **Flowers** solitary, sometimes yellow or green, 1½–2¾″, stalkless; fetid, petals ascending, ⅝–1½″, narrowly to broadly elliptic. Rich, moist woods.

1. Spreading pogonia

3″

2. Large twayblade

3. Southern twayblade

4. Prairie trillium

5. Purple trillium

4½″

6. Toad trillium

**LEAVES basal or
alternate
LEAVES simple
PETALS 3**

Wild Ginger, Heartleaves, Virginia Snakeroot, and related species
Dutchman's-pipe family

These species are low-growing herbs or vines. The flowers occasionally have fetid odors. They all appear to be pollinated by flies, either carrion flies or fungus gnats. Critical characters are flower and leaf shapes. See p. 168 for yellow-flowered species.

a. Flowers urn-shaped or cylindrical
  b. Leaves with a U-shaped or rounded basal notch

1. Wild ginger *Asarum canadense*
Plants hairy. **Leaves round or kidney-shaped, 1½–3″, with a deep, narrow, U-shaped, basal notch.** Flowers urn-shaped, ¾–1½″ long, with spreading or reflexed lobes. Rich woods.

2. Virginia heartleaf *Hexastylis virginica*
Plants hairless. Leaves round, 2–4″, with a deep, narrow, U-shaped, basal notch. Flowers urn-shaped, ⅜–1″ long, cylindric to slightly constricted above or slightly flaring; the lobes erect to spreading. Moist or dry woods.

3. Large-flowered heartleaf *Hexastylis shuttleworthii*
Plants hairless. Leaves round, 2–4″, with a deep, U-shaped basal notch. Flowers ½–1½″, often constricted above, with spreading lobes ¼–½″. Rich woods.

  b. Leaves with a V-shaped basal notch
4. Little-brown-jug *Hexastylis arifolia*
Plants hairless. **Leaves mostly triangular, 2–6″, with deep, V-shaped, basal notch.** Flowers urn-shaped, ½–1″ long, constricted toward tip, with erect to spreading lobes ⅛–¼″. Moist or dry woods. Pollinated by fungus gnats.

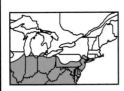

a. Flowers S-shaped
5. Virginia snakeroot *Aristolochia serpentaria*
Plants erect. Leaves oval, oblong, or linear, 2¼–4¾″, with a deep, rounded, basal notch. Flowers solitary, S-shaped, ⅜–¾″. Moist or dry, upland woods. The flower has a flap above the opening which allows flies in but not out until the stamens mature. Pipevine Swallowtail caterpillars feed on *Aristolochia*.

6. Dutchman's pipe *Aristolochia macrophylla*
Plants woody vines. Leaves heart-shaped, 4–16″, hairless or minutely hairy below. Flowers 1–few from axils, S-shaped, enlarged at base, flared at tip, 1½″long, 1″ wide. Rich, mountain woods.

1. Wild ginger

2. Virginia heartleaf

3. Large-flowered heartleaf

4. Little-brown-jug

5. Virginia snakeroot

6. Dutchman's pipe

LEAVES absent
PETALS 5 or irregular

Miscellaneous species without leaves
Several families
a. Petals 5, regular (equal shape and size)

1. Pine-drops *Pterospora andromedea*
Plants 12–36″, **unbranched**, hairy. Leaves scale-like. Flowers numerous on terminal racemes 4–12″ tall; flowers ¼″. Conifer woods. Heath family.

a. Petals irregular

*2. Lesser broom-rape *Orobanche minor*
Plants 6–20″, **unbranched**, sticky-hairy. Leaves absent. Flowers on spikes, ⅜–½″. Parasitic on a variety of plants, often on clover. See also p. 398. A serious pest in clover fields. Produces up to 100,000 seeds in a capsule. Broom-rape family.

3. Beech-drops *Epifagus virginiana*
Plants 4–20″, **branched**. Leaves scale-like, ≤⅛″. Flowers on terminal racemes, ¼–⅜″, white with brown-purple stripes. Parasitic on beech tree roots. The larger, upper flowers are male; the lower flowers are female and fertile. Broom-rape family.

4. Skunk-cabbage *Symplocarpus foetidus*
Plants ≤24″. Basal leaves oval, ≤24″, notched at base, not present at beginning of flowering. **Flowers on a thick, rounded spike, 3–6″ tall, surrounded by a tear-drop-shaped, mottled spathe; flowers minute**. Swamps and moist low areas. See page 250 for a discussion of the inflorescence of Arum family members. The spike produces heat early in the spring, often melting the snow around it and making it very visible. The crushed leaves give off a strong odor of skunk. Arum family.

5. Crane-fly orchid *Tipularia discolor*
Plants 4–12″. Basal leaves oval, ⅜–¾″, stalkless, absent at flowering time. Flowers on terminal racemes, 1–1½″, nodding, ephemeral; sepals and lateral petals narrowly oval, ½–¾″; lip ½–¾″, with 3 green ridges. Rich, moist woods; often on rotting wood. Pollinated by moths. Orchid family.

1. Pine-drops

*2. Lesser broom-rape

3 Beech-drops

4. Skunk-cabbage

5. Crane-fly orchid

LEAVES alternate and basal
LEAVES simple
PETALS 4

Miscellaneous alternate-leaved species
Several families
a. Petals absent; stamens very long (Boxwood family)

1. Allegheny pachysandra *Pachysandra procumbens*
Plants 6–24″. Leaves oval or round, 1¼–3¼″, toothed above, abruptly narrowed to leaf-stalk. Male and female flowers on separate plants, on **1–few spikes from near the base**, 2–4″ tall; petals absent; male flowers with 4 long stamens; female flowers with 3 recurved styles. Fruit a dry capsule. Rich, limey woods.

***2. Japanese pachysandra** *Pachysandra terminalis*
Plants 6–24″. Leaves oval, broader toward tip, 2–4″, tapering to base. Male and female flowers on separate plants, on **terminal spikes;** petals absent; male flowers with 4 long stamens; female flowers with 3 recurved styles. Fruit a berry. Garden escape.

a. Petals present; stamens shorter
  b. Fruit fleshy (Ruscus family)

3. Canada mayflower *Maianthemum canadense*
Plants 2–10′. **Leaves 2, heart-shaped,** 1¼–4″, pointed, notched at base, short-stalked or stalkless. Flowers 12–25 on erect racemes, 1–2″ tall; flowers ¼″. Moist woods. See also p. 286.

  b. Fruit dry (Mustard family)
    c. Fruit round or oblong, unwinged (cont. on next page)

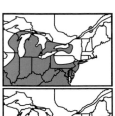

4. Whitlow-grass *Draba verna*
Plants 2–12″. **Leaves crowded at base,** narrowly oval, broader toward tip, or spoon-shaped, ⅜–¾″, hairy; **stem leaves absent.** Flowers on racemes, ⅛″; **petals notched.** Roadsides, fields.

5. Branched draba *Draba ramosissima*
Plants forming mats; flowering stems 3–18″, thinly hairy. Basal leaves narrowly oval, broader toward tip, ½–2″, minutely hairy; **stem leaves narrowly oval. Flowers on diffuse panicles;** flowers ½″. Fruit ⅛–½″, often twisted, thinly hairy. Calcareous cliffs, rocks.

6. Smooth draba *Draba glabella*
Plants matted, 2–16″. Basal leaves narrowly oval, broader toward tip, or spoon-shaped, ¼–2½″, sharp-toothed; **stem leaves oval or oblong,** ¼–2½″, blunt or rounded. **Flowers on loose racemes** 1–6″ tall; flowers ⅛–¼″. Fruit straight, ¼–½″. Rocks, gravel, cliffs.

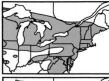

***7. Hoary alyssum** *Berteroa incana*
Plants stiffly erect, 12–40″, pale downy on stems and leaves. **Leaves at base and on stem,** narrowly oval, broader toward either end, ¾–2″, pointed, toothless. Flowers on racemes, ⅛″, **petals notched.** Fruit ⅛–¼″, hairy. Fields, weedy places.

***8. Sweet alyssum** *Lobularia maritima*
Plants 4–12″. **Leaves at base and on stem,** linear or narrowly oval, broader toward tip, ¾–2″, pointed. Flowers on racemes, ⅛″; **petals rounded.** Fruit ⅛″. Escape from cultivation.

1. Allegheny pachysandra

*2. Japanese pachysandra

3. Canada mayflower

5. Branched draba

6. Smooth draba

4. Whitlow-grass

*7. Hoary alyssum

*8. Sweet alyssum

LEAVES alternate and basal
LEAVES simple or lobed
PETALS 4

Mustards (1)
Mustard family
c. Fruit round, winged (cont. from previous page)
d. Stem leaves basally lobed or clasping

*1. Field penny-cress *Thlaspi arvense*
Plants 4–32″, hairless. Lower leaves oval, broader toward tip, stalked, stem leaves oblong or narrowly oval, with a few teeth or toothless, slightly clasping. Flowers on racemes; **flower-stalks ascending;** flowers ⅛″. **Fruit** spherical or broadly elliptic, ⅜–⅝″, **deeply notched, slightly winged; seeds 4.** Roadsides, weedy places, fields.

*A. Thoroughwort penny-cress *Microthlaspi perfoliatum* (not illus.)
Plants 4–12″, hairless. Lower leaves ⅜–1¼″; stem leaves oval or oblong, with a few teeth or toothless, basal lobes blunt or rounded. Flowers on racemes; **flower-stalks spreading;** flowers ⅛″. **Fruit** oval, broader toward tip, ⅜–¼″, **with an open notch, slightly winged; seeds 4.** Roadsides, weedy places, fields.

*2. Field peppergrass *Lepidium campestre*
Plants 8–18″, densely hairy, gray-green. Basal leaves narrowly oval, broader toward tip, entire or shallowly lobed, stalked; stem leaves narrowly oval or narrow oblong, 1–1½″, toothed or not, clasping. Flowers on racemes; flower-stalks divergent; flowers <¼″; stamens 6. **Fruit** oval, ¼″, **indented at tip, broadly winged; seeds 2.** Fields, roadsides, weedy places.

d. Stem leaves stalked
3. Wild peppergrass *Lepidium virginicum*
Plants 6–24″, smooth or minutely hairy. Basal leaves deeply lobed, narrowly oval, broader toward tip, sharply toothed; **stem leaves** smaller, linear or narrowly oval, broader toward tip, **stalked.** Flowers on racemes, ≤⅛″; stamens 2. Fruit broadly elliptic to spherical, widest below middle, ≤⅛″. Dry or moist sites, fields, gardens, roadsides, weedy places

B. Prairie-pepperweed *Lepidium densiflorum* (not illus.)
Similar to no. 3, but the petals are smaller. Roadsides, weedy places, pastures.

c. Fruit triangular, sides wingless (cont. on next page)

*4. Shepherd's purse *Capsella bursa-pastoris*
Plants 4–24″, smooth or hairy. Basal leaves oblong, lobed, 2–4″, dandelion-like; stem leaves much smaller, narrowly oval, toothless or with small teeth, clasping. Flowers on racemes, ≤¼″. **Fruit triangular, ⅛–¼″, notched at tip.** Roadsides, weedy areas, gardens. Principally pollinated by flies.

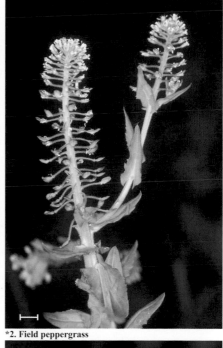

*1. Field penny-cress

*2. Field peppergrass

3. Wild peppergrass

*4. Shepherd's purse

Page 268
LEAVES alternate and basal
LEAVES simple or lobed
PETALS 4

Mustards (2)
Mustard family

c. Fruit linear (cont. from previous page)
d. Fruit erect to spreading

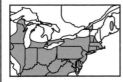

*1. Garlic-mustard *Alliaria petiolata*
Plants 12–36″, **smelling of garlic**. Lower leaves kidney-shaped, others triangular, 1¼–2½″, pointed, coarsely toothed. Flowers on racemes, ⅜–½″. Fruit 1–2½″, divergent. Roadsides, edges, open woods. **Leaves on first-year (non-flowering) plants are heart-shaped, close to the ground; these over-winter. The flowering stem has smaller, triangular leaves.**

2. Lyre-leaved rock-cress *Arabis lyrata*
Plants 4–12″. **Basal leaves deeply lobed, spoon-shaped**, 1–1½″; stem leaves linear, toothless, narrowed to base. Flowers on racemes, ¼–½″. **Fruit 1–2″**, widely ascending, flat. Dry woods, ledges, fields, especially sandy or gravely soil. See also p. 268.

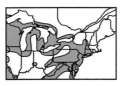

3. Spreading rock-cress *Arabis patens*
Plants 12–24″, hairy. **Basal leaves oval**, ½–2½″, **coarsely toothed**; stem leaves narrowly oval, 1–2″, clasping at base. Flowers on racemes, ½–¾″. **Fruit 1–1½″**, ascending to spreading, linear, flat. Moist, rocky woods.

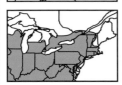

4. Spring-cress *Cardamine bulbosa*
Plants erect, 6–24″. **Leaves round**, ½–1½″, **toothless or toothed**; lower leaves long-stalked; upper leaves stalkless. Flowers on racemes; flowers often pink, ½–1″. **Fruit ½–1″**, ascending or spreading. Moist or wet woods, shallow waters, springs, meadows, bottomland woods. See also pp. 66, 300.

*5. Mouse-ear cress *Arabidopsis thaliana*
Plants 4–16″, hairy at base. **Basal leaves oblong or spoon-shaped**, ⅜–2″, **toothless**, rounded, hairy; stem leaves few, linear or narrowly oblong, pointed, stalkless. Flowers on open racemes, ¼″. **Fruit** ⅜–¾″. Dry fields, roadsides, weedy places.

d. Fruit nodding

6. Sickle-pod *Arabis canadensis*
Plants 12–36″, slightly hairy at base. **Basal leaves narrowly oval or oval, broader toward the tip**, 1–5″; stem leaves narrowly oval or elliptic, 1–4″, sharply and remotely toothed, somewhat hairy. Flowers on racemes, ½″, slightly nodding. **Fruit 2½–4″, nodding, sickle-shaped.** Rich woods, rocky places, thickets.

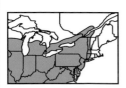

7. Smooth rock-cress *Arabis laevigata*
Plants 12–36″, whitish. **Basal leaves spoon-shaped**, 1–4″, toothed; stem leaves narrowly oval, 1–8″, toothed or not, clasping at base. Flowers on racemes, ¼–½″. **Fruit 2–4″, arching downward**. Rocky woods, ledges.

*1. Garlic-mustard

2. Lyre-leaved rock-cress

3. Spreading rock-cress

4. Spring-cress

*5. Mouse-ear cress

6. Sickle-pod

7. Smooth rock-cress

Bindweeds, Morning-glories, Jimson-weeds, and related species
Morning-glory and Nightshade families

a. Flowers axillary (Morning-glory family)
  b. Flowers >1″

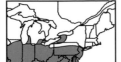

1. Wild potato-vine *Ipomoea pandurata*
Plants trailing or twining, ≤15′, hairless. Leaves oval, 2–6″, narrowly pointed, toothless, usually hairy below, deeply notched. **Flowers** 1–7 on axillary stalks, white with red-purple center, 2¾–4″; **stigma 1**. Dry woods, thickets. See also pp. 14, 70, 90.

2. Hedge bindweed *Calystegia sepium*
Plants twining or trailing, ≤9′. Leaves triangular or oblong, 2–4″, lobed at base, long-stalked. Flowers solitary in leaf-axils; flower-stalks 2–6″; **flowers** pink or white, 1½–2¾″; **stigmas 2**. Thickets, shores, disturbed sites.

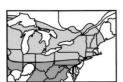

3. Low bindweed *Calystegia spithamaea*
Plants erect, 4–18″, or declining and ≤15′. Leaves oblong or oval, broader toward tip, 1¼–3¼″, broadly pointed or rounded, hairy; base notched or truncate, short-stalked. **Flowers** 1–4 in axils; flower-stalks ¾–3¼″, white or pink, 1½–2″; **stigmas 2**. Dry, rocky fields; sandy fields, open woods.

  b. Flowers ≤1″

*4. Field bindweed *Convolvulus arvensis*
Plants trailing or climbing, ≤3′, often forming mats. Leaves triangular or oblong, ½–2″; basal lobes spreading or descending, long-stalked. **Flowers** 1–2 in axils, ½–1″; **stigmas 2**. Fields, roadsides, weedy places.

5. White morning-glory *Ipomoea lacunosa*
Plants creeping or twining, 3–9′, hairless or nearly so. Leaves oval, 1¼–3¼″, sharply pointed, deeply notched, entire or 3-lobed. **Flowers** 1–5 on axillary stalks, ⅜–¾″; **stigma 1**. Moist fields, thickets, river banks.

6. Pickering's dawnflower *Stylisma pickeringii*
Plants prostrate or creeping, 3–6′. Leaves linear, 1¼–2¼″. **Flowers** 2–7 in axillary clusters, ½″; **stigmas 2**. Dry pine barrens.

a. Flowers terminal (Nightshade family)
*7. Jimson-weed *Datura stramonium*
Plants ≤4½′, hairless. **Leaves** coarsely toothed or lobed, ≤8″, **slightly hairy**. Flowers solitary, terminal, white or purple, 1¼–2″; **calyx winged**. Dry soil, waste places.

*8. Indian apple *Datura wrightii*
Plants ≤4½′, gray-hairy. **Leaves** broadly oval, ≤8″, rounded or notched at base, **velvety beneath**. Flowers solitary, terminal, ≤4¾″; **calyx not winged**. Weedy ground.

1. Wild potato-vine

2. Hedge bindweed

3. Low bindweed

*4. Field bindweed

5. White morning-glory

6. Pickering's dawnflower

*7. Jimson-weed

*8. Indian apple

LEAVES alternate
LEAVES simple
PETALS 4-5

Tearthumbs and Bindweeds
Smartweed family

Critical characters include shape of inflorescence; leaf lobing, shape, and size; and the margin of the tubular sheath surrounding the stem at the base of the leaf-stalk. See pink smartweeds p. 76.

a. Stems barbed (Tearthumbs)

1. Halberd-leaved tearthumb *Polygonum arifolium*
Plants 2–6'. Leaves triangular, 2–8"; **basal lobes** triangular, **divergent**, slender-stalked. Flowers on axillary head-like spikes; stalks barbed, <⅛"; petals 4. Marshes and swamps.

2. Arrow-leaved tearthumb *Polygonum sagittatum*
Plants 3–6'. Leaves arrow-shaped, 1–4"; **basal lobes pointing downward**. Flowers on short, head-like spikes, <½", flowers pink, white, or green, petals 5. Marshes, wet meadows. Much like no. 1, but basal lobes not spreading and stems less barbed.

***A. Mile-a-minute weed** *Polygonum perfoliatum* (not illus.)
Plants climbing, 2–6'. Leaves triangular or somewhat heart-shaped, 1¼–3", notched at base; **upper leaves only represented by expanded leaf-sheaths**. Flowers few on short racemes, ⅛–¼". Thickets, open woods, meadows, fields, roadsides. A serious, invasive species. Differs from nos. 1 and 2 in having upper leaf-sheaths expanded and blue-fruited.

a. Stems not barbed (Bindweeds)

***3. Black bindweed** *Polygonum convolvulus*
Plants ≤3'. Leaves triangular, ≤2½", basally notched or lobed; leaf-sheath smooth. Flowers on spikes 1–2½" long or in axillary clusters, green- or purple-tipped; **fruit ≤¼", wingless or sometimes minutely winged.** Roadsides.

4. Fringed bindweed *Polygonum cilinode*
Plants ≤6'. Leaves oval or triangular, 2–5", pointed; base notched; leaf-sheath oblique, with reflexed bristles at base. Flowers on spikes 1½–4" tall; flowers white or pink-tinged; **fruit ≤¼", wingless.** Dry, rocky woods; slopes.

5. Climbing false buckwheat *Polygonum scandens*
Plants ≤15'; stem reddish. Leaves oblong or oval, 1–5", pointed; base notched, long-stalked. Flowers on axillary racemes; flowers white; **fruit ≤½", winged.** Moist to dry sites.

1. Halberd-leaved tearthumb

2. Arrow-leaved tearthumb

*3. Black bindweed

4. Fringed bindweed

5. Climbing false buckwheat

LEAVES alternate
LEAVES simple
PETALS 5

Knotweeds (1)
Smartweed family

See other knotweeds on p. 76.
a. Flowers axillary (cont. on next page)
   b. Flowers reflexed

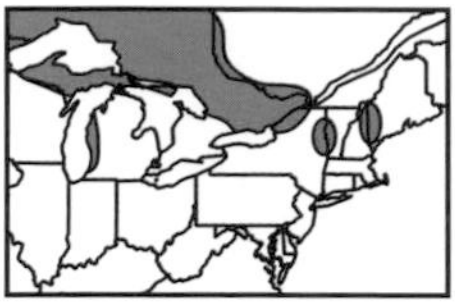

1. Douglas's knotweed *Polygonum douglasii*
Plants ½–5', branches ascending. Leaves linear or narrowly oval, ¾–2", very sharp. Flowers ⅛–¼", cleft nearly to the base; flowers reflexed after opening. Open, rocky areas, gravelly areas.

   b. Flowers erect
      c. Leaves on axillary shoots smaller than other leaves
*2. Field knotweed *Polygonum aviculare*
Plants sprawling or erect, ≤4". **Leaves blue-gray**, oval or narrowly oval, 1–2", those on axillary shoots ⅓ as large; leaf-sheath ¼–⅜", silvery, torn. Flowers ⅛", divided to below the middle. Roadsides, fields.

3. Erect knotweed *Polygonum erectum*
Plants erect, 4–40". **Leaves green or bright yellow-green**, oval or elliptic, 1–3"; leaves on axillary shoots ½ as large; leaf-sheath silvery, entire or slightly torn. Flowers greenish-white, ⅛"; petals cleft ¾ their length. Weedy, open areas.

      c. Leaves all relatively uniform in size
*4. Dooryard knotweed *Polygonum arenastrum*
Plants prostrate, mat-forming, 12–16" wide. **Leaves blue-green or gray-green**, broadly elliptic or oblong, ¼–¾", rounded or blunt; leaf-sheath ⅛–¼", notched and ragged. Flowers <⅛", divided to middle. Sidewalks, streets, dooryards.

A. Narrow-leaved knotweed *Polygonum bellardii* (not illus.)
Very similar to no. 4, but the leaves narrower and the plant more ascending. Beaches, dunes, and shores.

5. Seaside knotweed *Polygonum glaucum*
Plants prostrate, 1(–2)'. **Leaves whitish gray**, narrowly oval or narrowly oblong, ½–1"; leaf-sheath conspicuous, ¼–½", silvery. Flowers white or pink, ⅛", divided nearly to the base. Beaches, dunes.

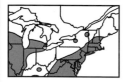

6. Slender knotweed *Polygonum tenue*
Plants 4–20", branches erect or ascending. **Leaves** linear or narrowly oval, ½–1", very sharp, **pleated**. Flowers ¼"; petals ⅛", deeply cleft, pale or roseate. Dry, open areas.

1. Douglas's knotweed

*2. Field knotweed

3. Erect knotweed

*4. Dooryard knotweed

5. Seaside knotweed

6. Slender knotweed

LEAVES alternate or basal
LEAVES simple
PETALS 4–5

Knotweeds (2)
Smartweed family

a. Flowers on terminal inflorescences (cont. from previous page)
  b. Leaves mostly on stem
    c. Leaves thread-like or broadly oval

1. Sand jointweed *Polygonella articulata*

Plants 4–20″, wiry and jointed. Leaves thread-like, ¼–1″. Flowers on erect, compound racemes, 1–1½″ tall; flowers pink, white, or green, <⅛″; fruits slightly larger. Dry, acid sands.

*2. Japanese knotweed *Polygonum cuspidatum*

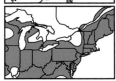

Plants 3–10′, bushy, grayish and often mottled. **Leaves broadly oval, 2–6″,** abruptly pointed; base truncate. Flowers on numerous racemes, 3–6″ long; flowers greenish-white, ¼–⅜″. Weedy areas. Note male and female flowers are on separate plants.

c. Leaves narrowly oval or elliptic

3. Virginia knotweed *Polygonum virginianum*

Plants erect, 1–4′. Leaves narrowly oval or elliptic, ≤6″, hairy; **leaf-sheaths hairy, fringed. Flowers on 1-several spikes with reflexed flowers;** spikes 4–16″ long; flowers greenish-white or pink, ¼″; **petals 4.** Rich woods, thickets.

*4. Common smartweed *Polygonum hydropiper*

Plants 8–24″. Leaves narrowly oval, ≤ 4″, hairless, wavy-margined, extremely peppery tasting; **leaf-sheaths brown or reddish, hairless, short-fringed. Flowers on drooping spikes 1–3″ tall;** flowers greenish- or red-tipped, ⅛″; **petals 4,** covered by glands. Wet shores.

5. Dotted smartweed *Polygonum punctatum*

Plants 1–36″, erect or ascending, smooth. Leaves narrowly oval or elliptic, ≤ 4″, dotted, hairless; **leaf-sheaths hairless or rough, fringed. Flowers on 1–several, much interrupted, erect or arching, spikes ≤ 6″ tall;** flowers greenish or white, <¼″; **petals 5,** covered with glandular dots. Swamps, shallow water, wet places.

6. Mild water-pepper *Polygonum hydropiperoides*

Plants decumbent, becoming erect, ½–3′. Leaves narrowly oval or linear, 2–4″; **leaf-sheaths short-hairy, with <¼″ bristles on top. Flowers on sparsely flowered spikes 1–3″ tall;** flowers ¼″; **petals 5.** Beaches, marshes, shallow water.

7. Dense-flowered smartweed *Polygonum densiflorum*

Plants ≤5′. Leaves narrowly oval, 4–10″, sharp-pointed, **leaf-sheaths cone-shaped, truncate, fringeless. Flowers on slender racemes to 4″;** flowers ⅛″; **petals 5.** Swamps, shallow water.

b. Leaves chiefly basal

*8. Bistort *Polygonum bistorta*

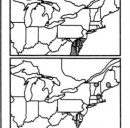

Plants ≤3′. **Basal leaves oblong or oval, 2½–8″,** rounded at base. Flowers on stout racemes 2–4″ tall; flowers ¼″. Fields, meadows, cultivated areas.

1. Sand jointweed

*2. Japanese knotweed

3. Virginia knotweed

*4. Common smartweed

5. Dotted smartweed

6. Mild water-pepper

7. Dense-flowered smartweed

*8. Bistort

LEAVES alternate
LEAVES simple
PETALS 5

Pennyworts and Coinleaf
Umbel family
Pennyworts are low, usually creeping, members of the Umbel family. They have the typical umbel inflorescence but it is often very small. Critical characters are leaf size and shape and inflorescence type.

a. Leaves attached to the stalk in the middle of the blade
1. Water pennywort *Hydrocotyle umbellata*
Plants with floating or creeping stems. Leaves round, ≤2¾″, toothed or shallowly lobed, attached to the stalk in the middle. **Flowers <10 usually on simple umbels;** flower-stalks spreading to reflexed ≤⅜″; flowers ⅛″. Pond-shores, ditches, low grounds.

2. Whorled pennywort *Hydrocotyle verticillata*
Plants with creeping stems. Leaves round, ≤2¼″, with 8–14 shallow lobes, attached to the stalk in the middle. **Flowers <7 per whorl, 2–several whorls;** flower-stalks absent or ≤⅜″, flowers ⅛″. Swamps, shores, low ground.

a. Leaves attached to the stalk at the base along leaf margin
3. Marsh pennywort *Hydrocotyle americana*
Plants with creeping stems. Leaves round to broadly oval, ⅜–2″, shallowly 6–10-lobed. Flowers 2–7 on short umbels; **umbels stalkless; flower-stalks absent or <⅛″;** flowers ⅛″. Moist, low ground.

4. Buttercup pennywort *Hydrocotyle ranunculoides*
Plants with floating or creeping stems. Leaves round or kidney-shaped, ≤2¾″, 5–6-lobed, deeply notched. Flowers 5–10 on umbels; **umbels stalked ½–2¼″; flower-stalks <⅛″;** flowers ⅛″. Shallow water, shores.

*5. Lawn pennywort *Hydrocotyle sibthorpioides*
Plants with creeping stems. Leaves round ≤⅜″, shallowly 7-lobed. Flowers 3–10 on umbels; **umbels stalked ¼–¾″; flowers stalkless,** ⅛″. Lawns.

A. Erect coinleaf *Centella erecta* (not illus.)
Plants with prostrate to creeping stems. Leaves oval or broadly oblong, ⅜–2¼″, blunt, without lobes, shallowly notched at base. Flowers 2–5 on umbels; **umbels stalked ½–4″; flowers stalkless or nearly so,** ⅛″. Marshes, wet soil.

1. Water pennywort

2. Whorled pennywort

3. Marsh pennywort

4. Buttercup pennywort

*5. Lawn pennywort

LEAVES alternate
LEAVES simple
PETALS 4–5

Miscellaneous alternate-leaved species (1)
Several families

a. Flowers solitary
 b. Petals not spreading (Heath family)

1. Creeping snowberry *Gaultheria hispidula*
Plants prostrate, 8–16″, bristly when young. **Leaves** elliptic or round, ¼–⅜″. Flowers bell-shaped, ≤¼″; **petals 4**. Fruit white, with wintergreen taste. Bogs, wet woods, often on decaying logs.

2. Wintergreen *Gaultheria procumbens*
Plants creeping with erect flowering stems, 2–6″. **Leaves crowded toward top of stem, elliptic, 1–2″,** toothless or toothed, smelling of wintergreen when crushed. Flowers solitary, nodding, ¼–⅜″; **petals 5**. Dry or moist woods, clearings.

3. Moss plant *Harrimanella hypnoides*
Plants 1–8″. **Leaves needle-like, ≤⅛″.** Flowers solitary, nodding, bell-like with red sepals, ½″; **petals 5**. Mossy, alpine summits.

 b. Petals spreading (Borage and Nightshade families)

*4. Corn gromwell *Buglossoides arvensis*
Plants 4–30″, minutely downy. **Leaves linear or narrowly oval, broader toward tip, ¾–2½″,** blunt or broadly pointed, 1-nerved. Flowers solitary, stalkless, ⅛–¼″; **petals 5**. Weedy places, sandy soil, roadsides, fields. Borage family.

*5. European gromwell *Lithospermum officinale*
Plants erect, 1½–3′, downy. **Leaves oblong or narrowly oval, broader toward either end, 1¼–3″,** tapering to tip, 2–3-nerved, nearly stalkless. Flowers solitary, ¼″; **petals 5**. Weedy places, roadsides, pastures, open places. Borage family.

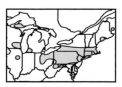

*6. Chinese-lantern plant *Physalis alkekengi*
Plants erect, 16–34″. **Leaves oval, 2–3″,** stalked. Flowers solitary, axillary, ½″; **petals 5;** mature fruit enclosed in a distinctive papery, orange, lantern-like calyx. Weedy places. See yellow-flowered Physalis p. 148. Nightshade family.

a. Flowers in small axillary clusters (Spurge family)
7. Rushfoil *Croton willdenowii*
Plants 4–12″. Leaves linear or narrowly oval, ⅜–1¼″, softly hairy. Flowers on spikes with male flowers above and female below; flowers ⅛″; **petals 4**. Sandy areas, rocky areas.

8. Tooth-leaved croton *Croton glandulosus*
Plants 8–24″, hairy and sticky. Leaves narrowly oblong or oval, 1¼–2¾″, coarsely toothed, with 1–2 glands at junction of leaf and leaf-stalk. Flowers on terminal spikes; male and female flowers on different plants; flowers ¼″; **petals 5** on male flowers, absent on female flowers. Dry or sandy woods, fields.

1. Creeping snowberry

2. Wintergreen

3. Moss plant

*4. Corn gromwell

*5. European gromwell

*6. Chinese-lantern plant

7. Rushfoil

8. Tooth-leaved croton

LEAVES alternate and basal
LEAVES simple
PETALS 4–5 or RAYS 5

Miscellaneous alternate-leaved species (2)
Several families

a. Flowers on umbels
1. Black nightshade *Solanum nigrum*
Plants 6–24″, hairless or minutely hairy. Leaves oval or triangular, ¾–3¼″, irregularly toothed, stalked. **Flowers 5–10 on umbels;** flowers ¼–½″, drooping; petals reflexed; stamens forming a yellow cone. Dry woods, thickets, shores, disturbed habitats. See also p. 14. Nightshade family.

a. Flowers on spikes
2. Lizard's-tail *Saururus cernuus*
Plants 1½ –5′. Leaves oval, 2½–6″, notched at base; stalk sheathing stem. Flowers on 1–2 terminal or apparently axillary spikes, 2½–12″ tall, nodding at tip; flowers ≤¼″; petals absent. Swamps, marshes, shallow water. Lizard's-tail family.

3. Slender cottonweed *Froelichia gracilis*
Plants erect to prostrate, ½–1½′. Leaves linear or narrowly oval, <3″, silky-hairy. Flowers on spikes ½–1¼″ tall; flowers ≤¼″, white-woolly. Dry, sandy soil. Amaranth family

4. Seneca-snakeroot *Polygala senega*
Plants 4–20″. Leaves narrowly oval, 1–3¼″, sharp-pointed. Flowers on dense, terminal racemes ½–1½″ tall; **flowers ⅛″, with broadly elliptic wings longer than petals.** Dry or moist woods, rocks, prairies. See also pp. 82, 136. Milkwort family.

a. Flowers on racemes or panicles
  b. Shrubs
5. Labrador-tea *Ledum groenlandicum*
Plants ≤3′, densely hairy twigs. **Leaves** narrowly oval or narrowly elliptic, ¾–2″, **densely white- or rusty-hairy below;** margins curled under. Flowers on panicles, ⅜″. Bogs, wet shores. Heath family.

6. Meadowsweet *Spiraea alba*
Plants ≤6′. **Leaves** oblong or narrowly to broadly oval, broader toward tip, 1¼–2¾″, toothed, **hairless or nearly so.** Flowers on panicles, ⅛–¼″. Wet meadows, swamps, shores, upland soil, rocky soil, overgrown fields. See also p. 70. Rose family.

b. Herbs with basal leaves (cont. on next page)
7. Eastern parthenium *Parthenium integrifolium*
Plants 1–3′, hairless or slightly hairy. **Basal leaves** narrowly to broadly oval, 2¾–8″, toothed, rough-hairy, long-stalked; stem leaves smaller, clasping. Heads numerous in flat-topped panicles, 2–10″ wide; heads ¼–½″; rays 5. Prairies, dry woods. Composite family.

8. Water-pimpernel *Samolus valerandi*
Plants 4–20″. **Basal and stem leaves** spoon-shaped or oval, broader toward tip, **1–2″,** blunt or rounded. Flowers on racemes 1–6″; flowers <⅛″. Brackish shores, mud, ditches. Theophrasta family.

1. Black nightshade

2. Lizard's-tail

3. Slender cottonweed

4. Seneca-snakeroot

5. Labrador-tea

6. Meadowsweet

7. Eastern parthenium

8. Water-pimpernel

LEAVES alternate and basal
LEAVES simple
PETALS 4–5

Miscellaneous alternate-leaved species (3)
Several families
b. Herbs with basal leaves (cont. from previous page)

1. Cliff stonecrop *Sedum glaucophyllum*
Plants tufted, spreading or prostrate, flowering stems erect, ≤4″.
Leaves spoon-shaped or narrowly oval, broader toward tip, ½″,
Flowers on 2–4 widely spreading, one-sided racemes; flowers ¼″;
petals 4. Rocky places. See also p. 150. Stonecrop family.

b. Herbs without basal leaves
c. Flowers on racemes

2. Spring forget-me-not *Myosotis verna*
Plants erect, 2–16″, hairy. Leaves oblong or narrowly oval, broader
toward tip, ½–2″, stalked; upper leaves narrower and stalkless. **Flow-ers** on racemes 1–7″, <½ of plant; flowers ⅛″. Upland woods, fields,
dry banks, prairies, openings. See also p. 8. Borage family.

3. Virginia stickseed *Hackelia virginiana*
Plants 3–4½′, finely hairy. Lower leaves 2–12″, round or heart-shaped, narrowed to stalk; middle and upper leaves oblong or oval,
2–10″, narrowed to both ends, stalkless. **Flowers** on spreading
racemes or loose panicles, 2–6″; flowers <⅛″. Upland woods; rich
woods, thickets. **Note the prickly fruit.** See blue-flowered stick-seeds p. 10. Borage family.

4. Pokeweed *Phytolacca americana*
Plants 4–10′. Leaves narrowly oval or oval, 4–12″; stalks ½–2″.
Flowers on bright pink racemes 4–8″ long; flowers greenish-white
or pink, ¼″. Fields, fence rows, damp woods. Pokeweed family.

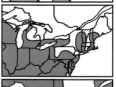

*5. White mullein *Verbascum lychnitis*
Plants erect, ≤4½′, white-woolly. Leaves oval or narrowly oval,
broader toward either end, ≤16″, pointed, obscurely toothed, hairy
beneath, sticky, stalkless. **Flowers** on panicles or racemes, yellow or
white, ⅜–¾″. Fields, weedy places. See also p. 146. Scroph family.

c. Flowers on panicles
6. Bastard-toadflax *Comandra umbellata*
Plants 4–16″. **Leaves narrowly oval or elliptic, 1–2″.** Flowers on
terminal, ellipsoid panicles, ⅛″. Prairies, shores, upland woods.
Sandalwood family.

7. Flowering spurge *Euphorbia corollata*
Plants erect, 1–3′, hairless or hairy. **Leaves oval, linear, or elliptic,
1–2½″,** opposite or upper whorled. Flowers on panicles; flowers ¼″.
Woods, fields, roadsides. See also p. 242. Spurge family.

8. Wood-nettle *Laportea canadensis*
Plants 1–4′, bristly with stinging hairs. **Leaves broadly oval, 3–8″,**
sharp-pointed, coarsely toothed, hairy. Male flowers in panicles from
lower axils; female flowers on loose panicles in upper axils; flowers
without petals, ⅛″. Rich, moist woods. Nettle family.

1. Cliff stonecrop

2. Spring forget-me-not

3. Virginia stickseed

4. Pokeweed

*5. White mullein

6. Bastard-toadflax

7. Flowering spurge

8. Wood-nettle

Miscellaneous alternate-leaved species (4)
Several families
False Solomon's seals have clusters of small flowers on panicles or racemes. Recently these plants have been moved to the same genus as Canada mayflower (p. 264). Critical characters are inflorescence type and number of flowers. Pollinated by halictid bees.

a. Flowers on racemes
1. Starry false Solomon's seal *Maianthemum stellatum*
Plants zigzag, 8–24″, finely hairy or hairless. **Leaves usually pleated**, narrowly oval or narrowly oblong, 2¼–6″, tapering, rounded at base, stalkless and clasping. **Flowers 6–15** on racemes 1–2″ tall; flowers ¼″. Moist, sandy woods; shores; prairies. Ruscus family.

2. Three-leaved Solomon's seal *Maianthemum trifolium*
Plants 4–16″ Leaves 1–4, usually 3, oval, oblong, or narrowly oval, 2¼–4¾″, pointed, stalkless. **Flowers 3–8** on racemes; flowers ¼″. Wet woods, peat bogs. Ruscus family.

a. Flowers on panicles
3. Narrow-leaved white-topped aster *Sericocarpus linifolius*
Plants 8–24′, hairless. **Leaves linear or narrowly oblong, ¾–3¼″**, stalkless. Heads on loose, flat-topped panicles, ½″, rays 3-6. Dry woods, open ground. Composite family.

4. False Solomon's seal *Maianthemum racemosum*
Plants usually curved, slightly zigzag, 16–32″, finely hairy. **Leaves** spreading horizontally in 2 rows, **elliptic, 2¾–6″**, pointed, blunt, or rounded at base, stalkless. Flowers 7–25 on panicles 1¼–6″ tall; flowers ⅛–¼″. Rich, deciduous woods. Ruscus family.

5. Moonseed *Menispermum canadense*
Plants climbing, 6–20′. **Leaves broadly oval or round, 4–6″**, shallowly 3-lobed or unlobed; leaf-stalk attached near the margin. Flowers on panicles, ⅛″; petals 6. Moist woods, thickets. Moonseed family.

a. Flowers in terminal or axillary clusters
6. Nodding mandarin *Disporum maculatum*
Plants 4–32″, hairy when young. Leaves oblong or oval, 1½–6″, pointed, rounded at base, sessile. **Flowers 1–3 in clusters at ends of branches; flowers ½–1″**, spotted with purple. Rich, mountain woods. See also p. 152. Colchicum family.

7. White twisted-stalk *Streptopus amplexifolius*
Plants 16–40″, hairless or with reddish hairs near base. Leaves oval or narrowly oval, 2¼–4¾″, pointed, notched at base, and clasping. **Flowers 1–2 in leaf axils, flowers deflexed and twisted, ⅜″**. Rich, moist woods. See also p. 70. Lily family.

1. Starry false Solomon's seal

2. Three-leaved Solomon's seal

3. Narrow-leaved white-topped aster

4. False Solomon's seal

5. Moonseed

6. Nodding mandarin

7. White twisted-stalk

LEAVES alternate
LEAVES simple
RAYS many

White-flowered asters
Composite family

a. Heads on flat-topped panicles

1. Common white heart-leaved aster *Eurybia divaricata*
Plants 8–40″, hairy. **Basal leaves oval**, 1½–4″, pointed, **notched at base**, stalked; stem leaves smaller. Heads on flat-topped panicles, becoming elongate, ¾–1″; involucral bracts firm, rounded, or broadly pointed, mostly whitish, with short, green tips, rays 5–16. Dry woods.

2. Toothed white-topped aster *Sericocarpus asteroides*
Plants 6–24′, hairy. **Leaves spoon-shaped, narrowly oval or oval, broader toward tip**, ½–4″, fringed and sometimes hairy, **toothed**, stalked or upper leaves stalkless. Heads on flat-topped panicles, ½″; involucral bracts broad, with short-spreading green tips; rays 4–8. Dry woods.

3. Tall flat-topped white aster *Doellingeria umbellata*
Plants 3–7′, hairless or minutely hairy. **Leaves narrowly oval or oval**, 2–6¼″, tapering at ends, **toothless**, stalkless. Heads many on flat-topped panicles, ½–¾″; involucral bracts hairless or minutely hairy, greenish, blunt or broadly pointed; rays 7–14. Moist, low places.

4. Appalachian flat-topped white aster *Doellingeria infirma*
Plants 16–44′, hairless. **Leaves elliptic or oval, broader toward tip**, 2¼–5″, pointed, **toothless**, hairless or rough above, stalkless or nearly so. Heads few to several on open, flat-topped panicles, 1″; involucral bracts firm; rays 5–9. Woods.

a. Heads on open panicles

5. Calico aster *Symphyotrichum lateriflorum*
Plants 1–5′, curly-hairy or hairless. Leaves linear, narrowly oval, or elliptic, 2–6″, tapering at ends, toothless or not, hairy only along midrib. Heads on wide panicles, white or purple, ¼–½″; **rays 9–14.** Various habitats, mostly in open woods, fields, beaches

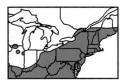

6. Small white aster *Symphyotrichum racemosum*
Plants 1–5′, hairless or hairy in lines. Leaves linear or narrowly oval, ≤4¼″, pointed, hairless or rough above, tapering to stalkless base. Heads many on open panicles, ¼–½″; **rays 15–30.** Moist, open places; flood plain woods.

a. Heads on one-sided branches

7. Many-flowered aster *Symphyotrichum ericoides*
Plants 1–4′, hairy. Leaves linear, ≤2¼″, toothless. Heads many on one-sided, spreading or recurved branches, white or pink, ¼–½″; involucral bracts rigid, bristle-tipped, with a broad, green band, spreading; **rays 8–20.** Dry, open places.

8. Awl aster *Symphyotrichum pilosum*
Plants ½–4½′. Leaves linear or elliptic, ≤4″, sharp-pointed, toothless or slightly toothed, sharp, stalkless. Heads on one-sided branches, ⅜–¾″; involucral bracts narrowly triangular, hardened, hairless, sharp; margins inrolled, often spreading; **rays 16–35.** Open, dry places.

1. Common white heart-leaved aster

2. Toothed white-topped aster

3. Tall flat-topped white aster

4. Appalachian flat-topped white aster

5. Calico aster

6. Small white aster

7. Many-flowered aster

8. Awl aster

LEAVES alternate
LEAVES simple
RAYS many

Fleabanes and miscellaneous composites (1)
Composite family

a. Rays 50 or more
1. Common fleabane *Erigeron philadelphicus*
Plants 8–28″, long-spreading, soft-hairy. Basal leaves narrowly oval or oval, broader toward tip, ≤6″, rounded, coarsely toothed or lobed; stem leaves oblong or oval, clasping. Heads 1–many, ½–1″; involucral bracts hairy; **rays 150–400.** A variety of habitats, often wet, often weedy.

2. Rough fleabane *Erigeron strigosus*
Flowers 12–28″, appressed-hairy. Basal leaves narrowly oval or elliptic, toothless or toothed, ≤6″; stem leaves linear or narrowly oval, mostly hairless. Heads several to very many, ≤½″; **rays 50–100;** involucral bracts slightly sticky and hairy. Disturbed places, weedy places.

3. Annual fleabane *Erigeron annuus*
Plants 24–60″, spreading-hairy. Basal leaves elliptic or round, ≤4″, toothed, stalked; stem leaves narrowly oval or oval, toothed. Heads several to many, ¼–¾″, **rays 80–125,** white or purple. Disturbed sites. **Similar to previous but generally with more leaves.**

a. Rays 25–50
4. Hyssop fleabane *Erigeron hyssopifolius*
Plants 6–12″, hairless to densely hairy. Leaves linear or nearly so, ≤1″, often with short, leafy, axillary shoots, 1¼″, broadly pointed, toothless. **Heads 1–5 on long stalks, <1″;** involucral bracts with long, sharp tips; rays 20-50. Mostly rocky shores, banks.

5. Horseweed *Conyza canadensis*
Plants 4–60″, foul-smelling, bristly. Leaves linear or narrowly oval, broader at either end, ≤3¼″, pointed, toothed or toothless. **Heads many on long, open panicles, ¼″;** involucral bracts with pale brown midrib and green sides, hairless; rays 25–40. Weedy places, overgrown fields.

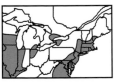

6. Boltonia *Boltonia asteroides*
Plants 1–4½′. Leaves linear, narrowly oval, or elliptic, 2–6″, broad-based, stalkless. **Heads 1–23 on loosely ascending stalks, ½–1″;** involucral bracts with green midrib; rays 25–35. Moist or wet places.

a. Rays 7–25
7. Upland white aster *Oligoneuron album*
Plants ½–2′, hairless or nearly so. Leaves linear or narrowly oval, 1¼–8″, rigid, toothless or nearly so, 3-nerved. **Heads 3–60 on open flat-topped panicles, ¾–1″;** involucral bracts firm with strongly thickened midrib, hairless or nearly so; rays 10–25. Prairies, open, dry, commonly calcareous places. See also p. 158.

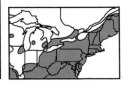

8. White goldenrod *Solidago bicolor*
Plants 4–40″, hairy. Leaves elliptic, narrowly oval, or oval, broader toward tip, 3¼–8″; toothed, hairy. **Heads on elongate spike-like panicles, ¼″;** involucral bracts white with a green band; rays 7–14. Dry woods, open, often rocky places. See also pp. 154–58.

1. Common fleabane

2. Rough fleabane

3. Annual fleabane

4. Hyssop fleabane

5. Horseweed

6. Boltonia

7. Upland white aster

8. White goldenrod

Miscellaneous composites (2)
Composite family

a. Rays present

*1. Ox-eye daisy *Leucanthemum vulgare*
Plants 8–32″, hairless or nearly so. Basal leaves spoon-shaped or narrowly oval, broader toward tip, 1½–6″, toothed and lobed; stem leaves oblong or narrowly oval, broader toward tip. **Heads solitary, 1½–2½″;** rays 15–35. Fields, roadsides, weedy places.

*2. Sneezeweed *Achillea ptarmica*
Plants 1¼–2½′. Leaves linear or narrowly oval, 1¼–4″. **Heads on flat-topped panicles, ⅜–¾″;** rays 8–10. Beaches, roadsides, waste places. See also p. 318.

a. Rays absent
    b. Involucral bracts in only 1 series, of equal length
    (cont. on next page)

3. Pilewort *Erechtites hieraciifolia*
Plants ¼–8′, strong smelling. **Leaves narrowly oval or oval, broader toward either end**, 2–8″, broadly pointed, sharply toothed, short-stalked; upper leaves clasping. Heads on flat-topped or elongate panicles, ½–¾″, swollen at the base, **many-flowered.** Various moist to dry habitats, often in burned areas.

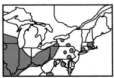

Indian-plantains have fairly large, often triangular leaves, and heads with one series of bracts of the same length. No. 4 has outer flowers without viable stamens; the others have all flowers with stamens. Critical characters are leaf shape and number of flowers per head. Visited, and possibly pollinated by butterflies.

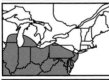

4. Sweet-scented Indian-plantain *Hasteola suaveolens*
Plants 3–8′. **Leaves arrowhead-shaped**, 2–8″, sharply toothed, stalked. Heads on flat-topped panicles, ¼–½″, **many-flowered.** Riverbanks; moist, low ground; woods.

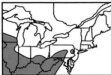

5. Pale Indian-plantain *Arnoglossum atriplicifolium*
Plants 3–9′, blue-green. Lower leaves triangular, 1½–6″, **veins spreading, branched,** toothed, **pale beneath**. Heads on terminal panicles, ¼–½″, **5-flowered.** Woods; moist to dry, open areas

6. Great Indian-plantain *Arnoglossum muehlenbergii*
Plants 3–9′. Lower leaves kidney-shaped, ≤32″; upper leaves smaller, oval, **veins spreading, branched,** toothed or shallowly lobed, **green on both sides**. Heads on many terminal panicles, ¼–⅜″, **5-flowered.** Open woods.

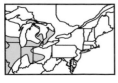

7. Tuberous Indian-plantain *Arnoglossum plantagineum*
Plants 2–5′. Leaves elliptic, narrowly oval, or oval, 2¼–8″, **veins converging toward tip,** toothless, tapering to the base, green below. Heads on terminal panicles, ¼–½″, **5-flowered.** Wet prairies, marshy or boggy places.

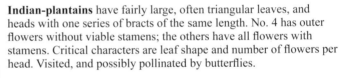

*1. Ox-eye daisy

*2. Sneezeweed

3. Pilewort

4. Sweet-scented Indian-plantain

5. Pale Indian-plantain

6. Great Indian-plantain

7.Tuberous Indian-plantain

Miscellaneous composites (3)
Composite family
b. Involucral bracts in several series or of different lengths
(cont. from previous page)
c. Heads nodding

1. Tall white lettuce *Prenanthes altissima*
Plants 1–6', hairless or spreading-hairy at base. Leaves highly variable, oval or triangular, 1¼–6", toothed, notched at base, sometimes deeply lobed, upper leaves smaller. Heads in axillary and terminal clusters, nodding; **involucral bracts hairless**, dark-tipped; heads ½–¾" long, **5–6-flowered**. Moist woods.

2. Lion's foot *Prenanthes serpentaria*
Plants 1½–5', hairless or rough-hairy in inflorescence. Leaves highly variable, pinnately few-lobed to merely toothed, ≤6½"; lobes rounded, hairless or nearly so. Heads on open panicles, nodding; **involucral bracts with a few long hairs**, often speckled with black; heads ½–¾" long, **10–11-flowered**. Woods, especially in sandy soil.

3. Gall-of-the-earth *Prenanthes trifoliolata*
Plants 1½–4', hairless. Leaves highly variable, lower leaves long-stalked, few-lobed, occasionally with basal lobes or only toothed; upper leaves smaller, hairless above, paler and slightly hairy beneath. Heads on elongate panicles, nodding; **involucral bracts hairless,** often purplish, with minute white waxy dots, sometimes also black spotted; heads ½–¾" long, **10–12-flowered**. Woods, especially in sandy soil.

c. Heads erect or spreading

4. Pearly everlasting *Anaphalis margaritacea*
Plants 1–3'; stems cottony. Leaves narrowly oval or linear, ≤4½", green above, white-woolly below, often with inrolled margins, stalkless. **Heads numerous and crowded in short, flat-topped panicles; involucral bracts pearly white;** heads ≤½", yellow surrounded by white bracts. Various dry and open habitats.

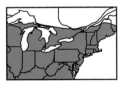

5. Sweet everlasting *Pseudognaphalium obtusifolium*
Plants 1–3', thinly white-woolly. Leaves linear or nearly so, ≤4", green above, white-woolly beneath. **Heads many on flat-topped or elongate panicles; involucral bracts tinged with yellow;** heads ¼". Dry fields.

6. False boneset *Brickellia eupatorioides*
Plants 1–5', minutely hairy. Leaves linear or narrowly oval, 1–4", slightly toothed or toothless, resin-dotted beneath, stalkless. **Heads in small, flat clusters at the ends of branches; involucral bracts green;** heads ⅜–½". Dry, open places, especially in sandy soil.

1. Tall white lettuce

2. Lion's foot

3. Gall-of-the-earth

4. Pearly everlasting

5. Sweet everlasting

6. False boneset

LEAVES alternate or basal
LEAVES simple
PETALS irregular

White Violets and miscellaneous species
Several families

a. Leaves heart-shaped
See other violets pp. 12, 58, 166, 216, 378.
1. Tall white violet *Viola canadensis*
Plants 8–16″, with scattered hairs. Leaves heart-shaped, 2–4″. **Flowers** solitary in leaf axils, ¼–½″, short-stalked, **yellow in center, purplish on backs of petals**. Rich, moist woods. The capsules of tall white violet, as well as others, can shoot their seeds up to 15 feet away from the parent plant. Violet family.

2. Creamy violet *Viola striata*
Plants 2½–12″. Leaves heart-shaped, 1–2½″, long-pointed; stipules large, sharply toothed, ½–1″. **Flowers** solitary in leaf axils, ½–1″, long-stalked. Sun or light shade, ditches, streams, often weedy. Violet family.

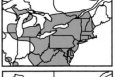

a. Leaves elliptic, narrowly oval or linear
***3. European field pansy** *Viola arvensis*
Plants ≤12″. Leaves elliptic, ½–2½″, toothed, tapered to base. **Flowers solitary in leaf axils**, ¼–1″. Fields, roadsides. Violet family.

4. Biennial gaura *Gaura biennis*
Plants ≤10′, downy. Basal leaves narrowly oval, broader toward tip, 4–12″; stem leaves narrowly oval or oblong, 1¼–4″, tapering at both ends, obscurely toothed. **Flowers on several terminal, wand-like spikes;** flowers ½–1″; sepals, reflexed. Open places, often disturbed habitats. Pollinated by long-tongued bees but visited by other insects, such as moths, for the nectar. Evening-primrose family.

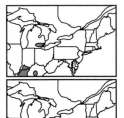

A. Threadstalk gaura *Gaura filipes* (not illus.)
Similar to no. 4, but the flowers are on racemes and the plants are perennial. Dry woods, fields. Evening-primrose family.

B. Small-flowered gaura *Gaura mollis* (not illus.)
Similar to no. 4, but the flowers smaller ¼″. Fields, pastures, streamsides. Pollinated by long-tongued bees. Evening-primrose family.

***5. Lesser toadflax** *Chaenorhinum minus*
Plants 2–16″, sticky-hairy. Leaves narrowly oval or linear, ⅜–1″. **Flowers on loose, leafy racemes**, ¼″; spur <⅛″. Weedy places, roadsides, railroad cinders. Almost exclusively along railroad tracks in our area. It is self-fertilizing and occasionally visited, and presumably cross-pollinated, by "sweat" bees. Plantain family.

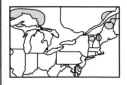

6. Northeastern paintbrush *Castilleja septentrionalis*
Plants 6–24″, hairless except woolly in inflorescence. Leaves linear or narrowly oval, 1¼–4″. **Flowers on compact spikes;** bracts pale yellow or white; flowers ½–1″. Damp, rocky soil, gravelly soil. See also p. 84. Broom-rape family.

1. Tall white violet

2. Creamy violet

*3. European field pansy

4. Biennial gaura

*5. Lesser toadflax

6. Northeastern paintbrush

LEAVES alternate
LEAVES simple
PETALS irregular

Miscellaneous Orchids
Orchid family

Orchids have a peculiar flower in which one of the petals is highly modified and called the lip.

a. Lip forming a pouch

1. White lady-slipper *Cypripedium candidum*
Plants 6–16″. Leaves 3–5, narrowly oval or elliptic, 3¼–6″. **Flowers** solitary, **1½–3″;** sepals and petals linear, often red-striped, ¾–1½″, petals often twisted; lip veined with violet, ½–1″. Calcareous bogs, open swamps, wet prairies. See also pp. 86, 132, 166.

2. Ivory lady-slipper *Cypripedium kentuckiense*
Plants 24–30″. Leaves 3-6, oval, narrowly oval or elliptic, 5–10″. **Flowers** 1-2, **6½–7″;** upper sepal 2¾–3¼″, the lower connate, 2–2¾″; lateral petals spreading, deflexed, 3¼–3½″; lip white or dull yellow, 2–2¼″. Ravines, low woods.

a. Lip fringed or lobed
See other Platantheras, pp. 36, 138, 248.

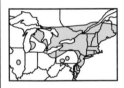

3. White fringed orchid *Platanthera blephariglottis*
Plants 3–48″. Leaves 2–several, linear or narrowly oval, 2–14″; upper leaves much smaller. Flowers on compact, oval spikes, 2–6″; **flowers** ½–1″; sepals round, ¼–½″, reflexed; petals shorter, linear-oblong, torn at tip; **lip reflexed, narrowly oval, broader toward tip,** ¼–½″, **fringed**. Acid swamps, bogs. Pollinated by sphinx moths.

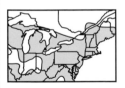

4. Ragged fringed orchid *Platanthera lacera*
Plants 12–30″. Leaves 1–4, narrowly oval or oval, 2½–9″, pointed or blunt. Flowers on compact spikes, 2–16″ tall; flowers ¼–½″; sepals broadly oval, ⅛–¼″, reflexed; petals linear, blunt; **lip descending,** ½″, **deeply 3-lobed, fringed**. Bogs, swamps, wet to dry, sunny places.

a. Lip not fringed, lobed or pouch-like

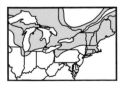

5. Tall white bog-orchid *Platanthera dilatata*
Plants 6–60″. Leaves narrowly oval, broader toward either end, 1½–13″. Flowers on compact spikes, 4–12″ tall; **flowers ½″;** sepals spreading to reflexed; petals oval or linear; lip descending, ¼″, blunt, narrowly oval. Bogs, wet woods. Note the clove-scented flowers.

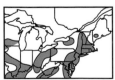

6. Nodding pogonia *Triphora trianthophora*
Plants 4–12″, nodding, straightening as they grow. Leaves 2–6, oval, ⅜–¾″, stalkless. Flowers 3-6 on racemes; **flowers** nodding to nearly erect, **1–1½″;** sepals and petals narrowly oval, broader at either tip, ½–¾″; lip oval, broader toward tip, ½–¾″, 3-lobed, white, often tinged with pink, with 3 prominent green ridges, wavy at tip. Rich, moist woods. A rarely seen plant which is only visible for about a month. Throughout much of the range the flowers open on the same day and only stay open for a day. Pollinated by bees.

1. White lady-slipper

2. Ivory lady-slipper

3. White fringed orchid

4. Ragged fringed orchid

5. Tall white bog-orchid

6. Nodding pogonia

Mustards
Mustard family
Mustards have 4 petals, 6 stamens, and fruit pods with a papery membrane between the two halves that remains after the seeds are released. Most of these are pollinated by small, native bees. Caterpillars of the West Virginia White butterfly feed on toothworts.

a. Leaves pinnately compound, stem leaves alternate

*1. Water-cress *Rorippa nasturtium-aquaticum*
Plants submersed, floating, or prostrate on mud. Leaves succulent, deeply lobed with 3–9 blunt lobes; lateral lobes oval or round; terminal lobe larger, round, peppery tasting. Flowers on racemes, ⅛–¼″; petals twice as long as sepals. Fruit ⅜–1″. Clear, quiet waters See also p. 142.

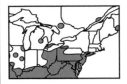

*2. Hoary bitter-cress *Cardamine hirsuta*
Plants 2–16″, hairless. **Basal leaves numerous**, 1¼–3″, with 1–3 pairs of round leaflets and a round or kidney-shaped terminal leaflet; **stem leaves 2–6;** leaflets 4–6, narrow. Flowers on racemes, <¼″. Fruit ½–1″. Sandy soil, roadsides, overgrown fields. See also pp. 66, 268.

3. Pennsylvania bitter-cress *Cardamine pensylvanica*
Plants 2–24″, minutely hairy. **Basal leaves few;** leaflets 2–12, elliptic; terminal leaflet round, ½–¾″ wide; **stem leaves 4–10,** 1½–3″; leaflets 2–12, narrow; terminal leaflet oval, broader toward tip. Flowers on raceme, <¼″. Fruit ½–1″. Swamps, wet woods, meadows, springs.

a. Leaves palmately cleft, stem leaves opposite

4. Broad-leaved toothwort *Cardamine diphylla*
Plants 6–16″, smooth. Leaves 2, opposite; **leaflets 3, oval, 2–4″, coarsely toothed**, rounded at base. Flowers on racemes, 1–1½″. Fruit 1–1½″. Rich, moist woods.

5. Cut-leaved toothwort *Cardamine concatenata*
Plants 8–15″, downy with spreading hairs. **Leaves in whorls of 3,** lobes 5, narrow, slightly toothed. Flowers on racemes, 1–1½″. Fruit 1–1½″. Moist, rich woods, wooded bottomlands, rocky banks.

6. Slender toothwort *Cardamine angustata*
Plants 8–16″. Basal leaves with 3 leaflets, oval, ½–3″, coarsely toothed; stem leaves 2, opposite; **leaflets 3, narrow, 6–30″, sharply toothed**. Flowers on racemes, 1″. Fruit 1–1½″. Rich woods.

A. Dissected toothwort *Cardamine dissecta* (not illus.)
Plants 8–16″. Basal leaves resembling stem leaves; stem leaves 2, opposite; **leaflets 6, linear, 2–3″, toothless.** Flowers on racemes, ¾–1¼″. Fruit ¾–1½″. Rich, moist woods.

*1. Water-cress

*2. Hoary bitter-cress

3. Pennsylvania bitter-cress

4. Broad-leaved toothwort

5 Cut-leaved toothwort

6. Slender toothwort

LEAVES alternate and basal
LEAVES compound
PETALS 4–5

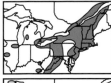

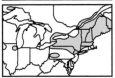

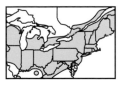

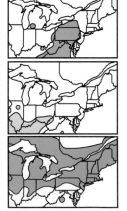

Miscellaneous alternate, compound-leaved species
Rose family

a. Flowers numerous on panicles or spikes

1. Goat's beard *Aruncus dioicus*
Plants 3–7′. **Leaves 2–3 times compound,** ≤20″; leaflets oval or oblong, 2–6″, narrow-pointed, toothed, rounded or notched at base. Flowers on pyramidal panicles of spikes, 4–8″ tall; flowers <¼″; sepals 5. Rich woods, ravines, banks.

2. American burnet *Sanguisorba canadensis*
Plants 1–6′. **Leaves pinnately compound:** lower ≤20″; upper smaller; leaflets 7–17, oval, oblong, or elliptic, 1–4″, blunt, sharply toothed, stalked. Flowers on 1–several cylindrical spikes, 1–8″ tall; flowers ¼″; **sepals 4; petals absent.** Wet areas. See also p. 92.

***3. Queen-of-the-meadow** *Filipendula ulmaria*
Plants 3–6′. **Leaves pinnately compound;** leaflets 5–11; terminal leaflet round, deeply divided into 3–5 oval or oblong, toothed lobes; lateral leaflets 1–3″, oblong or oval, toothed. Flowers on panicles, ¼–½″; **sepals 5; petals 5.** Roadsides, thickets. See also p. 90.

a. Flowers few on racemes or solitary

4. White avens *Geum canadense*
Plants 1–4′, hairless below. **Basal leaves trifoliate;** leaflets oval, broader toward tip, long-stalked; stem leaves short-stalked; leaflets diamond-shaped or narrowly oval, broader toward tip. **Flowers on leafy racemes;** flower-stalks hairless, ½″; **sepals shorter than petals.** Dry or moist woods, open areas. See also pp. 90, 174.

5. Rough avens *Geum laciniatum*
Plants 1–3′, **hairy. Lower leaves trifoliate;** leaflets pinnately lobed, toothed; upper leaves trifoliate; leaflets oval, broader toward tip, or 3-lobed. **Flowers on leafy racemes, ¼–¾″; sepals longer than petals.** Thickets, meadows, roadsides. Differs from no. 4 in having hairy flower-stalks and sepals longer than petals.

6. Mountain Indian-physic *Porteranthus trifoliatus*
Plants 1½–3′, hairless or nearly so. **Leaves trifoliate;** stipules linear, ¼–⅜″; leaflets narrowly oval, broader toward tip, 1½–4″, broadly pointed at both ends, sharply toothed. Flowers on loose panicles; **flowers 1–1¾″; tube ¼–⅜″ long.** Dry or moist upland woods.

7. Midwestern Indian-physic *Porteranthus stipulatus*
Plants 1–3′, sparsely hairy. **Leaves trifoliate;** leaflets narrowly oval, 2–3″, tapering at both ends, sharply toothed; stipules leaf-like; **leaflets therefore appearing to number 5. Flowers** on loose panicles, ¾–1; **tube ⅛–¼″ long.** Dry or moist upland woods; thickets; rocky slopes.

8. Tall cinquefoil *Potentilla arguta*
Plants 1–3′, **sticky-hairy. Basal leaves pinnately compound; leaflets 7–11,** 2¾″, downy beneath. Flowers on cymes; **flowers ½–¾″;** sepals equal petals. Woods, prairies, fields. See also pp. 172, 222.

1. Goat's beard

2. American burnet

*3. Queen-of-the-meadow

4. White avens

5. Rough avens

6. Mountain Indian-physic

7. Midwestern Indian-physic

8. Tall cinquefoil

Cloudberry, Blackberries, Dewberries, and Rose
Rose family

Blackberries and dewberries are very difficult to distinguish because of extreme variation. See also p. 88.

a. Leaves lobed

1. Cloudberry *Rubus chamaemorus*
Plants erect, 4–12″. **Leaves** 2-3, round or kidney-shaped, 1½–3½″, **shallowly 5–7 lobed**, toothed. Flowers solitary, ¾–1¼″; sepals reddish, petals oval, broader toward tip. Fruit usually orange. Peat bogs, heaths.

a. Leaves compound
  b. Plants trailing or low-arching

2. Swamp dewberry *Rubus hispidus*
Plants trailing or low-arching, **bristly, sometimes with sticky hairs intermingled**, ≤8′. Leaves of non-flowering stems evergreen; leaflets 3–5; terminal leaflet oval, broader toward either end, 1–2¾″, short-stalked. **Flowers several to many on racemes;** flowers ½–¾″. Moist or dry habitats.

3. Coastal-plain dewberry *Rubus trivialis*
Plants trailing, **bristly; bristles tipped with round, sticky glands.** Leaves of non-flowering stems evergreen; leaflets 5; terminal leaflet elliptic to oblong, ¾–4″, toothed, long-stalked. **Flowers 1–few on erect stalks;** flowers ½–¾″. Dry, sandy areas.

4. Common dewberry *Rubus flagellaris*
Plants prostrate or low-arching, **armed with broad-based prickles.** Leaves of non-flowering stems hairless, trifoliate; terminal leaflet oblong, narrowly oval, or oval, broader toward tip, narrowly pointed with straight sides below the middle. **Flowers 1–few,** 1–1¼″. Dry or moist woods; usually open, oak-hickory woods.

  b. Plants erect or viney

5. Sand blackberry *Rubus cuneifolius*
Plants erect or nearly so, 1–3′, heavily armed with thorns. **Leaves** on non-flowering stems **trifoliate;** terminal leaflets <2″, diamond-shaped or oval, broader toward tip, truncate or rounded, toothed, densely white-hairy beneath, tapering to base. **Flowers 1–3;** flower-stalks ascending; flowers ¾–1″. Dry, sandy or rocky soil.

A. Common blackberry *Rubus alleghensiensis* (not illus.)
Similar to no. 5, but with sticky hairs on flower-stalks, and leaves usually with **5 leaflets**. Meadows, weedy areas.

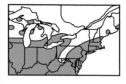

*6. Multiflora rose *Rosa multiflora*
Plants climbing or scrambling, <10′. Leaves pinnately compound; **leaflets 5–11**, elliptic or oval, broader toward tip. Flowers many, ¾–1¼″, sometimes pink. Many habitats.

1. Cloudberry

2. Swamp dewberry

3. Coastal-plain dewberry

4. Common dewberry

5. Sand blackberry

*6. Multiflora rose

LEAVES alternate and basal
LEAVES compound
PETALS 4–5

Meadow-rues
Crowfoot family

See also p. 382.
a. Leaflets generally with 4 or more lobes
1. Early meadow-rue *Thalictrum dioicum*
Plants 1–2′. Basal and stem leaves compound, stalked; leaflets kidney-shaped or oval, broader toward tip, 5–9-lobed, ½–2″ wide. Flowers many on terminal and axillary panicles; flowers drooping, purplish-brown or greenish-white, **male and female flowers on separate plants; sepals 5; filaments yellow or greenish-yellow,** longer than anthers. Moist or rocky woods, ravines, alluvial terraces.

2. Mountain meadow-rue *Thalictrum clavatum*
Plants 1–2½′. Basal and stem leaves compound, the lower stalked, upper stalkless; leaflets kidney-shaped or oval, broader toward tip, 4–7-lobed, 1–1½″ wide. Flowers few on lax, terminal panicles; flowers erect, **female and male parts in the same flower; sepals 4; filaments white,** much longer than anthers. Moist woods, cliffs.

A. Northern meadow-rue *Thalictrum venulosum* (not illus.)
Plants 1–3′. Basal and stem leaves compound, stalked; leaflets round or oval, broader toward tip, 3–5-lobed, ¼–¾″ wide. Flowers many on terminal panicles; flowers erect, greenish-white, **male and female flowers separate; sepals 5; filaments yellowish,** longer than anthers. Moist or rocky woods, ravines, alluvial terraces.

a. Leaflets unlobed or with only 2–3 lobes
3. Purple meadow-rue *Thalictrum dasycarpum*
Plants 3–6′, **stems often purple**, sometimes whitish. **Leaves** chiefly on stem, compound, lower stalked, upper stalkless; leaflets oval, broader at either end, undivided or usually 2–3 lobed, ½–2¼″ wide; **margins often rolled under, short-hairy beneath;** veins prominent. Flowers many on pointed, pyramidal panicles; flowers erect, whitish, often tinged purplish or brown; male and female flowers usually on separate plants; filaments white, longer than the anthers. Various wet areas.

4. Skunk meadow-rue *Thalictrum revolutum*
Plants malodorous, 2–6′. **Leaves** on stem only, compound, lower stalked, upper stalkless; leaflets narrowly oval, elliptic, kidney-shaped, or oval, broader at either end, undivided or 2–3-lobed, ¼–2″ wide; **margin rolled under, finely waxy, hairless beneath.** Flowers many on elongate racemes or panicles; flowers drooping; sepals 5; filaments white, longer than anthers. Dry woods, prairies, thickets.

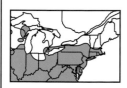

5. Tall meadow-rue *Thalictrum pubescens*
Plants 3–8′. Basal and stem **leaves** compound, lower stalked, upper stalkless; leaflets heart-shaped or round, undivided or 2–3 lobed, ¼–2¾″ wide; **margins flat, hairless or more often minutely hairy beneath.** Flowers many on rounded panicles; flowers erect, white, rarely purplish; filaments white, longer than the anthers, club-shaped, forming a starry burst of stamens. Rich woods, wet meadows, stream banks, swamps.

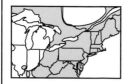

1. Early meadow-rue

2. Mountain meadow-rue

3. Purple meadow-rue

4. Skunk meadow-rue

5. Tall meadow-rue

LEAVES alternate or basal
LEAVES compound
PETALS 5

Miscellaneous umbels (1)
Umbel family
Umbels have a distinctive inflorescence that looks like the spokes of an umbrella. Critical characters are primarily in the fruits.

a. Leaflets generally 3
1. Honewort *Cryptotaenia canadensis*
Plants 1–3′, hairless. Leaves trifoliate; leaflets oval, 1¼–6″, toothed. Flowers on irregular umbels; spokes 2–5 of unequal length; flowers ¼″. Rich woods, thickets.

2. Cow-parsnip *Heracleum maximum*
Plants 3–10′, woolly. Leaves trifoliate; leaflets oval, 4–12″ or the lower ≤24″, lobed and coarsely toothed. Flowers on umbels 4–8″ wide; spokes 15–30, flowers ¼–⅜″. Rich, damp soil

a. Leaflets >3, usually branched in 3's (cont. on next page)
  b. Leaflets <½″
3. Spreading chervil *Chaerophyllum procumbens*
Plants 10–24″, hairless or nearly so. Leaves dissected; leaflets oblong to oval, ⅛–¼″, hairless. Flowers on umbels ¾–1½″ wide; spokes 1–3; flowers ¼″. Moist woods, alluvial soils.

4. Southern chervil *Chaerophyllum tainturieri*
Plants 10–28″, hairy. Leaves dissected; leaflets elliptic to narrowly oblong, ⅛–¼″, hairy. Flowers on umbels ¾–1½″ wide; spokes 1–4; flowers ¼″. Open roads, weedy places.

  b. Leaflets >1″
5. Sweet cicely *Osmorhiza claytonii*
Plants 16–32″. Leaves compound; **leaflets numerous,** oval or narrowly oval, 1½–2¾″. Flowers on compound umbels, 2½–4″; spokes 3–5; flowers ¼″. **Moist woods, slopes.**

6. Anise-root *Osmorhiza longistylis*
Plants 1–4″. Leaves compound; **leaflets numerous,** oval or narrowly oval, ½–2¼″. Flowers on compound umbels, 2½–4″; spokes 3–5; flowers ¼″. **Moist woods. Very similar to no. 5, but anise-scented and long-styled.**

7. Scotch lovage *Ligusticum scoticum*
Plants 1–2′. Leaves compound; **leaflets 9,** diamond-shaped or oval, 1¼–4″. Flowers on compound umbels 2½–4″; spokes 10–20; flowers ¼″. **Salt-marshes, sandy or rocky seashores.**

*8. Goutweed *Aegopodium podagraria*
Plants 1–3′. Leaves compound; **leaflets 9,** oval, 1–3″. Flowers on compound umbels 2¼–4½″ wide; spokes 15–25; flowers ¼″. **Mostly in shady, moist places; escapes from cultivation.** Note the variegated leaves of this cultivated form.

1. Honewort

2. Cow-parsnip

3. Spreading chervil

4. Southern chervil

5. Sweet cicely

6. Anise-root

7. Scotch lovage

*8. Goutweed

LEAVES alternate
LEAVES compound
PETALS 5

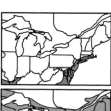

Miscellaneous umbels (2)
Umbel family

a. Leaflets >3, usually pinnate (cont. from previous page)
    b. Upper leaves without blades, only sheaths
1. Purple-stem angelica *Angelica atropurpurea*
Plants ≤ 6'. Leaves pinnately compound; **leaflets** oval or narrowly oval, 1½–4", **broadly pointed; margins fringeless.** Flowers on umbels 4–8" wide; spokes 20–45; flowers ¼". Swamps, wet woods.

2. Mountain angelica *Angelica triquinata*
Plants ≤5'. Leaves pinnately compound; **leaflets** narrowly oval or oblong, 1¼–3¼", **narrowly pointed; margins fringed.** Flowers on umbels 2¼–6" wide; spokes 13–25; flowers ¼". Woods, thickets.

    b. Upper leaves with blades
        c. Leaves pinnately compound; leaflets fewer than 25
3. Water-parsnip *Sium suave*
Plants 1–6', **hairless.** Leaves pinnately compound; leaflets 5–17, linear or narrowly oval, 1½–6"; **teeth numerous.** Flowers on compound umbels 1¼–4¾" wide; spokes 10–25; flowers ¼". **Sunny, wet meadows; swamps; wet banks.**

4. Cowbane *Oxypolis rigidior*
Plants 1½–6', **hairless.** Leaves pinnately compound; leaflets 5–15, oblong, narrowly oval, or oval, broader toward either end, 1–6"; **teeth 0–5 per side.** Flowers on compound umbels 2½–6" wide; spokes 12–25; flowers ¼". **Swamps, marshes, ditches, wet prairies.**

*5. Burnet-saxifrage *Pimpinella saxifraga*
Plants 1–3', **hairless.** Leaves near the base, pinnately compound; leaflets 8–16, oblong or oval, broader toward tip, 2½–5½"; **teeth numerous, coarse.** Flowers on compound umbels, ¾–2"; spokes 8–20; flowers white or pink, ¼". **Roadsides, fields, shores.**

*6. Hog-weed *Heracleum sphondylium*
Plants 3–10', **spreading-hairy,** rank-smelling. Leaves pinnately compound; leaflets 3–7, oval or rounded, 2–4", coarsely toothed. Flowers on compound umbels 4–8" wide; spokes numerous; flowers ¼". Fields, roadsides, weedy places.

        c. Leaves 2–3 times pinnately compound; leaflets > 25
            d. Leaflets thread-like (cont. on next page)
7. Atlantic mock bishop-weed *Ptiliminium capillaceum*
Plants 1–6', hairless. Leaves pinnately compound; leaflets numerous, thread-like, ⅛–¼", toothless. **Flowers on compound umbels ¾–2" wide;** spokes 2–16; flowers <⅛". Freshwater or brackish swamps, marshes.

*8. Caraway *Carum carvi*
Plants 8–30', hairless. Leaves pinnately compound; leaflets thread-like, ¼–½". **Flowers on compound umbels, 2–4";** spokes 7–14; flowers occasionally pink or purple, ¼". Weedy sites.

1. Purple-stem angelica

2. Mountain angelica

3. Water-parsnip

4. Cowbane

*5. Burnet-saxifrage

*6. Hog-weed

7. Atlantic mock bishop-weed

*8. Caraway

LEAVES alternate
LEAVES compound
PETALS 5

Miscellaneous umbels (3)
Umbel family
d. Leaflets not thread-like, broader (cont. from previous page)
 e. Fruit prickly or barbed
*1. Queen Anne's lace *Daucus carota*
Plants 1–3′, bristly. Leaves pinnately compound; leaflets linear, narrowly oval, or oblong, ⅛–½″. Flowers on flat-topped, compound **umbels, 2¼–4″ wide; spokes 10–60 or more;** flowers white; central flower often red-purple, ¼″. In fruit, the inflorescence curls up like a bird's nest. Dry fields, waste places.

*2. Japanese hedge-parsley *Torilis japonica*
Plants ≤ 3′. Leaves pinnately compound; leaflets ¼–2½″. Flowers on **umbels, ⅜–2″; spokes 2–10, hairy;** flowers ¼″. Margins of woods, fields, roadsides, weedy areas.

*3. Burr chervil *Anthriscus caucalis*
Plants ≤3′. Leaves pinnately compound; leaflets oval, broader toward tip, ⅛–¼″, sparsely hairy. Flowers on compound **umbels** opposite the upper leaves, ¾–1½″ **wide; spokes 3–5, hairless,** ¼″. Weedy places.

 e. Fruit smooth
  f. Leaflets firm, not deeply dissected into segments
4. Common water-hemlock *Cicuta maculata*
Plants 2–6′, **often purple mottled.** Leaves pinnately compound; leaflets narrowly oval, 1¼–4″, toothed. **Flowers on compound umbels 2–4¾″ wide; spokes 15–30 or more;** flowers ¼″. Swamps, marshes, prairies, ditches.

5. Bulbiferous water-hemlock *Cicuta bulbifera*
Plants 1–3′, **green.** Leaves pinnately compound; leaflets linear to narrowly oval, ⅜–3″ wide. **Flowers on compound umbels ¾–3″ wide, these often absent or not bearing fruit;** spokes 5–7; flowers ¼″. Swamps, marshes. **Note the bulbils in leaf axils.**

  f. Leaflets limp, deeply dissected into segments
*6. Poison hemlock *Conium maculatum*
Plants ≤10′, **usually with purple blotches.** Leaves pinnately compound; leaflets oval, ⅜–1¼″. Flowers on umbels 1½–2¼″ wide; spokes 10–15; flowers ¼″. Weedy places.

*7. Fool's parsley *Aethusa cynapium*
Plants 10–28″, foul smelling. Leaves pinnately compound; **leaflets** oval or narrowly oval, **narrowly pointed,** ⅛–½″. Flowers on compound umbels ¾–2″ wide; spokes 10–20; flowers ¼″. Roadsides, weedy places. **Note the drooping bracts.**

*8. Wild chervil *Anthriscus sylvestris*
Plants ≤3′, hairy at base. Leaves pinnately compound; **leaflets** pinnately lobed, **blunt,** ⅛–¼″. Flowers on compound umbels ≤3″; spokes 6–15; flowers ¼″. Fields, weedy places.

*1. Queen Anne's lace

*2. Japanese hedge-parsely

*3. Burr chervil

4. Common water-hemlock

5. Bulbiferous water-hemlock

*6. Poison hemlock

*7. Fool's parsley

*8. Wild chervil

LEAVES alternate
LEAVES lobed or
compound
PETALS 3–6

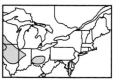

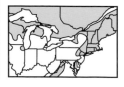

Miscellaneous alternate-, compound- or lobed-leaved species
Several families
a. Leaves palmately lobed
1. Golden seal *Hydrastis canadensis*
Plants 8–20″, hairy. Basal leaf 1; stem leaves 2 near the top, broadly heart-shaped, 5-lobed, 1¼–4″ wide, later ≤10″; lobes sharp-pointed, toothed. **Flowers solitary, terminal, ½″; sepals 3, soon falling; petals absent; stamens showy**. Deep, rich woods. Crowfoot family.

2. Maple-leaved waterleaf *Hydrophyllum canadense*
Plants 6–24″; hairs scattered or absent. Leaves palmately 5–7-lobed, 1½–10″, rounded, notched at base; lobes sharp-pointed, toothed. **Flowers on compact panicles, ⅜–½″; sepals 5; petals 5**. Rich, moist woods. See also p. XXX. Borage family.

3. False bugbane *Trautvetteria caroliniensis*
Plants 1½–5′. Basal leaves kidney-shaped, 5–11-lobed, 4–16″ wide, long-stalked; lobes tapering, irregularly toothed or cleft; stem leaves smaller, nearly stalkless. **Flowers long-stalked in flattish clusters, ½–¾″; sepals 4, soon falling; petals absent; stamens showy**. Mountain woods, stream banks, bluffs, prairies. Crowfoot family.

4. Umbrella-leaf *Diphylleia cymosa*
Plants 16–40″. Basal leaves 2-cleft, many-lobed, ¾–2″ wide, toothed, long-stalked; stem leaves smaller; stalk attached near leaf margin. **Flowers many in clusters** 2–4″ wide; flowers ½″ (photo shows bud); **sepals 6; petals 6**. Mountain woods. Barberry family.

a. Leaves pinnately lobed
5. Blue-Ridge phacelia *Phacelia fimbriata*
Plants 4–14″. Leaves pinnately 5–9-lobed, <2″; lobes oblong or narrowly oval. Flowers 6–20 on racemes, ¼–½″; **petals fringed;** stamens < petals. Mountain woods. See also p. XXX. Borage family.

6. Glade-mallow *Napaea dioica*
Plants 3–10′. Leaves deeply 5–9-lobed, 4–12″; lobes round, toothed. Flowers many on terminal panicles; **male flowers ¼–¾″; female flowers much smaller; petals fringeless; stamens shorter than petals**. Moist, alluvial woods; roadsides. Mallow family.

7. Hairy waterleaf *Hydrophyllum macrophyllum*
Plants 12–28″, densely gray-hairy. Leaves deeply pinnately 7–13-lobed, 4–14″, oblong; lobes oval, pointed or blunt, toothed. Flowers on dense panicles, ⅜–½″; petals fringeless; stamens shorter than petals. Rich, moist woods. Borage family.

a. Leaves trifoliate
8. Mountain white cinquefoil *Sibbaldiopsis tridentata*
Plants woody at base, 4–12″. **Leaves mostly near the base, trifoliate;** leaflets narrowly oval, broader toward tip, ½–1″, blunt, shallowly toothed, slightly hairy below. Flowers several on flat panicles, ⅜–½″. Rocky shores, open summits, alpine areas. Rose family.

1. Golden seal

2. Maple-leaved waterleaf

3. False bugbane

4. Umbrella-leaf

5. Blue-Ridge phacelia

6. Glade-mallow

7. Hairy waterleaf

8. Mountain white cinquefoil

LEAVES alternate
LEAVES lobed or
compound
PETALS 5–6

Aralias, Snakeroots and Cucumbers
Several families

a. Herbs
   b. Flowers on umbels (Ginseng family)
1. Bristly sarsaparilla *Aralia hispida*
Plants ¾–3′, bristly. Leaves 2 times pinnately compound; leaflets oblong, oval, or narrowly oval, ≤4″, toothed. **Flowers on several terminal, rounded umbels** stalked above the leaves; flowers ¼″. Dry, rocky or sandy open woods; clearings. See also p. 388.

2. Spikenard *Aralia racemosa*
Plants 1½–10′, smooth. Leaves pinnately compound; leaflets 6–21, oval, ≤2½″, basally notched. **Flowers on panicles;** flowers ¼″. Rich woods.

   b. Flowers in heads (Umbel family)
3. Black snakeroot *Sanicula marilandica*
Plants 1–4′. Leaves palmately compound, 1½–4″; **leaflets 5,** elliptic or oval, broader toward tip. **Flowers stalkless in small heads;** heads on 2–several ascending branches; flowers ⅛″. Thickets, shores, meadows, open woods. See also p. 176.

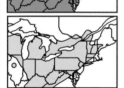

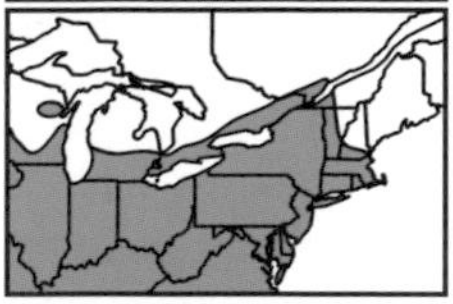

4. Canada snakeroot *Sanicula canadensis*
Plants ¼–6½′. Leaves 3-lobed or leaflets 3, ¾–2½″; **lateral leaflets deeply cleft. Flowers short-stalked in heads;** heads on few branches; flowers ⅛″. Dry, open woods.

5. Beaked snakeroot *Sanicula trifoliata*
Plants 1–2½′. Leaves 3-lobed or leaflets 3, 2½–4½″; **lateral leaflets shallowly cleft. Flowers short-stalked in heads;** heads on few branches; flowers ⅛″. Rich woods.

   b. Flowers on racemes (Mignonette family)
***6. White upright mignonette** *Reseda alba*
Plants erect, ≤32″. Leaves pinnately lobed, 1–1¼″; lobes linear or oblong, pointed, Flowers on dense racemes, greenish-white, ⅜″; petals 5–6. Weedy places. See also p. 176.

a. Vines (Gourd family)
7. Bur-cucumber *Sicyos angulatus*
Plants <7′, sticky-hairy. Leaves round, shallowly 3-5-lobed; lobes narrowly pointed, toothed, leaf notched at base; forked tendrils in leaf axils. Flowers in leaf axils; male flowers on stalks 2–3″; female flowers longer-stalked, ¼–⅜″; **petals 5. Fruit firmly spiny.** Damp thickets, riverbanks.

8. Wild cucumber *Echinocystis lobata*
Plants <7′, nearly hairless. Leaves round, 5-lobed; lobes sharp, 3-forked tendrils in leaf axils; leaves long-stalked. Flowers in leaf axils; male flowers on erect racemes 1–1¼′ long; female flowers solitary or few, ⅜–½″, short-stalked; **petals 6. Fruit weakly spiny.** Stream banks.

1. Bristly sarsaparilla

2. Spikenard

3. Black snakeroot

4. Canadan snakeroot

5. Beaked snakeroot

*6. White upright mignonette

7. Bur-cucumber

female flowers

male flower

8. Wild cucumber

Miscellaneous Composites with dissected leaves
Composite family

a. Flowers on panicles
1. Common yarrow *Achillea millefolium*
Plants 8–40″, **woolly-hairy to hairless**. Leaves pinnately dissected, 1¼–6″. **Heads on flat or rounded panicles; ⅜″; disk ⅛″; rays 5, ⅛″.** Various habitats. See also. p. 292.

The yarrows are a complex group, represented in this range by introduced forms from Europe and by native forms. The most common is native *Achillea millefolium* var. *occidentalis* with leaves having a 3-dimensional aspect. The European form, *Achillea millefolium* var. *millefolium*, has flat leaves and pale involucral bracts surrounding the heads; it is rarely found outside gardens. It may be found in various colors. A more northern form has flat leaves and black-tipped involucral bracts; it is called *Achillea millefolium* var. *nigrescens*.

*2. Feverfew *Tanacetum parthenium*
Plants 12–32″, **upper parts short-hairy.** Leaves pinnately lobed, >3″; lobes rounded. **Heads on flat-topped panicles, ⅝–1″; rays 10–21, ⅛–¼″; disk ¼–⅜″.** Weedy places. See also p. 182.

a. Flowers solitary
*3. Stinking chamomile *Anthemis cotula*
Plants 4–36″, foul-smelling, **nearly hairless**. Leaves 2–3 times pinnately dissected, ¾–2¼″. Heads solitary, ½–1″; **rays 10–16**, ¼–½″. Fields, waste places. The foul odor repels bees but is attractive to flies, which pollinate this species.

*4. Corn-chamomile *Anthemis arvensis*
Plants 12–28″, **hairy**. Leaves 2–3 times pinnately dissected, ¾–2¼″. Heads solitary ½–1″; **rays 10–16**, ¼–½″. Fields, weedy places.

5. Dune-thistle *Cirsium pitcheri*
Plants 20–40″, **white-hairy**. Leaves deeply pinnately lobed, linear, 3″; lobes weakly spine-tipped. Heads 1-few in axils, 1–1¼″; **rays absent**. Sand dunes. The flowers are visited by a variety of insects including bees, butterflies, and moths. See also pp. 24, 94, 180.

1. Common yarrow

*2. Feverfew

*3. Stinking chamomile

*4. Corn-chamomile

5. Dune-thistle

LEAVES alternate
LEAVES compound
PETALS irregular

Clovers, Vetches, and other beans
Bean family

a. Leaflets 3
  b. Leaves toothed

*1. White clover *Trifolium repens*
Plants creeping, 4–10″, hairless; flowers and leaves on separate stalks. **Leaflets 3, broadly elliptic to oval, broader toward tip,** ⅜–¾″, rounded or notched at tip, notched at base. **Flowers in round heads,** long-stalked, ½–1¼″ wide; flowers, ¼–⅜″. Lawns, roadsides. Pollinated by bees. See also pp. 100, 184.

2. Shale-barren clover *Trifolium virginicum*
Plants prostrate, 4–8″, hairy. **Leaflets 3, linear to narrowly oval, broader toward tip,** ¾–2¾″, hairy, long-stalked. **Flowers in round heads,** ¾–1¼″ wide; stalks 1½–4″; flowers ⅜–½″. Shale-barrens.

*3. White sweet-clover *Melilotus alba*
Plants 2–8′. Leaflets 3, narrowly oblong or narrowly oval, broader toward either end, ⅜–1″. **Flowers numerous on slender, tapering racemes,** 2–8″ tall; flowers ⅛–¼″. Roadsides, waste places. Flowers are visited by a wide range of insects, particularly honey-bees. See also p. 184.

  b. Leaves toothless

4. Round-headed bush-clover *Lespedeza capitata*
Plants erect, 2–5′. Leaflets 3, elliptic, 1–1½″, hairy, stalked ⅛″. **Flowers in numerous round heads,** ½–1″; **flowers white with red spot,** ¼–½″. Open, dry woods; sand dunes; prairies. See also p. 98.

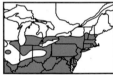

5. Hairy bush-clover *Lespedeza hirta*
Plants erect, 2–5′. Leaflets 3, round, elliptic, or linear, ½–1½″, stalked ¼–1″. **Flowers in numerous short-cylindrical spikes,** ½–1½″ long; **flowers ¼″.** Dry, open places; roadsides.

6. White false indigo *Baptisia alba*
Plants 1–7′, hairless. Leaflets 3, oblong or narrowly oval, broader toward tip, 2–6″; stalks ¼–½″. **Flowers on terminal, long-stalked racemes** 8–24″; **flowers ½–1″.** Dry, sandy woods; prairies; open, upland woods. See also pp. 36, 188.

a. Leaflets more than 3

7. White prairie-clover *Dalea candida*
Plants erect, 1–3′. Leaflets 5–9, linear, narrowly oval, or oblong, flat or folded, ⅛–1¼″. **Flowers on 1-few, very dense spikes, the terminal spike cylindrical,** ½–4″ tall; flowers ¼″. Dry prairies; dry, upland woods. See also p. 34.

8. Pale vetch *Vicia caroliniana*
Plants creeping or climbing, 1–5′. Leaflets 10–24, elliptic, oblong, or narrowly oval, ⅜–¾″, rounded or notched. **Flowers 7–20 on axillary, loose racemes,** 2½–4″; flowers ¼–½″. Moist woods, thickets, shores. See also pp. 34, 102, 188.

*1. White clover

2. Shale-barren clover

*3. White sweet-clover

4. Round-headed bush-clover

5. Hairy bush-clover

6 White false indigo

7. White prairie-clover

8. Pale vetch

Miscellaneous opposite-leaved species
Several families

a. Flowers solitary

1. Pearlwort *Sagina procumbens*
Plants prostrate to ascending, ¾–6″, hairless. **Leaves linear, ⅛–¾″,** sharp-pointed. **Flowers** solitary, ¼″; sepals oval or round; petals much shorter than sepals or absent. Damp, open soil; shores; rocky places; pavement. Pink family.

2. Virginia buttonweed *Diodia virginiana*
Plants 8–32″, hairy on the stem edges. **Leaves narrowly elliptic or narrowly oval, 1¼–2½″,** stalkless, tip pointed, base pointed or blunt, stipules linear, ⅛–¼″. **Flowers** solitary, stalkless in axils, ¼–½″; sepals 2, linear or narrowly oval. Wet ground. See also p. 38. Madder family.

3. Clustered mille graines *Oldenlandia uniflora*
Plants erect to spreading, ¾–20″, hairy. **Leaves narrowly oval or oval, ½–1″,** stalkless, hairy on veins and margin. **Flowers** solitary or in terminal and axillary clusters, <⅛″; sepals narrowly oval or oval. Wet soil, damp sands. Madder family.

4. American marsh willow-herb *Epilobium leptophyllum*
Plants 1–4′, hairy. **Leaves numerous in axillary clusters, linear, ½–2½″,** tapered to base. **Flowers** solitary in upper leaf axils, pink or white, ⅛–¼″. Wet meadows, marshes. **Note white hairs on seeds.** See also p. 112. Evening-primrose family.

5. Eastern willow-herb *Epilobium coloratum*
Plants erect, 4–12″, stem purple, hairy. **Leaves narrowly oval, broader toward either end, 1½–6″,** toothed, purple marked, hairy. **Flowers** solitary in upper axils, white or pink, ¼″, often nodding. Swamps, moist thickets, springy slopes. **Differs from no. 4 in having purple stems and brown seed hairs.** Evening-primrose family.

6. Virginia whitehair leather-flower *Clematis coactilis*
Plants 8–20″, densely hairy. **Leaves narrowly oval or oval, 2½–4½″,** densely hairy below. **Flowers solitary, ½–1¼″.** Shale-barrens. See also pp. 112, 346. Crowfoot family.

a. Flowers paired
7. Partridge-berry *Mitchella repens*
Plants 4–12″, forming mats. Leaves round or oval, ½–1″, stalked. Flowers paired, joined at base, terminal, ⅜–½″. Woods. Madder family.

a. Flowers in various clusters
8. Frogfruit *Phyla lanceolata*
Plants creeping, with ascending stems ≤24″. Leaves narrowly oval or elliptic, ¾–2½″, broadly pointed, toothed, short-stalked. Flowers on axillary spikes, white or pink, <⅛″. Moist ground, mud flats. Verbena family.

1. Pearlwort

2. Virginia buttonweed

3. Clustered mille graines

4. American marsh willow-herb

5. Eastern willow-herb

6. Virginia whitehair leather-flower

7. Partridge-berry

8. Frogfruit

LEAVES opposite or whorled
LEAVES simple
PETALS 5

Sandworts, Pearlworts, and Spurreys
Pink family

Pinks typically have opposite leaves and small white or pink flowers with 5 petals. In some species the petals are so deeply lobed as to appear to number 10. Critical characters are in the flowers and fruit.

a. Petals unlobed or merely notched at tip
  b. Leaves linear (cont. on next page)
    c. Leaves opposite

1. Pine-barren sandwort *Minuartia caroliniana*
Plants forming dense mats, 4–12″. **Leaves firm,** linear, ⅛–½″, pointed. **Flowers on terminal panicles,** ¼–½″; sepals 5, separate, blunt; petals 5, not notched. Dry, sandy pine barrens.

2. Rock sandwort *Minuartia michauxii*
Plants prostrate or erect, 2–12″. **Leaves stiff,** linear, ¼–1″, pointed, in-rolled, **with tufts of smaller leaves in leaf axils. Flowers 3–30, open panicles,** ¼–½″; sepals 5, separate, pointed; petals longer than sepals. Rocky and gravelly areas, particularly calcareous.

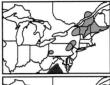

3. Mountain sandwort *Minuartia groenlandica*
Plants matted, 2–8″. **Leaves not stiff,** linear, ¼–1″, blunt. **Flowers 1–50 on terminal panicles,** ½″; sepals blunt; petals shallowly notched. Mountain ledges, gravel.

4. Trailing pearlwort *Sagina decumbens*
Plants ascending or prostrate, 1–6″. **Leaves not stiff,** linear, ⅛–½″, sharp-pointed. **Flowers solitary, terminal, flower-stalks hairless,** <¼″; sepals (4)5, separate, blunt; petals shorter than or equalling sepals, rarely absent. Moist or dryish sandy places, fields, lawns.

***5. Japanese pearlwort** *Sagina japonica*
Plants ascending or prostrate, 1–6″. **Leaves slightly succulent,** linear, ⅛–½″, sharp-pointed. **Flowers solitary, terminal; flower-stalks sticky-hairy,** < ¼″; sepals 5, separate, blunt; petals absent or shorter than sepals. Lawns, compacted soils, sidewalks.

    c. Leaves whorled
***6. Corn-spurrey** *Spergula arvensis*
Plants erect, 6–18″, sticky-hairy. Leaves in whorls of 6–8, linear or narrowly triangular, 1–2″. Flowers in terminal, loosely branching clusters, ¼″; sepals 5, separate; petals shorter than sepals. Cultivated fields, waste areas. **Note the channels beneath the leaves.**

***7. Spurrey** *Spergula morisonii*
Plants erect, 6–18″, sticky-hairy. Leaves in whorls of 6–8, linear or narrowly triangular, 1–2″. Flowers in terminal, loosely branching clusters, ¼″; sepals 5; separate, petals shorter than sepals. Sandy, weedy areas; dredge spoils. **Similar to no. 6, but without channels beneath the leaves.**

1. Pine-barren sandwort

2. Rock sandwort

3. Mountain sandwort

4. Trailing pearlwort

*5. Japanese pearlwort

*6. Corn-spurrey

*7. Spurrey

LEAVES opposite or whorled
LEAVES simple
PETALS 5 or 8–12

Sandworts and related species
Pink family
b. Leaves oval or oblong (cont. from previous page)

1. Seabeach sandwort *Honckenya peploides*
Plants matted and spreading, 4–20″, 3–6′ wide. **Leaves succulent,** elliptic or oval, broader toward either end, ¼–2″. Flowers on terminal, leafy racemes or solitary in forks of stem, ¼″; sepals 5, separate, **pointed; petals 5,** longer than sepals in male flowers, shorter than sepals in female flowers. Sea beaches, sand dunes.

2. Thyme-leaved sandwort *Arenaria serpyllifolia*
Plants diffuse and delicate, clumped, 2–12″. **Leaves** oval, ⅛–¼″, pointed. Flowers on leafy panicles, ⅛″; **sepals 5, separate, pointed; petals 5.** Dry, sandy, or stony places.

3. Grove sandwort *Moehringia lateriflora*
Plants weakly erect, 2–16″. **Leaves** oval or oblong, ½–1″, blunt, hairy. Flowers 1–5, ¼–½″; **sepals 5, separate, blunt or pointed; petals 5.** Open woods, gravelly areas, shores.

4. Large-leaved sandwort *Moehringia macrophylla*
Plants 2–6″. **Leaves** narrowly oval, ½–2½″, pointed, hairless. Flowers 1–few, terminal or axillary, ¼–½″; **sepals 5, separate, pointed; petals 5.** Rocky woods, shores, plains.

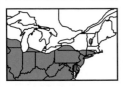

a. Petals deeply lobed (>¼ total length), fringed, or petals >5
b. Leaves linear (cont. on next page)
5. Starry campion *Silene stellata*
Plants 1–3½′, hairy. Leaves in whorls of 4, linear or oval, 1–4″. Flowers on loose panicles, ¾″; sepals 5, fused, bell-shaped, inflated, unnerved; **petals 8–12, lobed or fringed.** Open woods, clearings. See also p. 108.

6. Long-leaved stichwort *Stellaria longifolia*
Plants weak, 5–22″, hairless, or sometimes rough along edges. Leaves linear, narrowly elliptic, or narrowly oval, ¾–2″. Flowers few on terminal and axillary panicles; flower-stalks spreading; flowers ½–¾″; sepals 5, separate; **petals 5, more or less equal to sepal length, 2-cleft.** Moist, grassy places; damp woods.

*****7. Common stichwort** *Stellaria graminea*
Plants prostrate or ascending, 8–18″, hairless or rough. Leaves linear or narrowly oval, ½–2″. Flowers many on terminal, diffuse panicles, flower-stalks spreading or reflexed; flowers ¼–½″; sepals 5, separate; **petals 5, longer than sepals, 2-cleft.** Grassy places. See also p. 249.

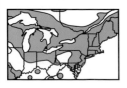

8. Field chickweed *Cerastium arvense*
Plants often matted at base; flowering stalks branching or erect, 4–16″. Leaves linear or narrowly oval, 1–2½″. Flowers in axillary clusters, ½″; sepals 5, separate; **petals 5, much longer than sepals, lobed.** Grassy and rocky, open areas.

1. Seabeach sandwort

2. Thyme-leaved sandwort

3. Grove sandwort

4. Large-leaved sandwort

5. Starry campion

6. Long-leaved stitchwort

*7. Common stitchwort

8. Field chickweed

Stitchworts and Chickweeds
Pink family
b. Leaves oval or elliptic (cont. from previous page)
 c. Sepals separate

*1. Greater stitchwort *Stellaria holostea*
Plants 6–22″. Leaves narrowly oval, ¾–3″, stalkless. Flowers on open panicles, ½–¾″; **petals longer than sepals, notched more than half way.** Garden escape.

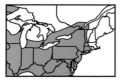

2. Nodding chickweed *Cerastium nutans*
Plants 4–24″, sticky-hairy. Leaves narrowly oblong or narrowly oval, broader toward tip, ½–3″. Flowers 1–41 on open panicles or solitary, flowers ¼–½″, **nodding; sepals without long hairs from tip; petals slightly shorter to much longer than sepals or absent, slightly notched.** Woods, open places.

*3. Mouse-ear chickweed *Cerastium fontanum*
Plants 4–25″. Leaves oval or oblong, ¼–1½″, long-hairy, stalkless. Flowers 3–60 on compact panicles, becoming open; flowers ¼″, **erect; sepals with long hairs from tip; petals equaling sepal length, deeply lobed.** Roadsides, fields, weedy areas.

*4. Gray chickweed *Cerastrium brachypetalum*
Plants 2–12″. Leaves oval, broader at either end, ⅜–1″. Flowers on open panicles, ⅜″, **erect; sepals with long hairs from tip; petals shorter than or equaling sepals.** Roadsides, weedy areas. Similar to no. 3, but with more open panicles and longer flower-stalks.

*5. Common chickweed *Stellaria media*
Plants matted, 4–16″. **Leaves** oval, broader toward either end, ¼–1½″, **long-stalked or upper leaves stalkless.** Flowers solitary or on small panicles; flowers ¼–½″; **petals shorter than sepals or absent, 2-lobed.** Various habitats including woods, meadows, yards.

6. Star chickweed *Stellaria pubera*
Plants erect or ascending, 6–16″. **Leaves** elliptic, narrowly oval, or oval, broader toward either end, 1–4″, **stalkless.** Flowers on open panicles; flowers ½″; **petals shorter than sepals, sometimes cleft nearly to the base.** Woods, rocky slopes.

 c. Sepals fused
*7. White campion *Silene latifolia*
Plants 1–4′, downy, sticky. Leaves narrowly oval or broadly elliptic, 1–4″. Flowers on much branched panicles, 1″, night blooming, male and female separate; sepals 5, fused, becoming much inflated in female flowers, veiny; petals deeply lobed. Weedy areas.

*8. Bladder campion *Silene vulgaris*
Plants ½–2′, hairless. Leaves somewhat succulent, narrowly oval, broader toward either end, 1–3″, often clasping. Flowers few to many on open panicles. ½–¾″; sepals 5, fused, inflated, papery, veiny; petals deeply lobed. Weedy areas.

*1. Greater stitchwort

2. Nodding chickweed

*3. Mouse-ear chickweed

*4. Gray chickweed

*5. Common chickweed

6. Star chickweed

*7. White campion

Male flower

Female flowers

*8. Bladder campion

LEAVES opposite or whorled
LEAVES simple
PETALS 5–6, or 9

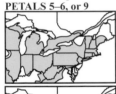

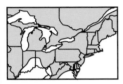

Miscellaneous opposite-leaved species
Several families

a. Flowers on umbels

1. Tall milkweed *Asclepias exaltata*
Plants 2–6′. **Leaves opposite,** oval or narrowly oval, the larger leaves 4¾–12″, tapering at both ends, hairy beneath. Flowers few on umbels, ⅜–¾″, nodding. Moist, upland woods; rich woods; borders. See also pp. 110, 136, 246, 354. Dogbane family.

2. Four-leaved milkweed *Asclepias quadrifolia*
Plants 1–2½′. **Leaves opposite or whorled, normally the middle nodes with whorls of 4 leaves,** narrowly oval, 2½–4¾″, narrowly pointed, tapering to base. Flowers on 1–3 terminal or nearly terminal umbels, ≤½″. Dry, upland woods; slopes. Dogbane family.

a. Flowers on few-flowered panicles

3. Indian hemp *Apocynum cannabinum*
Plants 1½–5′. Leaves narrowly oval, oval, or broadly elliptic, 2–4¼″, short-stalked or stalkless. **Flowers erect, white, ⅛–¼″; lobes erect or slightly spreading.** Thickets, borders of woods, stream banks, open areas. Dogbane family.

4. Spreading dogbane *Apocynum androsaemifolium*
Plants 4–32″. **Leaves drooping,** narrowly oval or oval, 1¼–3″, **stalked. Flowers** on terminal panicles, pink or white, ¼–⅜″, nodding; **lobes spreading or recurving.** Upland woods, fields, roadsides. Dogbane family.

5. Lance-leaved rose gentian *Sabatia difformis*
Plants 1–3′. **Leaves** narrowly oval, linear, elliptic, or oblong, 1–2″, narrowly pointed, rounded at base, **stalkless. Flowers** on loose terminal panicles, ¼–1″; **lobes spreading to reflexed.** Bogs, sandy swamps, peaty shores, wet pine barrens. See also p. 104. Gentian family.

a. Flowers on spike-like racemes

6. Two-leaved mitrewort *Mitella diphylla*
Plants 4–18″, sticky above. Basal leaves oval, shallowly 3–5-lobed, toothed, hairy, notched at base, long-stalked, stem leaves 2, mostly 3-lobed, stalked. **Flowers on racemes** 2–6″ tall; flowers <¼″. Rich, loamy woods; rocky woods. See also p. 244. Saxifrage family.

a. Flowers axillary (cont. on next page)

7. Narrow-leaved loosestrife *Lythrum lineare*
Plants erect, 1–5′, hairless. Leaves linear, ⅜–1¼″, narrowed to stalkless base; upper leaves smaller, ¼–½″. **Flowers axillary,** pale purple or white, ⅛″. Brackish marshes. Loosestrife family.

8. Mayapple *Podophyllum peltatum*
Plants 1–1½′. Leaves round, 12–16″, deeply 3–7-lobed; stalk attached near middle. Flowers solitary in fork between leaves, flowers 1½–2″, petals 6 or 9. Rich woods, thickets, and pastures. Barberry family.

1. Tall milkweed

2. Four-leaved milkweed

3. Indian hemp

4. Spreading dogbane

5. Lance-leaved rose gentian

6. Two-leaved mitrewort

7. Narrow-leaved loosestrife

8. Mayapple

LEAVES opposite and basal
LEAVES simple or compound
PETALS 5–6

Miscellaneous opposite-leaved species
Several families

a. **Flowers on many-flowered panicles (cont. from previous page)**
 b. **Basal leaves differing from stem leaves (Valerian family)**
 c. **Stem leaves pinnately lobed**
1. Taprooted valerian *Valeriana edulis*

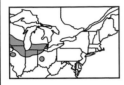

Plants 1–4′. **Basal leaves spoon-shaped or narrowly oval, broader toward tip, 4–12″,** toothless or rarely with basal lobes, tapering to short stalk; stem leaves pinnately compound; leaflets few. Flowers on elongate panicles, becoming diffuse; **flowers ⅛″ wide, ≤⅛″ long.** Prairies; sloughs; hillsides; swamps; wet, open soil. See also p. 124.

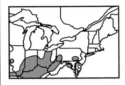

2. Long-tube valerian *Valeriana pauciflora*
Plants 1–3′. **Basal leaves heart-shaped, 2–3″, unlobed or with 2 leaflets;** stem leaves pinnately lobed; lobes 3–7, terminal lobe broadly oval or triangular, much larger than others. Flowers on small, dense, flat-topped panicles, becoming diffusely pyramidal; **flowers ¼″ wide, ½–⅝″ long.** Moist, rich soil; alluvium.

c. **Stem leaves unlobed**
3. Great Lakes corn-salad *Valerianella chenopodiifolia*

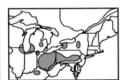

Plants 4–24″. Basal leaves oblong, spoon-shaped, broadly rounded, ¾–2½″; **stem leaves narrowly oval,** broadly pointed. Flowers on compact panicles; **flowers ⅛″.** Meadows, bottoms, fields.

4. Corn-salad *Valerianella radiata*

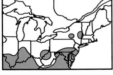

Plants 4–24″, slightly hairy. Basal leaves spoon-shaped, ¾–2½″, **stem leaves oblong,** broadly rounded. Flowers on compact panicles, ¼–½″; **flowers <⅛″.** Damp or dry woods, meadows, fields, roadsides.

b. **Basal leaves absent or like stem leaves (Nettle and Verbena families)**
5. Clearweed *Pilea pumila*

Plants 4–28″, succulent, **hairless.** Leaves oval, 1–6″, long-pointed, coarsely toothed, broadly tapered or rounded at base, hairless, long-stalked. **Flowers many on axillary and terminal panicles;** flowers <⅛″. Moist, rich, shaded soil. Nettle family.

6. White vervain *Verbena urticifolia*

Plants 1–5′, **hairy.** Leaves narrowly oval or oval, 4–6″, narrowly pointed, coarsely toothed, hairy or nearly hairless, stalked 2–8″. **Flowers on erect panicles of slender spikes, more or less interrupted;** flowers ⅜″. Rich thickets, borders of woods, weedy places. See also p. 44. Verbena family.

1. Taprooted valerian

2. Long-tube valerian

3. Great Lakes corn-salad

4. Corn-salad

5. Clearweed

6. White vervain

Page 334
LEAVES opposite or
alternate
LEAVES simple
RAYS 8 or more

Snakeroots, Bonesets, and Thoroughworts
Composite family

Eupatoriums have small, few-flowered heads without disks. Critical characters are leaf size and shape, and involucral bract shape. See also pp. 128, 356.

a. Leaves stalked

1. Small-leaved white snakeroot *Ageratina aromatica*
Plants ≤2.5'. **Leaves** opposite or the upper leaves often alternate, narrowly oval or oval, **1½–3″, blunt or broadly pointed,** toothed, rounded at base, leaf stalk < ½″. Heads on rounded panicles; **involucral bracts** blunt or rounded, **in one row;** heads ¼″; flowers 10–19. Dry woods, thickets, clearings, especially in sand.

2. White snakeroot *Ageratina altissima*
Plants 1–5'. **Leaves** narrowly oval, oval, or heart-shaped, **2½–7″, mostly narrowly pointed,** sharply toothed, leaf stalk > ¾″. Heads on rounded panicles; **involucral bracts** narrowly pointed or blunt, **in one row;** heads ⅛–¼″; flowers 12–30. Woods, thickets. Poisonous.

A. Rock-house white snakeroot
Ageratina luciae-brauniae (not illus.)
Similar to no. 2, but **leaves thinner, triangular, or heart-shaped, 1¼–2½″,** long-stalked. Sandstone cliffs.

3. Late-flowering thoroughwort *Eupatorium serotinum*
Plants 1–6'. **Leaves** opposite or the upper leaves often alternate, narrowly oval or oval, **2–8″, mostly narrowly pointed,** sharply toothed, 3-veined; leaf stalk > ⅜″. Heads on rounded panicles; **involucral bracts** broadly rounded to blunt, in **2–3 rows;** heads ¼″; flowers 8–15. Bottomlands, moist woods, sometimes in drier or more open places.

a. Leaves stalkless (cont. on next page)
  b. Leaves clasping

4. Boneset *Eupatorium perfoliatum*
Plants 2–5', hairy. **Leaves** narrowly oval, 2¾–8″, narrowly pointed, toothed, **broadly based and fused around the stem.** Heads on flat-topped panicles; involucral bracts blunt; heads ¼″; flowers 9–40. Moist or wet, low ground; thickets; swamps.

5. Pine-barren eupatorium *Eupatorium resinosum*
Plants 1–3', sticky-hairy. **Leaves** narrowly oval, 2–5″, narrowly pointed, toothed; **base clasping but not fused.** Heads on flat-topped panicles; involucral bracts broadly rounded to pointed; heads ⅛–¼″; flowers 9–14. Pocosins, bogs, wet areas, often in pine barrens. Note the leaves are much narrower than in no. 4.

1. Small-leaved white snakeroot 2. White snakeroot 3. Late-flowering thoroughwort

4. Boneset 5. Pine-barren eupatorium

LEAVES opposite or alternate
LEAVES simple
RAYS 5 or more

Bonesets, Thoroughworts, and Eupatoriums
Composite family
b. Leaves not clasping or perfoliate (cont. from previous page)
c. Leaves tapering to base

1. White boneset *Eupatorium album*
Plants 1½–3′. **Leaves elliptic or narrowly oval, broader at either end,** 1½–5″, coarsely toothed, veiny. Heads on flat-topped panicles; **involucral bracts** sharply pointed, **hairless, whitened above the middle;** heads ½″; flowers 5. Dry, open woods; especially sandy pinelands.

2. White-bracted boneset *Eupatorium leucolepis*
Plants 1½–3′. **Leaves linear or narrowly oval, broader toward tip, often folded,** 1–3″, with a few scattered teeth or toothless, often with smaller leaves in the axils of the larger leaves, with 1 nerve or 3 obscure nerves. Heads on flat-topped panicles; **involucral bracts** sharply pointed, **white-hairy;** heads ¼″; flowers 5. Wet meadows, margins of pine-barren ponds, especially on sandy soil.

3. Tall boneset *Eupatorium altissimum*
Plants 2½–6′. **Leaves narrowly oval or elliptic,** 2–5″, broadly or narrowly pointed, narrowed to base. Heads on flat-topped panicles; **involucral bracts** rounded to blunt, **green;** heads ¼″; flowers 5. Woods, thickets, savannas, glades, clearings.

A. Mohr's thoroughwort *Eupatorium mohrii* (not illus.)
Similar to no. 3, but with thickened rhizomes and deflexed leaves. Pond margins; ditches; shores; low, moist areas.

c. Leaves broad-based
4. Upland boneset *Eupatorium sessilifolium*
Plants 2–6′, **hairless.** Leaves narrowly oval, 2¾–7″, narrowly pointed, toothed, rounded at base, 1-veined. Heads on flat-topped panicles; involucral bracts rounded or obtuse; heads ¼″; flowers 5. Upland woods, especially sandy, acid soils.

B. Godfrey's thoroughwort *Eupatorium godfreyanum* (not illus.)
Similar to no. 4, but plants hairy and net-veined. Nos. 5 and 6 have 3-veins from base. Woods, open areas.

5. Round-leaved boneset *Eupatorium rotundifolium*
Plants 2–5′, **hairy.** Leaves opposite, oval, oblong, or round, 1–4½″, stalkless, rarely clasping. Heads on flat-topped panicles; involucral bracts sharply pointed or blunt; heads ¼″; flowers 5. Woods; in dry, or seldom wet, soil.

6. Rough boneset *Eupatorium pilosum*
Plants 3–5′, **hairy.** Leaves opposite or upper leaves alternate, narrowly oval, oval, or elliptic, 1–4″, broadly pointed, coarsely and unevenly toothed, rounded at base. Heads on flat-topped panicles; heads ¼″; involucral bracts sharply pointed or blunt; flowers 5. Wet soils; bogs; wet, sandy soils.

1. White boneset

2. White-bracted boneset

3. Tall boneset

4. Upland boneset

5. Round-leaved boneset

6. Rough boneset

LEAVES opposite
LEAVES simple
PETALS irregular

Miscellaneous mints
Mint family

a. Flowers in spikes and axillary whorls
1. Hairy wood-mint *Blephilia hirsuta*
Plants 16–40″, hairy. Leaves narrowly oval or oval, 1½–3″; base
blunt or rounded and notched. Flowers on terminal spikes; flowers
spotted with purple, ¼″; **stamens 2.** Moist, shaded places. See also
p. 48.

*2. Catnip *Nepeta catarica*
Plants 1–3′, gray-downy. Leaves triangular or oval, 1¼–3″, white-
hairy below; base notched or truncate. Flowers on terminal spikes,
¾–2¼″ tall; flowers spotted with pink or purple, ¼–½″; **stamens 4.**
Yards, roadsides, weedy places.

a. Flowers in axillary whorls
*3. Horehound-motherwort *Chaiturus marrubiastrum*
Plants 40–60″, finely hairy. Leaves narrowly oval or oval, ¾–2″,
finely hairy, **narrowed to base.** Flowers on interrupted spikes; **flowers ¼″.** Weedy places. **Note the long, stiff, spiny sepal-lobes.**

*4. Common balm *Melissa officinalis*
Plants 16–32″, hairy, lemon-scented. Leaves oval or triangular, 1½–
2¾″; **base blunt or truncate**, long-stalked. Flowers few on axillary
whorls; **flowers ¼–½″.** Roadsides, open woods, weedy places.

*5. White dead nettle *Lamium album*
Plants 8–20″, hairy. Leaves oval or triangular, 1¼–4″, stalked. Flowers in several whorls; **flowers ¾–1¼″.** Gardens, roadsides, weedy
places. See also p. 120.

a. Flowers in dense heads
  b. Flowers ½–1½″ long
6. Basil bee-balm *Monarda clinopodia*
Plants 1½–4′, hairy. **Leaves** narrowly triangular or narrowly oval,
2¼–4″; base rounded or truncate; **stalks ⅜–½″.** Flowers in heads
⅜–1½″, surrounded by whitish bracts; flowers ½–1½″. Moist woods,
thickets, ravines, banks. See also pp. 118, 206.

7. Red-purple bee-balm *Monarda russeliana*
Plants 8–24″, hairless or nearly so. **Leaves** oval, triangular, or narrowly oval, 2–4″; base rounded or notched, **stalkless or nearly so.**
Flowers in heads ½–1¼″, surrounded by pinkish bracts; flowers
rose-purple or white, 1–1½″. Woods, bluffs, ravines, thickets.

  b. Flowers ¼″ long
8. Hoary mountain-mint *Pycnanthemum incanum*
Plants 1–3′, hairy. **Leaves oval, narrowly oval, or oblong**, 1¾–4½″,
hairy; stalks ¼–½″. Flowers in loose heads, ½–1½″; **bracts** elliptic,
whitened, hairy; flowers ¼–⅜″; calyx hairy; teeth broadly pointed.
Upland woods, thickets.

1. Hairy wood-mint

*2. Catnip

*3. Horehound-motherwort

*4. Common balm

*5. White dead nettle

6. Basil bee-balm

7. Red-purple bee-balm

8. Hoary mountain-mint

LEAVES opposite
LEAVES simple
PETALS irregular

Mountain-mints
Mint family

a. Leaves narrow, >3× longer than broad, <¾″ wide

1. Basil mountain-mint *Pycnanthemum clinopodioides*
Plants ≤3′, hairy. **Leaves narrowly oval, 1¾–3¾″**; hairs scattered; leaf-stalks ⅛–¼″. Flowers in loose, open heads, ⅝″; **bracts** narrowly oval, **green;** flowers ¼″; calyx short-hairy; bristle-tipped. Dry to moist wooded slopes; thickets; shores. **Similar to hoary mountain-mint (p. 338), but leaves narrower, bracts not whitened.**

2. Whorled mountain-mint *Pycnanthemum verticillatum*
Plants ≤5′, hairy. Leaves narrowly oval, 1¼–2″, **hairy above**, tapering to base. Flowers in dense heads, ¼–½″; flowers ¼″; calyx hairy; **teeth narrowly pointed.** Woods, thickets, wet places.

3. Virginia mountain-mint *Pycnanthemum virginianum*
Plants 1–3′, hairy only on the 4 edges of stems. Leaves narrowly oval, 1¼–2½″, **hairless above, hairy on veins below;** base rounded, **stalkless.** Flowers in dense heads, ¼–½″; flowers ¼″; calyx densely hairy; **teeth broadly pointed.** Woods, prairies, shores, thickets.

4. Torrey's mountain-mint *Pycnanthemum torrei*
Plants 1–3′, finely hairy. **Leaves** narrowly oval or linear, 1½–2½″, **hairless, short-stalked.** Flowers in dense heads, ¼–½″; flowers ¼″; calyx hairy; **teeth long-pointed.** Dry woods, thickets. Similar to no. 3, but **calyx teeth more narrowly pointed and leaves stalked.**

5. Narrow-leaved mountain-mint *Pycnanthemum tenuifolium*
Plants 1–3′, hairless. Leaves linear, 1–2″, **hairless,** stalkless. Flowers in dense heads, ⅛–⅜″; flowers ⅛″; calyx hairy; **teeth narrowly pointed.** Dry to moist woods, prairies, thickets, bogs.

a. Leaves broad, <3× longer than broad, >¾″ wide

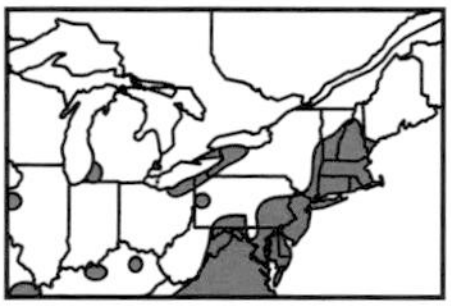

6. Short-toothed mountain-mint *Pycnanthemum muticum*
Plants ≤3′, hairy. **Leaves oval or narrowly oval, 1½–2¾″;** veins hairy below; stalks ≤⅛″. Flowers in dense, hemispherical heads, ⅜–½″; bracts **oval, velvety-hairy above;** flowers ¼″; calyx short-hairy; teeth pointed. Dry to moist woods, meadows, thickets, clearings.

7. Thin-leaved mountain-mint *Pycnanthemum montanum*
Plants 20–45″, minutely hairy or hairless. **Leaves narrowly oval or narrowly elliptic, 2½–4¾″,** hairless; stalks ⅛–½″. Flowers in dense, round heads, ½–¾″; **bracts linear, hairless, fringed;** flowers ¼″; calyx with scattered long hairs; teeth bristle-tipped. Mountain woods.

8. Awned mountain-mint *Pycnanthemum setosum*
Plants ≤3′, minutely hairy. **Leaves oval or narrowly oval, ½–2″,** hairless; stalks ≤⅛″. Flowers in dense, hemispherical heads, ½–¾″; bracts **narrowly oval, minutely hairy;** flowers ¼″; calyx short-hairy; teeth pointed or bristle-tipped. Dry fields, woods.

1. Basil mountain-mint

2. Whorled mountain-mint

3. Virginia mountain-mint

4. Torrey's mountain-mint

5. Narrow-leaved mountain-mint

6. Short-toothed mountain-mint

7. Thin-leaved mountain-mint

8. Awned mountain-mint

LEAVES opposite
LEAVES simple or lobed
PETALS 4–5

Water-horehounds
Mint family

a. Sepals broadly triangular, blunt or broadly pointed
1. Virginia water-horehound *Lycopus virginicus*
Plants 6–30″, densely hairy. Leaves narrowly oval, elliptic, oval, or narrowly diamond-shaped, 3–4½″, coarsely toothed; base narrowly tapered. Flowers in axillary whorls, ⅛″; sepals 4, oval or triangular; petals 4, longer than sepals. Rich, moist soil.

2. Northern water-horehound *Lycopus uniflorus*
Plants 8–40″, hairless or nearly so. Leaves narrowly oval or oblong, ¾–4¼″; teeth few; base broadly tapering or rounded. Flowers in axillary whorls, ⅛″; sepals 5, broadly triangular; petals 5, longer than sepals. Low ground.

a. Sepal lobes narrowly triangular or sharp-tipped
    b. Leaf teeth or lobes ≥¼″

3. American water-horehound *Lycopus americanus*
Plants 6–36″, hairless or nearly so. **Leaves** narrowly oval, narrowly oblong, or linear, 1¼–3″, coarsely and irregularly incised, **hairless or rough above;** base tapering, short-stalked. Flowers in axillary whorls, ⅛″; sepals 5, linear; petals 4 due to fusion of upper 2 lobes, twice as long as sepals. Low ground; wet, sandy or peaty swamps; wet woods; swales.

*4. European water-horehound *Lycopus europaeus*
Plants 16–40″, hairy. Leaves narrowly oval or oval, 1½–4½″, toothed becoming lobed toward base, **hairy above**. Flowers in axillary whorls, ⅛″; sepals 5, narrowly oval, pointed; petals 5, same length as sepals. Weedy places, roadsides.

    b. Leaf teeth ≤⅛″
5. Sessile-leaved water-horehound *Lycopus amplectens*
Plants 6–50″, hairless or nearly so. Leaves oblong, oval, or narrowly oval, 1¼–4″, 4–6 teeth per side; **base blunt or rounded,** stalkless. Flowers in axillary whorls, ⅛″; sepals 5, linear; petals 5, twice the sepal length. Damp sand, peat.

6. Stalked water-horehound *Lycopus rubellus*
Plants 16–50″, hairless to densely hairy. Leaves narrowly oval, elliptic or linear, 2–4½″, toothed to nearly toothless; **base narrowly tapering,** stalked or stalkless. Flowers in axillary whorls, ⅛″; sepals 5, narrow triangular; petals 5, twice the calyx length. Damp soils in woods, thickets.

1. Virginia water-horehound

2. Northern water-horehound

3. American water-horehound

*4. European water-horehound

5. Sessile-leaved water-horehound

6. Stalked water-horehound

Whitlow-worts, Forked nailworts, and Spurges
Pink and Spurge families

a. Leaf base equal on both sides (Pink family)
   b. Leaves linear or nearly so

1. Silver whitlow-wort *Paronychia argyrocoma*
Plants forming mats or tufts, 2–12″, **silky-hairy**. Leaves linear or narrowly oval, ⅜–1¼″, narrowly pointed; stipules silvery, narrowly oval, broadly pointed. **Flowers in dense clusters subtended by silvery bracts;** flowers ⅛–¼″. Rocky slopes, ridges, ledges.

A. Appalachian whitlow-wort *Paronychia virginica* (not illus.)
Plants prostrate or tips erect, 4–16″, **hairless or rough, short-hairy.**
Leaves narrowly linear, ¾–1¼″, bristle-tipped; stipules narrowly oval, papery, narrowly pointed. **Flowers on open panicles;** flowers ⅛″. Crevices, ledges, rocky places, woods.

   b. Leaves elliptic or oval

2. Smooth forked nailwort *Paronychia canadensis*
Plants erect, 3–16″, **hairless.** Leaves elliptic or oval, ¼–1¼″, usually dotted. Flowers in diffuse panicles, <⅛″. Sandy or rocky woods, open places.

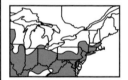

3. Hairy forked nailwort *Paronychia fastigiata*
Plants erect or diffusely spreading, 4–10″, **hairy.** Leaves narrowly oval, broader toward either end, ¼–¾″, hairless or nearly so, white-dotted. Flowers on diffuse panicles, ≤⅛″. Sandy or rocky woods, openings.

a. Leaf base usually unequal (Spurge family)

4. Spotted spurge *Chamaesyce maculata*
Plants prostrate, often forming circular mats, 4–36″, hairy. Leaves oblong, oval, or linear, ⅛–⅝″, often red-spotted. Flowers on short lateral branches, cleft on one side, <⅛″; **ovary and fruit hairy.**
Lawns, gardens, roadsides, weedy places. See also p. 242.

5. Hairy spurge *Chamaesyce vermiculata*
Plants prostrate or ascending, ≤16″, hairy. Leaves oval, ¼–¾″, toothed. Flowers solitary or in small clusters, axillary, <⅛″; **ovary and fruit hairless.** Fields, roadsides, weedy places.

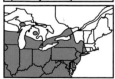

6. Eyebane *Chamaesyce nutans*
Plants erect or ascending, 4–36″, hairy. Leaves oblong or oval, ⅜–1½″, toothed. Flowers solitary or clustered; stalks ≤ ¼″; flowers <⅛″. Dry or moist fields, roadsides.

1. Silver whitlow-wort

2. Smooth forked nailwort

3. Hairy forked nailwort

4. Spotted spurge

5. Hairy spurge

6. Eyebane

LEAVES opposite
LEAVES simple or compound
PETALS or RAYS various

Miscellaneous opposite-leaved species
Several families

a. Herbs (Composite family)
1. Pale-flowered leaf-cup *Polymnia canadensis*
Plants 2–6½′, sticky-hairy. Lower leaves deeply 3–5-lobed, 4–12″, upper broadly oblong or oval, ≤12″. **Heads in panicles; heads 1″;** rays 5 or absent. Moist woods and ravines.

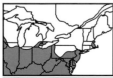

2. Yerba-de-tajo *Eclipta prostrata*
Plants prostrate or ascending, 8–36″, finely hairy. Leaves narrowly oval, elliptic or linear, ¾–4″, broadly pointed, stalkless. **Heads in axillary clusters, ¼″;** rays many. Bottomlands, muddy places.

*3. Common quickweed *Galinsoga quadriradiata*
Plants 8–28″, hairy. **Leaves** oval, 1–2¾″, toothed. **Heads on open, leafy panicles, ¼″;** rays 4–5, 3-toothed. Weedy places, gardens, yards.

*4. Common cosmos *Cosmos bipinnatus*
Plants 2–8′, hairless or rough. Leaves pinnately lobed, 2½–4¼″; lobes linear or thread-like. **Heads numerous, 2½–4″;** disk ⅜–½″; rays 8. Garden escape.

a. Vines (several families)
5. Common hops *Humulus lupulus*
Plants vines, ≤40′. **Leaves 3-lobed,** 1¼–6″. Female flowers on spikes ⅜″ long, becoming 1¼–2¼″ tall in fruit; flowers ¼″. Moist sites. See another photo of this species, p. 238. Hemp family.

6. Climbing hempweed *Mikania scandens*
Plants climbing ≤16′. **Leaves triangular,** 1–5½″, slightly toothed; base notched. **Heads** in branching clusters, white or pink, ¼″; flowers 4. Wet thickets, swamps. Composite family.

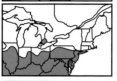

*7. Yam-leaved clematis *Clematis terniflora*
Plants climbing 6–10′. **Leaflets 5,** triangular or oval, ≤2½″, blunt or pointed, **toothless,** rounded or notched at base. Flowers numerous, ¾–1¼″. Roadsides, thickets, edges of woods. See also pp. 112, 322. Crowfoot family.

8. Virgin's bower *Clematis virginiana*
Plants climbing, 6–10′. **Leaflets 3,** oval, ¾–4″, narrowly pointed, **toothed,** notched at base. Flowers on panicles, ⅜–⅝″. Moist areas. Crowfoot family.

1. Pale-flowered leaf-cup

2. Yerba-de-tajo

*3. Common quickweed

*4. Common cosmos

3″

5. Common hops

6. Climbing hempweed

*7. Yam-leaved clematis

8. Virgin's bower

LEAVES opposite and basal
LEAVES simple
PETALS irregular

Miscellaneous opposite-leaved species
Several families

a. Flowers on racemes (Evening-primrose and Plantain families)

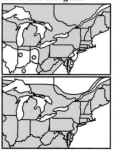

1. Alpine enchanter's-nightshade *Circaea alpina*
Plants weak, 1–12″. **Leaves** oval or triangular, **1–2½″,** broadly pointed, **sharply toothed.** Flowers ≤15 on racemes, ½–2¾″ tall; flowers ≤⅛″. Moist or wet woods, mossy bogs. Evening-primrose family.

2. Common enchanter's-nightshade *Circaea lutetiana*
Plants erect, 8–36″. **Leaves** oblong or oval, **2½–6″,** narrowly pointed, **shallowly toothed.** Flowers many on racemes 3–10″ tall; flowers ⅛–¼″. Moist woods. Evening-primrose family.

3. Purslane speedwell *Veronica peregrina*
Plants erect or depressed, 2–12″, hairless or sticky-hairy. **Leaves** oblong, linear, or narrowly oval, broader toward tip, ¼–1¼″, **toothless or irregularly toothed.** Flowers on terminal, elongate, leafy racemes; flowers <⅛″. Damp, open soil. See also pp. 40–42. Plantain family.

a. Flowers on spikes (Plantain family)
4. White turtlehead *Chelone glabra*
Plants erect, 1–3′. Leaves linear or narrowly oval, ≤6″, narrowly pointed, toothed. **Flowers on spikes 1¼–3″ tall;** flowers 1–1½″. Wet woods, swales, thickets. See also p. 122.

a. Flowers on panicles (Plantain family)
5. Eastern white beard-tongue *Penstemon pallidus*
Plants 12–30″, hairy. Leaves narrowly oval or oblong,<2½″, ⅜–¾″ wide, toothless or nearly so. Flowers on panicles, 4–10″ tall, sticky-hairy; flowers ½–1″; **corolla tube not strongly swollen.** Dry woods, openings, fields. See also p. 54.

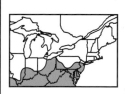

6. Tall beard-tongue *Penstemon digitalis*
Plants 2–4′, hairless or hairy. Basal leaves elliptic or oval, broader toward tip; stem leaves narrowly oblong to narrowly triangular, 2½–8″, toothed or toothless. Flowers on panicles 4–12″ tall, often sticky; flowers ½–1¼″; **corolla tube markedly swollen in middle.** Moist, open woods; prairies; meadows; fields; clearings.

a. Flowers terminal or in leaf axils (Gentian & Plantain families)
7. Pennywort *Obolaria virginica*
Plants fleshy, 3–6″. **Lower leaves scale-like; upper leaves oval, broader toward tip,** ¼–½″, tapered to base. Flowers 1–3 in axillary clusters and a single terminal flower; flowers ⅜–½″; sepals ≤ corolla. Moist woods, thickets. Gentian family.

8. Clammy hedge-hyssop *Gratiola neglecta*
Plants 3–16″, sticky-hairy. **Leaves narrowly oval, broader toward either end,** ⅜–2¼″; base narrowed; leaf-stalk ⅜–¾″. Flowers solitary in leaf axils; flowers ⅜–½″; sepals narrowly oval. Wet or muddy places. See also p. 196. Plantain family.

1. Alpine enchanter's-nightshade

2. Common enchanter's-nightshade

3. Purslane speedwell

4. White turtlehead

5. Eastern white beard-tongue

6. Tall beard-tongue

7. Pennywort

8. Clammy hedge-hyssop

White trilliums
Bunch-flower family

Trilliums are woodland herbs with three leaves subtending a single, 3-petaled flower. See other trilliums pp. 64, 126, 210, 258.

a. Leaves without stalks
  b. Flowers horizontal or bent downward

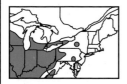

1. Nodding trillium *Trillium cernuum*
Plants 8–20″. Leaves oval or diamond-shaped, 2¼–4″, sharply pointed, narrowed to the base. Flowers curved down below the leaves; flower-stalk ½–2″; flowers 1–2″; **petals recurved,** narrowly oval or oval, blunt or rounded; **anthers pink**. Moist or wet woods; often acidic soils.

2. Bent trillium *Trillium flexipes*
Plants 8–16″. Leaves broadly diamond-shaped, 3¼–6″, sharply pointed, narrowed to base. Flowers horizontal or drooping; flower-stalk 1½–4½″; flowers 1½–2″; **petals spreading**, narrowly oval, or ovate, blunt; **anthers white**. Moist or wet woods, often in calcareous soils.

  b. Flowers erect or ascending

3. White trillium *Trillium grandiflorum*
Plants 8–18″. **Leaves oval, diamond-shaped, or round, 3¼–4¾″,** pointed, tapering to base, stalkless. Flowers erect; **flower-stalks 2–3¼″;** flowers 2–4″; petals turning pink with age, erect, elliptic or oval, broader toward tip, pointed. Rich, moist woods; thickets.

4. Least trillium *Trillium pusillum*
Plants 3–12″. **Leaves elliptic or narrowly oval, 1–2¼″,** pointed or blunt, stalkless. Flowers erect; **flower-stalks absent or nearly so;** flowers 1–2½″; petals white, turning pink or purple with age, spreading-ascending, wavy, oval or narrowly oval. Damp to dry woods, acidic soils.

a. Leaves stalked

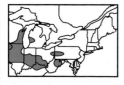

5. Painted trillium *Trillium undulatum*
Plants 8–20″. **Leaves oval, 2–4″, sharply pointed,** rounded base; leaf-stalk ¼–½″. Flowers erect; flower-stalks ¾–2″; flowers 1½–2½″; petals spreading, white with a red-violet inverted V-shaped marking, narrowly oval or oval, broader toward tip, wavy, pointed. Moist or wet woods, stream banks, mountains, acidic soils.

6. Snow trillium *Trillium nivale*
Plants 1–6″ at flowering, then enlarging to 6″. **Leaves** elliptic, narrowly oval, or oval, **1¼–2″, usually blunt;** leaf-stalk ¼–½″. Flowers erect; flower-stalk ½–1″; recurved in fruit; flowers 2–3″; petals recurved or spreading, elliptic or oval, broader toward tip, blunt. Rich, moist woods; clearings; shaded ledges.

1. Nodding trillium

2. Bent trillium

3. White trillium

4. Least trillium

5. Painted trillium

6. Snow trillium

LEAVES whorled or opposite
LEAVES simple
PETALS 3–4

Bedstraws, Madders, and Cleavers
Madder family

Bedstraws have whorled leaves and small, 3–4-lobed flowers. Many have rasping barbs on the stems and leaves. Critical characters are leaf number, shape, as well as fruit characters. See also pp. 126, 210, 244.

a. Fruit smooth

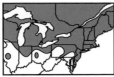

1. Northern three-lobed bedstraw *Galium trifidum*
Plants forming dense mats, rasping. **Leaves in whorls of 4,** linear or narrowly elliptic, ¼–¾″, blunt, **1-nerved,** rasping on margin. Flowers solitary (or 2) in small clusters of stalks from axils or at branch ends, <⅛″; **petals 3.** Fruit without bristles. Swamps, wet shores.

2. Southern three-lobed bedstraw *Galium tinctorium*
Very similar to no. 1, but usually with at least some whorls of 5 or 6 leaves, and 2–3 flowers per stalk. Moist places.

3. Marsh bedstraw *Galium palustre*
Plants 8–24″, rasping. **Leaves in whorls of 2–6,** linear or narrowly oval, ¼–¾″, **1-nerved,** blunt. Flowers many on terminal panicles, <⅛″; **petals 4.** Fruit without bristles. Wet meadows, banks.

4. Rough bedstraw *Galium asprellum*
Plants reclining, 1½–6′, rasping. **Leaves in whorls of 6 or 4–5 on branches,** narrowly elliptic or narrowly oval, broader toward tip, ½–¾″, **3-nerved,** sharply pointed, prickly margins and midrib. Flowers few, terminal and upper axillary; flowers ⅛″; **petals 4.** Fruit without bristles. Wet woods, thickets.

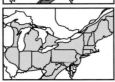

***5. Wild madder** *Galium mollugo*
Plants erect, 1–3′, hairless. **Leaves in whorls of 6 or 8,** linear or narrowly oval, broader toward tip, ½–1½″, **1-nerved.** Flowers many in terminal or axillary clusters, <⅛″; **petals 4.** Fruit without bristles. Roadsides, fields.

a. Fruit bristly

6. Blunt-leaved bedstraw *Galium obtusum*
Plants erect or matted, 8–32″, densely hairy at nodes, otherwise hairless. **Leaves in whorls of 4,** linear or narrowly oval, broader toward either tip, ⅜–1¼″, blunt, **1-nerved,** rasping on margins. Flowers 3–5 on short, terminal panicles, ⅛″; petals 4. Fruit bristly. Swampy thickets, woods, swamps.

7. Fragrant bedstraw *Galium triflorum*
Plants prostrate or scrambling, 8–32″, rasping. **Leaves in whorls of 6 or 4 on branches,** vanilla-scented, narrowly elliptic or narrowly oval, broader toward tip, ½–3″, **1-nerved,** bristle-tipped. Flowers terminal and axillary, ⅛″; petals 4. Fruit bristly. Woods, thickets.

8. Cleavers *Galium aparine*
Plants weak, 4–40″, rasping. **Leaves in whorls of 8,** narrowly oval, 1–3″, **1-nerved,** bristle-tipped. Flowers 3–5, <⅛″; petals 4. Fruit bristly. Rich woods, thickets, shores, waste ground.

1. Northern three-lobed bedstraw

2. Southern three-lobed bedstraw

3. Marsh bedstraw

4. Rough bedstraw

*5. Wild madder

6. Blunt-leaved bedstraw

7. Fragrant bedstraw

8. Cleavers

Miscellaneous whorled-leaved species (1)
Several families

a. Leaves simple (cont. on next page)
b. Petals 4

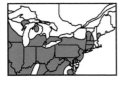

1. Culver's root *Veronicastrum virginicum*
Plants erect, 2–7′, hairless or nearly so. **Leaves in whorls of 3–7,** narrowly oblong or narrowly oval, broader toward either end, 1½–6″, narrowly pointed, toothed; leaf stalks ⅛–⅜″. Flowers on erect spikes, 2–6″ tall, flowers white or purplish, ¼–⅜″, petals 4. Moist or dry, upland woods; thickets; meadows; prairies. Plantain family.

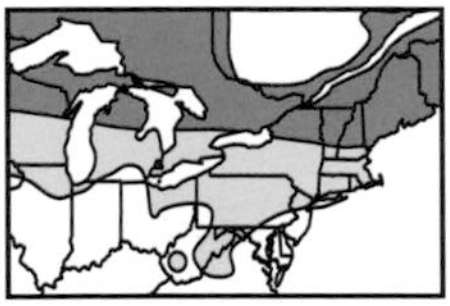

2. Bunchberry *Cornus canadensis*
Plants erect, 2–12″. Leaves in whorls of 6, clustered at top of stem, narrowly oval, broader toward either end, ¾–3½″, pointed at both ends, short-stalked. Flowers small, in a terminal cluster surrounded by 4 petal-like bracts, 1″. Moist woods, thickets, bogs. Named for the bunches of red berries it produces. Dogwood family.

3. Woodland stonecrop *Sedum ternatum*
Plants 4–8″. Leaves alternate, opposite, or in whorls of 3, oval, broader toward tip, ⅜–¾″, toothless, tapering to base. Flowers on 2–4 divergent panicles, ¼–⅜″; petals 4. Rocks, cliffs, woods. See also pp. 150, 284. Stonecrop family.

b. Petals 5

4. Whorled milkweed *Asclepias verticillata*
Plants 1–2′. Leaves in whorls of 3–6, narrowly linear, ½–2″, inrolled. **Flowers on small umbels from the upper nodes; flowers** ¼″; petals 5; horns very narrowly triangular, much surpassing the hoods. Dry slopes, open woods, fields, roadsides, prairies. See also pp. 110, 136, 246, 330. Dogbane family.

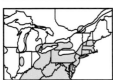

*5. Carpet-weed *Mollugo verticillata*
Plants forming mats ≤16″ wide. Leaves in whorls of 3–8, narrowly oval, broader toward tip, ½–1¼″, tapering to stalk. **Flowers 2–5 in axillary clusters,** pale green or white, ⅛″; **sepals 5; petals absent.** Roadsides, banks, fields, dunes. Carpet-weed family.

6. Spotted wintergreen *Chimaphila maculata*
Plants 4–10″. Leaves narrowly oval, 1–2¾″, broadly pointed, remotely toothed, **with white stripes along veins. Flowers in terminal clusters of 2–5;** flowers nodding, ½–⅝″; petals 5; anthers opening by round pores at tip. Dry woods, especially sandy soil. Heath family.

7. Pipsissewa *Chimaphila umbellata*
Plants spreading, fertile stems 4–12″. Leaves narrowly oval, broader toward tip, 1¼–2½″, broadly pointed or bristle-tipped, sharp-toothed, **without white veins. Flowers on umbels or clusters of 4–8;** flowers white or pink, nodding, ½–⅝″; petals 5; anthers opening by round pores at tip. Dry woods, especially sandy soil. Heath family.

1. Culver's root

2. Bunchberry

3. Woodland stonecrop

4. Whorled milkweed

*5. Carpet-weed

6. Spotted wintergreen

7. Pipsissewa

Page 356
LEAVES whorled
LEAVES simple or
compound
PETALS 5, 7 or RAYS
5–many

Miscellaneous whorled-leaved species (2)
Several families

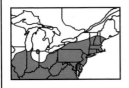

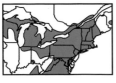

a. Leaves simple (cont. from previous page)
  b. Petals 7 (Myrsine family)
1. Starflower *Trientalis borealis*
Plants 1¼–10″. Leaves in irregular whorls near top, narrowly oval, 1½–4″, narrowly pointed. Flowers 1–few from axils, ¼–½″; **petals 7**. Rich woods and bogs.

  b. Rays many (Composite family)
2. Whorled aster *Oclemena acuminata*
Plants 8–32″, hairy and sticky, **green. Leaves** in irregular whorls, elliptic or oval, broader toward tip, 2¼–6¾″, narrowly pointed, toothed. Heads several to many on flat-topped panicles; heads white to pink, ¾–1″; rays 10–21. Woods. See also pp. 22, 288.

3. Hyssop-leaved boneset *Eupatorium hyssopifolium*
Plants 1–5′, finely hairy, **green;** stems solid. **Leaves** in whorls of 3 or usually 4, linear or narrowly oval, **1–3″**, toothed or toothless, tapering to the base. Heads on dense, flat-topped panicles; involucral bracts rounded to acute; heads ¼″; flowers 5. Fields, open places, especially dry, sandy soil. See also pp. 128, 334–36.

4. Hollow-stemmed Joe-Pye weed *Eupatorium fistulosum*
Plants 1½–5′, hairless, **gray suffused with purple throughout; stems hollow. Leaves** in whorls of 4–7, narrowly oval, **4–10″**, tapering to base and apex, toothed. Heads on round-topped panicles, 4–20″ tall; heads purple, white, or lilac-pink, ¼″; flowers 5–8. Damp thickets, meadows.

5. Purple-node Joe-Pye weed *Eupatorium purpureum*
Plants 2–7′, hairless, **purple only at nodes; stems solid. Leaves** mostly in whorls of 3 or 4, narrowly oval, oval, or elliptic, **3–12″**. Heads on panicles; heads ¼″; flowers white or lavender, 4–7. Thickets, open woods, **often in drier habitats than other Joe-Pye weeds.**

a. Leaves compound (Ginseng family)
6. American ginseng *Panax quinquefolius*
Plants 8–24″. Leaves usually in whorls of 3–4; **leaflets 5**, oval, broader toward tip, 2½–6″, narrowly pointed, toothed. Flowers on umbels, greenish-white, ⅜–¾″ (photo is in fruit). Rich woods. Now becoming rare because of over-collecting.

7. Dwarf ginseng *Panax trifolius*
Plants 4–8″. Leaves in whorls of 3; **leaflets 3–5**, elliptic or narrowly oval, broader toward either end, **1½–3″**, blunt, stalkless. Flowers on terminal umbels; flowers ⅛″. Rich woods, bottomlands.

1. Starflower

2. Whorled aster

3. Hyssop-leaved bonset

4. Hollow-stemmed Joe-Pye weed

5. Purple-node Joe-Pye weed

6. American ginseng

7. Dwarf ginseng

Arrowheads, Sagittarias, and Water-plantains
Water-plantain family

Arrowheads and water-plantains have numerous pistils and stamens. There are several additional species similar to no. 1 that are not treated here, and that can be definitively separated only by the fruits. Critical characters are in the leaf shape and in the fruit.

a. Inflorescence a raceme with 1-12 whorls
    b. Leaves arrowhead-shaped

1. Common arrowhead *Sagittaria latifolia*
Plants ≤2½'. Leaves arrowhead-shaped, ½–12", very variable. Flowers on racemes of 3–9 whorls; flowers ≤1½"; sepals recurved to spreading; flower-stalks spreading. Wet ditches, pools, margins of streams and lakes.

2. Acid-water arrowhead *Sagittaria engelmanniana*
Plants ≤2'. Leaves arrowhead-shaped, 1½–4". Flowers on racemes of 2–4 whorls, ≤1¼"; sepals recurved to spreading; flower-stalks spreading. Acid waters of ponds, lakes, bogs, streams. **Differs from no. 1 in having very narrow leaves.**

    b. Leaves quill-like, elliptic, or narrowly oval

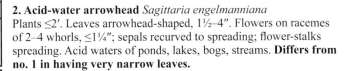

3. Sessile-fruited arrowhead *Sagittaria rigida*
Plants ≤4'. **Leaves linear to elliptic**, 2–6", rarely basally lobed. Flowers on racemes of 2–8 whorls, ≤1¼"; sepals recurved; **flower-stalks absent**. Calcareous or brackish shallow water; shores of ponds, swamps, rivers.

4. Grass-leaved sagittaria *Sagittaria graminea*
Plants 3'. **Leaves linear or narrowly oval**, 1–10". Flowers on racemes of 1–12 whorls, ≤1"; sepals recurved to spreading; **flower-stalks spreading**. Streams, lakes, swamps, freshwater tidal areas.

5. Quill-leaved sagittaria *Sagittaria teres*
Plants 2'. **Leaves quill-like**, ≤24". Flowers on racemes of 1–4 whorls, ≤½"; sepals recurved; **flower-stalks ascending**. Sandy pond shores, swamps of acid waters.

a. Inflorescence a panicle of many flowers
6. Southern water-plantain *Alisma subcordatum*
Plants ≤2'. **Leaves oval or elliptic,** 6". Flowers on diffuse, terminal panicles 3' tall; **flowers ⅛–¼".** Shallow ponds, stream margins, marshes, ditches. Differs from no. 7 by smaller flowers and petals.

7. Northern water-plantain *Alisma triviale*
Plants ≤3'. **Leaves oval or elliptic,** 6–14", stalked. Flowers on diffuse, terminal panicles 3' tall; **flowers ¼–⅜".** Shallow ponds; streams; margins; marshes; ditches; shallow, muddy shores.

A. Grass-leaved water-plantain *Alisma gramineum* (not illus.)
Similar to nos. 6 and 7, but **leaves submersed, floating, or emersed, ribbon-like or linear; or emergent leaves linear, narrowly oval, or narrowly elliptic.** Shallow, fresh or brackish water or shores.

1. Common arrowhead

2. Acid-water arrowhead

3. Sessile-fruited arrowhead

4. Grass-leaved sagittaria

5. Quill-leaved sagittaria

6. Southern water-plantain

7. Northern water-plantain

LEAVES basal
LEAVES simple
PETALS 4

Plantains
Plantain family

Plantains are herbs with small, 4-petaled flowers on spikes. Recent research shows plantains to be related to many former scrophs with showy flowers. Critical characters are leaf shape, bract shape, and in the fruit.

a. Leaves broad
1. American plantain *Plantago rugelii*
Plants 2–10″. Leaves broadly elliptic or broadly oval, 1½–7″, hairless; **veins 3, parallel with margin;** base contracted, reddish. Flowers on a terminal spike, 2–12″ tall; **bracts hairless;** flowers greenish or purplish, ⅛″. Lawns, gardens, roadsides, weedy places.

2. King-root *Plantago cordata*
Plants ≤40″. Leaves oval, 4½–10″, **veins 3, not parallel with margin,** base deeply notched. Flowers on spikes ≤40″, **bracts hairless,** flowers ⅛″. Marshes, streams.

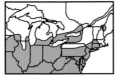

3. Virginia plantain *Plantago virginica*
Plants ≤8″. Leaves narrowly oval or oval, broader toward tip, 2–4″, blunt. Flowers in spikes 1¼–4″ tall; **bracts hairy; flowers ⅛″. Dry or sandy soil.**

a. Leaves narrow
  b. Plants of seasides.
4. Seaside plantain *Plantago maritima*
Plants 2–8″. Leaves fleshy, linear, ≤6″. Flowers on loose or dense spikes ¾–4″ tall, flowers <⅛″. **Salt marshes, beaches, coastal rocks.**

*5. Buck's-horn plantain *Plantago coronopus*
Plants 2–8″. Leaves pinnately lobed, ¾–2½″. Flowers on round or cylindrical spikes, ¼–2½″; flowers ⅛″. **Weedy areas.**

  b. Plants of lawns, roadsides, plains, prairies.
*6. English plantain *Plantago lanceolata*
Plants 6–24″. **Leaves ascending, narrowly elliptic or narrowly oval,** 4–16″; nerves 3–many, parallel with margin. Flowers on dense oval spikes, ½–3″ tall; flowers ⅛″. Lawns, roadsides.

7. Buckhorn *Plantago aristata*
Plants 4–10″. **Leaves linear,** ≤7″. Flowers on cylindrical spikes 1¼–2½″ tall; bracts long, linear; flowers ⅛″. Disturbed sites, roadsides.

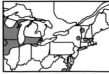

*8. Woolly plantain *Plantago patagonica*
Very similar to no. 7, but spikes woollier and bracts not long. Dry prairies, plains.

1. American plantain

2. King-root

3. Virginia plantain

4. Seaside plantain

*5. Buck's-horn plantain

*6. English plantain

7. Buckhorn

*8. Woolly plantain

LEAVES basal or alternate
LEAVES simple or lobed
PETALS 5

Grass-of-Parnassus and Saxifrages
Grass-of-Parnassus and Saxifrage families
Grass-of-Parnassus have distinctive green-lined flowers and occur in calcareous areas. Their flowers have 5 fertile stamens and 5 sterile stamens, each with 3 branches and topped by glistening green glands. Saxifrages are highly variable but usually have 5 petals, 5 sepals, 10 stamens, and generally basal leaves.

a. Petals green-lined (Grass-of-Parnassus family)
1. American grass-of-Parnassus *Parnassia glauca*
Plants 8–24″. **Basal leaves** leathery, oval or round, **1¼–2″,** rounded or notched at base; stem leaves usually present, at or below the middle, stalkless, smaller than basal leaves. Flowers solitary, ¾–1″; petals with 5 or more veins. Calcareous bogs, shores, wet meadows.

A. Big-leaved grass-of-Parnassus *Parnassia grandifolia* (not illus.)
Plants 8–16″. **Basal leaves** leathery, oval or round, **2–3¼″,** notched at base; stem leaves near the middle of the stem, stalkless, smaller than basal leaves. Flowers solitary, 1–2″; petals with 7–9 veins. Wet, calcareous soil.

a. Petals not lined (Saxifrage family)
   b. Stem leaves present
2. White alpine saxifrage *Saxifraga paniculata*
Plants 2–20″. **Leaves** stiff, oval, broader toward tip, ½–1¼″, **un-lobed,** sharply toothed, stalkless, edges encrusted with lime. Flowers on panicles, ⅛″, white, rarely with red dots. Exposed gravels, rocks.

3. Alpine-brook saxifrage *Saxifraga rivularis*
Plants ½–6″. Lower **leaves** kidney-shaped, ⅛–½″, **3–7-lobed,** long-stalked; upper leaves smaller, less cleft on shorter stalks. Flowers 1–5 on long stalks, ¼–½″; petals unspotted. Wet, alpine areas.

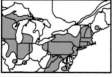

   b. Stem leaves absent
4. Swamp saxifrage *Saxifraga pensylvanica*
Plants 1–4′. Leaves narrowly oval, 2–12″, toothed or toothless, somewhat clasping. Flowers on long panicles, ⅛–¼″; **petals unspotted, sometimes reddish; sepals reflexed.** Wet meadows, swamps, bogs, prairies.

5. Lettuce saxifrage *Saxifraga micranthidifolia*
Plants 1–3′. Leaves oblong or narrowly oval, broader toward tip, 4–12″, coarsely toothed, tapering to base. Flowers on large, elongate panicles, ¼″; **petals with a yellow spot at the base; sepals reflexed.** Wet cliffs, rocks, banks.

6. Early saxifrage *Saxifraga virginiensis*
Plants 4″, elongating to 16″ later, sticky-hairy. Leaves oval, ½–3″, toothed or toothless, tapered and stalked. Flowers on loosely spreading panicles, ¼–½″; **petals unspotted; sepals erect or ascending.** Moist or dry, open woods; ledges.

1. American grass-of-Parnassus

2. White alpine saxifrage

3. Alpine-brook saxifrage

4. Swamp saxifrage

5. Lettuce saxifrage

6. Early saxifrage

LEAVES basal
LEAVES simple
PETALS 5

Shinleafs
Heath family

Shinleafs have small, white or pink flowers either on racemes or solitary. Critical characters are leaf shape and style type. See also pink-flowered shinleaf p. 128.

a. Flower solitary

1. One-flowered shinleaf *Moneses uniflora*
Plants 2–5″. Leaves round, ½–1″, toothed or not; stalks ¼–½″. Flowers solitary, nodding, ½–¾″; style protruding. Damp, mossy woods; bogs.

a. Flowers on racemes
  b. Inflorescence 1-sided

2. One-sided shinleaf *Orthilia secunda*
Plants 4–8″. Leaves elliptic to nearly round, ½–2½″, blunt or rounded at ends, toothed or not, short-stalked. Flowers crowded on one-sided racemes, ¼–⅜″; style protruding. Moist or dry woods, mossy bogs.

  b. Inflorescence not 1-sided
    c. Styles protruding

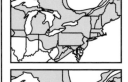

3. Greenish-flowered shinleaf *Pyrola chlorantha*
Plants 4–12″. **Leaves not shiny,** broadly elliptic, ½–1½″, rounded at ends, long-stalked, generally longer than the blade. Flowers usually <10 on racemes, ⅜–⅝″; petals green-veined; style protruding. Dry, conifer woods; thickets.

4. Common shinleaf *Pyrola elliptica*
Plants 4–12″. **Leaves not shiny,** broadly elliptic, 1–3″, rounded or pointed at the ends, short-stalked, shorter than the blade. Flowers often >10 on racemes, nodding, ⅜–½″; petals green-veined; style protruding. Dry, upland woods.

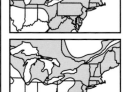

5. Rounded shinleaf *Pyrola americana*
Plants 4–12″. **Leaves shiny,** broadly elliptic or round, 1–3″, rounded at ends; stalks no longer than blade. Flowers 3–13 on racemes, ½–1″; style protuding. Dry or moist woods, bogs.

c. Styles not protruding

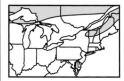

A. Little shinleaf *Pyrola minor* (not illus.)
Plants 4–8″. **Leaves** elliptic or round, ½–1¾″, rounded at ends. Flowers 5–15 on racemes, nodding, ⅜″; style not protruding. Cool, moist woods; thickets.

1. One-flowered shinleaf

2. One-sided shinleaf

3. Greenish-flowered shinleaf

4. Common shinleaf

5. Rounded shinleaf

Miscellaneous basal-leaved species
Several families
a. Flowers on spikes or coiled racemes
b. Leaves with red, sticky hairs
Sundews have sticky hairs that trap insects. The leaves secrete enzymes that help digest the insect, so that the plant can absorb its nutrients. See also p. 130. Sundew family.

1. Spatulate-leaved sundew *Drosera intermedia*
Plants 2–8″. **Leaves erect, spoon-shaped, or oval, broader toward tip**, 3¼–8″, stalked ¾–2″, covered with reddish sticky hairs. Flowers 1–20 on one-sided coiled racemes; flowers white or pink, ¼″; usually 1–2 open at a time. Wet places, shallow water, mostly acidic.

2. Round-leaved sundew *Drosera rotundifolia*
Plants 2¾–14″. **Leaves spreading, nearly round, ⅛–¾″,** stalked ½–2¼″, covered with reddish, sticky hairs. Flowers 3–15 on terminal, one-sided coiled racemes; flowers ⅛–¼″. Bogs, swamps.

b. Leaves without red, sticky hairs
3. Beetleweed *Galax urceolata*
Plants 8–24″. **Leaves glossy,** round or broadly oval, 1½–6″, toothed, deeply notched at base, stalked 3¼–8″. Flowers on terminal, dense spikes, 2–4″, **flowers ⅛″.** Moist or dry woods. Diapensia family.

4. Trailing arbutus *Epigaea repens*
Plants prostrate; stems 8–16″ long, hairy. **Leaves leathery**, oval or oblong, ¾–4″, rounded or notched at base, hairy, long-stalked. Flowers on spikes ¾–2″; **flowers ½″,** tube ¼–½″, white, often tinged with pink, fragrant. Sandy or rocky, acid soils. Heath family

a. Flowers solitary
5. Pyxie *Pyxidanthera barbulata*
Plants prostrate, forming mats ≤3′ wide. **Leaves narrowly oval, broader toward tip,** ⅛–¼″, pointed, usually hairy at base. **Flowers** solitary, stalkless, ⅛–¼″. Sandy pine barrens. Diapensia family.

6. Diapensia *Diapensia lapponica*
Plants forming dense cushions, 2–4″ high. **Leaves** opposite, s**poon-shaped, ¼–½″.** Flowers solitary, stalked ⅜–1½″; **flowers ½–¾″.** Alpine summits, bare gravel, ledges. Diapensia family.

7. Oconee bells *Shortia galacifolia*
Plants 4–8″. **Leaves round or broadly elliptic, ¾–2¾″,** toothed, often notched at base, stalked 1¼–5½″. **Flowers** solitary, nodding, ½–1″. Shady stream banks; moist, wooded ravines. Diapensia family.

8. Dewdrop *Dalibarda repens*
Plants 2–4″. **Leaves heart-shaped, 1¼–2″,** sparsely hairy. **Flowers** solitary, ¾–2″, often on recurved stalks. Swamps, moist woods. Rose family.

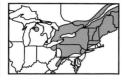

1. Spatulate-leaved sundew

2. Round-leaved sundew

3. Beetleweed

4. Trailing arbutus

5. Pyxie

6. Diapensia

7. Oconee bells

8. Dewdrop

Miscellaneous 6-petaled species (1)
Several families

The species on the next few pages with 6 petals were once placed in the lily family, but that family has since been split up, based on DNA research showing that these species are not all closely related to each other.

a. Flowers on umbels or clusters
*1. Summer snowflake *Leucojum aestivum*
Plants 8–24″. Leaves linear, 12–20″, blunt. **Flowers 2–8 in umbels, nodding, 1–2″**; petals white with green spots at tips. Roadsides, fields, weedy places, garden escape. Amaryllis family.

2. White clintonia *Clintonia umbellulata*
Plants 8–24″. Leaves oblong to oval, broader toward tip, 7–12″. **Flowers 10–24 in umbel-like clusters, erect, ½″**; petals spotted with purple. Rich hardwoods. See also p. 214. Lily family.

3. False garlic *Nothoscordum bivalve*
Plants 8–16″. Leaves thread-like or linear, ≤12″, truncate. **Flowers 3–6 on umbels, ¾–1″**; petals white or greenish-white. Open woods, prairies, barrens. Onion family.

a. Flowers on panicles (cont. on next page)
b. Flowers <2″
4. Virginia bunch-flower *Melanthium virginicum*
Plants 1½–6′. Leaves linear, 12–24″, pointed. Flowers on oval panicles 5–30″ tall, hairy; **flowers** becoming dull green or purple, **½–1″; petals** spreading, oblong or oval, broader toward either end, **abruptly contracted at base, blunt at tip**. Wet woods, meadows. See also p. 368. Bunch-flower family.

5. Broad-leaved bunch-flower *Melanthium latifolium*
Plants 1½–5′. Leaves narrowly oval, broader toward tip, 10–22″, pointed. Flowers on oval panicles, 7–28″ tall, hairy; **flowers ¼–½″; petals** spreading, triangular, oval or round, **abruptly contracted at base, pointed and rolled inward at the tip, crinkled on margin**. Rocky, wooded slopes. Bunch-flower family.

6. White camas *Zigadenus elegans*
Plants 8–24″. Leaves linear, 8–16″. Flowers on panicles 4–12″ tall; **flowers ½–1″**, white or greenish-yellow with a dark gland on upper surface; **petals oval, broader toward tip**. Beaches, bogs. Bunch-flower family.

A. Pine-barren death camas *Zigadenus densus* (not illus.)
Plants 1½–6′. Leaves linear, 4–20″. Flowers 40–100 on pyramidal panicles, 1½–6″ tall; **flowers ⅛–⅜″**; petals oval or elliptic, **narrowed to base but not abruptly so**. Pine-barren bogs, wet woods. Bunch-flower family.

*1. Summer snowflake

2. White clintonia

3. False garlic

4. Virginia bunch-flower

5. Broad-leaved bunch-flower

6. White camas

LEAVES basal
LEAVES simple
PETALS 6

Miscellaneous 6-petaled species (2)
Several families
a. Flowers on panicles (cont. from previous page)
  b. Flowers >2″
1. Adam's needle *Yucca filamentosa*
Plants 3–15′. Leaves erect or reflexed in middle, linear, 20–30″, with long curling threads along margin, sharp-pointed. Flowers on oval panicles with nodding flowers, 30–60″ tall; flowers 4–5½″; petals oval, pointed. Sandy soil, dunes; not native in the northern part of the range. Agave family.

a. Flowers on racemes or spikes (cont. on next page)
  b. Leaves linear

2. Sticky false-asphodel *Tofieldia glutinosa*
Plants 2–20″, **very sticky upward**. Leaves sword-shaped, 3–12″. **Flowers 3–30 in spike-like heads,** ½–2½″ tall; flower stalks sticky; flowers ¼″; petals narrowly oval, broader toward tip. Marshes, shores, wet meadows. Tofieldia family.

3. Scotch false-asphodel *Tofieldia pusilla*
Plants 2–8″, **hairless**. Basal leaves numerous, linear, ⅜–2½″; stem leaves sometimes 2, smaller than basal leaves. **Flowers on dense racemes,** ¼–¾″ tall; flowers ¼″; petals narrowly oval, broader toward tip. Wet rocks, alpine meadows. Tofieldia family.

A. Coastal false-asphodel *Tofieldia racemosa* (not illus.)
Plants 12–32″, **short-hairy, not sticky**. Basal leaves linear, 8–16″, one bract-like stem leaf. **Flowers on dense racemes,** 2–6″ tall; flowers ¼″; petals oblong. Wet pinelands, bogs. Tofieldia family.

4. Turkey-beard *Xerophyllum asphodeloides*
Plants 2–5′. Leaves linear, evergreen, 12–20″. **Flowers on terminal racemes** 6–12″ tall; **flower stalks with bracts well above the base;** flowers ½″; petals oblong or oval. Pine barrens; dry, mountain woods. Bunch-flower family.

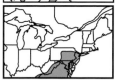

5. Fly-poison *Amianthium muscitoxicum*
Plants 1–3′. Leaves linear ≤16″. Flowers on terminal, conical or cylindrical racemes 1¼–4½″; flowers ⅜″, oblong or oval, broader toward tip. Open woods. **Similar to no. 4, but leaves not evergreen and flower stalks with bracts at base.** Bunch-flower family.

***6. Star-of-Bethlehem** *Ornithogalum umbellatum*
Plants 8–12″. **Leaves linear, 8–12″, white-striped. Flowers 8–20, upright on flat-topped racemes, 1¼″,** opening midday, closing at night; petals narrowly oval or oblong, white above, green striped below. Roadsides, yucca woods, weedy places. Hyacinth family.

***7. Nodding star-of-Bethlehem** *Ornithogallum nutans*
Plants 8–20″. **Leaves linear, 8–20″, white-striped. Flowers 3–12, nodding on elongate racemes, 1¼–2½″,** petals white above, green-striped below. Escape from gardens. Hyacinth family.

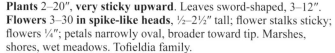

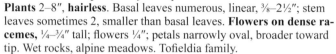

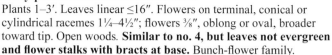

1. Adam's needle

2. Sticky false-asphodel

3. Scotch false-asphodel

4. Turkey-beard

5. Fly-poison

*6. Star-of-Bethlehem

*7. Nodding star-of-Bethlehem

Miscellaneous 6-petaled species (3)
Several families

a. **Flowers on racemes or spikes (cont. from previous page)**
    b. **Leaves oval, elliptic or spoon-shaped**

1. Colic-root *Aletris farinosa*
Plants 1½–4′. **Leaves narrowly oval, broader toward either end,
1½–8″**, pointed. Flowers on terminal spikes, 4–8″; flowers ¼–⅜″;
petals triangular, with a mealy texture. Sandy soil, open woods, barrens. Bog-asphodel family.

2. Devil's-bit *Chamaelirium luteum*
Plants 1–4′, the female plants smaller. **Leaves spoon-shaped or
oval, broader toward tip, 3¼–6″;** stem leaves progressively
smaller. Flowers on terminal, dense racemes or spikes; female-
flowered racemes or spikes erect, 1¼–4½″ tall; flowers divergent;
male-flowered racemes or spikes arching, ≤12″ tall with ascending
flowers, ¼″. Moist woods, bogs. Bunch-flower family.

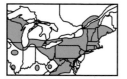

3. Lily-of-the-valley *Convallaria majalis*
Plants 4–8″. **Leaves 2–3, narrowly elliptic, 6–20″**. Flowers on
loose, one-sided racemes, flowers nodding, ¼–⅜″, very fragrant.
Disturbed, open areas. Often cultivated; a native form grows in the
Appalachians on montane slopes and sandy woods. Ruscus family.

*4. Fragrant plantain-lily *Hosta plantaginea*
Plants 1½–2½′. **Leaves oval, 6–10″**. Flowers on terminal racemes;
flowers 4–5″ long, very fragrant. Garden escape. Agave family.

a. **Flowers solitary**
*5. Atamasco-lily *Zephyranthes atamasca*
Plants 8–16″. Leaves linear, ≤16″. **Flowers solitary, erect, 2½–3½″;**
petals broad, pointed. Rich, moist woods; meadows. Amaryllis
family.

6. White trout-lily *Erythronium albidum*
Plants 4–8″. Leaves 2, narrowly oval, ovate or elliptic, 3–9″, irregu-
larly mottled. **Flowers solitary, nodding, ¾–1½″;** petals reflexed.
Moist woods, thickets. See also p. 214. Lily family.

1. Colic-root

2. Devil's-bit

female plant

male plant

3. Lily-of-the-valley

*4. Fragrant plantain-lily

5. Atamasco-lily

3"

6. White trout-lily

Pussytoes, Daisies and Rattlesnake-masters
Composite and Umbel families

Pussytoes are a small group of difficult species. The male and female parts are produced on different types of heads on different plants. The female heads are often fuzzier (see inserts). They are somewhat related to cudweeds. Critical characters are leaf shape, venation, and head number. See also p. 132.

a. Heads small, fuzzy (Composite family)
 b. Leaves with 1 vein

1. Field pussytoes *Antennaria neglecta*
Plants 4–14″, often forming mats. Basal **leaves** spoon-shaped or narrowly oval, broader toward tip, **1–2″**, 1-nerved, densely hairy below, sparsely hairy on top; stem leaves 3–10, smaller. Heads several, terminal, ¼–⅜″. Woods, dry fields, open slopes. A highly variable species.

2. Shale-barren pussytoes *Antennaria virginica*
Plants 4–16″. Basal **leaves** narrowly oval, broader toward tip, **<1″**, 1-nerved, densely woolly below, sparsely hairy on top; stem leaves 4–8, smaller. Heads several, terminal, ¼″. Shale-barrens.

 b. Leaves with 3-5 veins

3. Plantain-leaved pussytoes *Antennaria plantaginifolia*
Plants 3–16″; stem woolly. Basal leaves elliptic or oval, broader toward either end, 1½–3″, 3–5-nerved, densely hairy below, hairless or nearly so above; stem leaves smaller. **Heads several in terminal clusters**, ¼–⅜″; bracts white-tipped. Open woods, dry ground.

4. Solitary pussytoes *Antennaria solitaria*
Plants 4–10″. Leaves mostly at the base, oblong or oval, broader toward tip, 1½–3″, 3–5-nerved, densely hairy below, hairless or nearly so above; stem leaves smaller. **Heads solitary**, ⅜″; bracts white-tipped. Woods.

a. Heads with rays and disk (Composite family)

*5. English daisy *Bellis perennis*
Plants 2–6″, spreading hairy. Leaves elliptic, oval or round, ≤1½″, toothed, narrowed to a winged stalk ≤1½″. Heads solitary, white to pink, ¾″; rays numerous; disk ¼–⅜″, yellow. Weedy places, lawns.

a. Heads round (Umbel family)
6. Rattlesnake-master *Eryngium yuccifolium*
Plants 2–6′. Leaves linear, the lower 6–36″, the upper gradually reduced, remotely spiny on margin. Heads on terminal, flat-topped panicles, round or button-shaped, ⅜–¾″. Moist or dry, sandy soil; open woods; prairies. See other Eryngiums p. 16.

1. Field pussytoes

2. Shale-barren pussytoes

female   male

3. Plantain-leaved pussytoes

4. Solitary pussytoes

*5. English daisy

6. Rattlesnake-master

Miscellaneous basal-leaved species with heads
Several families

a. Heads solitary
  b. Head subtended by large white bract (Arum family)
1. Wild-calla *Calla palustris*
Plants 4–12″. Leaves oval or round, 2–4″, short-pointed, notched at base; stalk 2–6″. Flowers minute in heads, surrounded by a single white bract 1¼–2½″, prolonged into a curled tip; head cream-colored, short-cylindrical, ½–1″. Swamps, bogs, shallow water.

  b. Heads not subtended by a bract (Pipewort family)
2. Seven-angle pipewort *Eriocaulon aquaticum*
Plants 1½–8″, sometimes submersed. Leaves linear, ½–4″, very sharply pointed. Heads solitary on 5–7-ribbed stalks <⅛″ wide; **heads round, ⅛–⅜″, soft and easily compressed**. Sandy, peaty, or sphagnous shores; bogs; shallow water.

A. Estuary pipewort *Eriocaulon parkeri* (not illus.)
Similar to no. 1, but heads smaller on 4–5-ribbed stalks. Freshwater, upper edges of freshwater tidal marshes.

3. Ten-angle pipewort *Eriocaulon decangulare*
Plants 12–42″. Leaves linear, 4–20″; apex sharp. Heads on multi-ribbed stalks <⅛″ wide; **heads hemispherical or round, ¼–½″, hard**. Moist or wet sands, peats, ditches, savannahs.

4. Flattened pipewort *Eriocaulon compressum*
Plants 8–28″. Leaves linear, 2–12″, apex very sharp. Heads on multi-ribbed stalks <⅛″ wide; **heads hemispherical or round, ⅜–¾″, soft and easily compressed**. Sands, peats, shallow pineland ponds, seeps, ditches.

a. Heads more than 1 (Bur-reed family)
5. Floating bur-reed *Sparganium fluctuans*
Plants ≤5′. Leaves floating, ≤2–3′, flat in cross-section. Inflorescence floating; heads 2–4 on branches, ½–¾″; **stigma 1**. Mud or shallow water, swamps, ponds.

6. Branched bur-reed *Sparganium androcladum*
Plants 1½–4′. Leaves erect, emergent, ≤4′, triangular in cross-section. Heads 2–4 ascending, on stem; heads rarely on branches, 1–1½″; **stigma 1**. Muddy shores, shallow water.

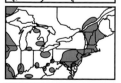

7. Giant bur-reed *Sparganium eurycarpum*
Plants 1½–4′. Leaves erect, emergent, ≤3′, nearly flat in cross-section. Heads 2–6, erect, ½–2″; **stigmas 2**. Mud, shallow water.

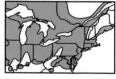

1. Wild-calla

2. Seven-angle pipewort

3. Ten-angle pipewort

4. Flattened plpewort

5. Floating bur-reed

6. Branched bur-reed

7. Giant bur-reed

LEAVES basal
LEAVES simple
PETALS irregular

Orchids and Violets
Orchid and Violet families

a. Leaves marked with white veins or markings (Orchid family)
1. Downy rattlesnake-plantain *Goodyera pubescens*
Plants 8–16″, woolly. **Leaves** oval or narrowly oval, 1¼–2½″,
usually finely white-reticulate with white stripes along midrib.
Flowers on cylindrical spikes, 2½–4″ tall; flowers ¼–⅜″; hood
¼″; lip ⅛″, round. Dry woods.

2. Checkered rattlesnake-plantain *Goodyera tesselata*
Plants 4–14″. **Leaves** oval or oblong, ¾–2¾″, **usually white-
reticulate. Flowers on loose, spiralling spikes, 1¼–6¼″ tall;**
flowers ½″; hood ¼″; lip ⅛–¼″, with a shallow pouch. Dry or moist,
conifer or hardwood forests. Note old flower in inset is turning brown.

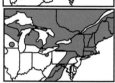

3. Lesser rattlesnake-plantain *Goodyera repens*
Plants 4–8″. **Leaves** oval or oblong, ⅜–1¼″, **usually white-reticulate. Flowers on loose, one-sided, spiralling spikes, 1¼–2½″ tall;**
flowers ¼″; hood ⅛″; lip ⅛″, with a deep pouch. Dry or moist, cold
woods, especially under conifers.

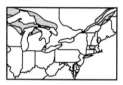

4. Green-leaved rattlesnake-plantain *Goodyera oblongifolia*
Plants 8–16″. **Leaves** narrowly oval or narrowly elliptic, 1¼–4″, **usually white only along the midrib. Flowers on loose or tight, spiral-
ing spikes, 2¼–7″; flowers ½″;** hood ¼–⅜″; lip ¼″, deeply concave
tapering to a boat-shaped tip. Dry or moist, conifer or mixed woods.

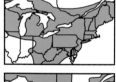

a. Leaves green throughout (Violet family)
5. Sweet white violet *Viola blanda*
Plants 1–10″. **Leaves heart-shaped, 1½–3″, broadly pointed,
deeply, narrowly notched at base.** Flowers solitary, ½″, fragrant;
upper petals twisted backwards; lateral petals angled forward. See
also pp. 12, 58, 166, 216, 296. Rich woods, thickets, openings, cool
ravines, moist slopes.

6. Wild white violet *Viola macloskeyi*
Plants 1–5″. **Leaves heart-shaped or round, ½–3″, blunt or
rounded at apex, deeply notched at base, not as narrow as no.
5.** Flowers solitary, ¼–½″, sometimes fragrant; upper petals not
twisted; lateral petals not angled forward. Wet woods, cold streams,
glades; often in shallow water. **Similar to no. 5, but with smaller
leaves; upper petals are not twisted; found in wetter sites.**

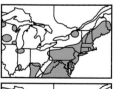

7. Primrose-leaved violet *Viola ×primulifolia*
Plants 2–10″. **Leaves oblong or oval, 1–5″, blunt or rounded at
tip, abruptly narrowed or slightly notched to the base.** Flowers
solitary, ¼–½″. Moist, open meadows; bogs; swamps; stream-banks;
especially in sandy soil. Intermediate between nos. 5 or 6 and no. 8.

8. Lance-leaved violet *Viola lanceolata*
Plants 2–6″. **Leaves narrowly oval, 1–4½″, pointed, tapering to
base.** Flowers solitary, ¼–½″. Bogs, swamps, along streams, pond
edges; especially in sandy soil.

1. Downy rattlesnake-plantain

2. Checkered rattlesnake-plantain

3. Lesser rattlesnake-plantain

4. Green-leaved rattlesnake-plantain

5. Sweet white violet

6. Wild white violet

7. Primrose-leaved violet

8. Lance-leaved violet

LEAVES basal
LEAVES simple
PETALS irregular

Ladies'-tresses orchids
Orchid family

a. Spikes with a single spiral of flowers

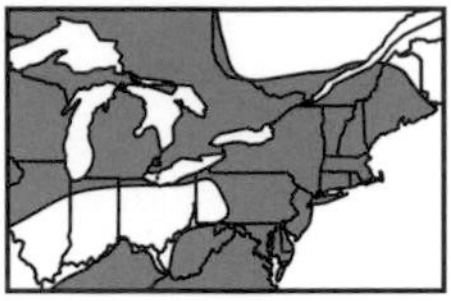

1. Slender ladies'-tresses *Spiranthes lacera*

Plants 6–26″. **Leaves** oval, broader toward tip, ¾–2″. Flowers on hairless (or nearly so) spikes 1¼–6″ tall; flowers spreading or declining, ½″; sepals spreading, elliptic, ¼″; petals narrowly oval, ¼″; **lip with bright green spot**, oval to oblong, ¼″; apex dilated, toothed. Dry to moist meadows, open woods, prairies, fields, roadsides.

2. Little ladies'-tresses *Spiranthes tuberosa*

Plants 2–12″. **Leaves** narrowly oval, broader toward tip, ¾–2¼″, usually absent at flowering time. Flowers on hairless spikes ¾–3″ tall; flowers spreading, ½″; sepals slightly spreading, ¼″; petals linear, ¼″, broadly pointed or blunt; **lip** oval to oblong, ¼″; **white**; apex dilated with wavy, finely torn margin. Dry, open woods; outcrops; fields; roadsides; cemeteries.

3. Spring ladies'-tresses *Spiranthes vernalis*

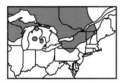

Plants 8–26″. **Leaves** narrowly oval, **2–10″**. Flowers on densely hairy spikes 1¼–6″ tall; flowers nodding to ascending, ½–¾″; sepals spreading, narrowly oval, ¼–⅜″; petals oblong, ¼–⅜″, blunt; **lip** oval, ¼″, **with a yellow center**. Dry to moist, open areas.

A. Case's ladies'-tresses *Spiranthes casei* (not illus.)
Somewhat similar to no. 3, but the spikes not as densely hairy and the **lips yellow, greenish-yellow or green toward the base.** Dry to moist, open areas.

a. Spikes with a double spiral of flowers

4. Nodding ladies'-tresses *Spiranthes cernua*

Plants 4–24″. **Leaves dull,** elliptic, oval, or narrowly oval, broader toward either end, ≤10″. **Flowers on hairy spikes;** flowers spreading or declining, ½–1″; sepals ¼–½″; petals linear or narrowly oval, ¼–½″; **lip oval to oblong,** ¼–⅜″, wavy at tip, **yellow in center.** Open, wet to dry, often sandy places.

5. Shining ladies'-tresses *Spiranthes lucida*

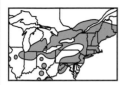

Plants 1½–15½″. **Leaves shiny,** elliptic or narrowly oval, 1¼–4¾″. **Flowers on hairy spikes;** flowers spreading or nodding, ½″; sepals projected forward, linear-oblong, ⅛–¼″; petals linear-oblong, ¼″, blunt; **lip oblong,** ¼″, toothed, **yellow in center with green lines.** Rocky and sandy riverbanks, seeps, fens, limey areas.

6. Hooded ladies'-tresses *Spiranthes romanzoffiana*

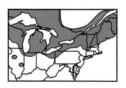

Plants 3–22″. Leaves linear, elliptic, or narrowly oval, broader toward tip, ≤10″. **Flowers on hairless (or nearly so) spikes;** flowers ascending, ½″; **sepals and petals fused, forming a hood;** sepals ¼″; petals linear to oval, blunt; **lip fiddle-shaped,** ⅛–⅜″; apex broad, **white**. Moist to wet meadows, marshes, fens, prairies, stream banks, coastal bluffs, dunes.

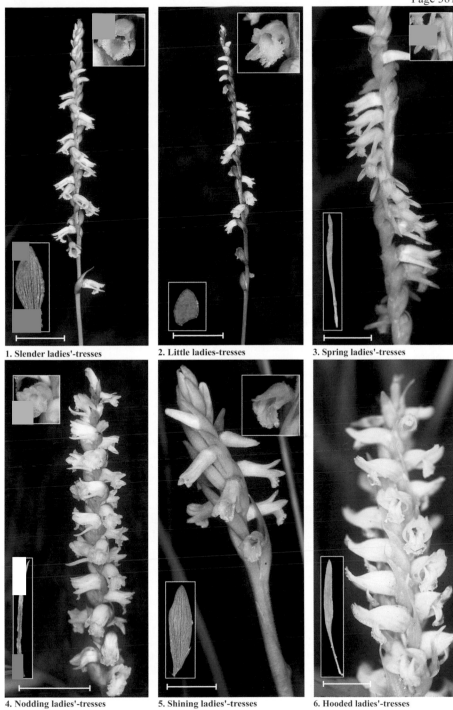

1. Slender ladies'-tresses

2. Little ladies-tresses

3. Spring ladies'-tresses

4. Nodding ladies'-tresses

5. Shining ladies'-tresses

6. Hooded ladies'-tresses

Anemones and Hepaticas
Crowfoot family

a. Stem leaves present
  b. Stem leaves sessile

1. Canadian anemone *Anemone canadensis*
Plants ½–2 ½'. Basal leaves round, 1½–4", pointed, toothed, or usually 3-lobed; stalk 3–8"; stem leaves in 2 whorls of 3, and, like basal leaves, 3-lobed. **Flowers 1, 1–2".** Damp or wet areas.

  b. Stem leaves stalked

2. Wood anemone *Anemone quinquefolia*
Plants 2–12". Basal leaves trifoliate; terminal leaflet diamond-shaped or narrowly oval, broader toward tip, ½–2", pointed to blunt, toothed; lateral leaflets lobed, usually stalked 1½–10"; **stem leaves in 1 whorl of 3, trifoliate or usually 5-lobed. Flowers solitary,** white or rarely pink, ½–1". Moist or wet areas.

3. Lance-leaved anemone *Anemone lancifolia*
Plants 4–16". Basal leaves trifoliate; terminal leaflet oval or narrowly oval, broader toward either end, ½–2¾", pointed; lateral leaflets unlobed, stalked ⅜–1"; **stem leaves in 1 whorl of 3 leaflets. Flowers solitary, 1–1½".** Damp, rich woods.

4. Long-fruited anemone *Anemone cylindrica*
Plants 1–2'. Basal leaves trifoliate; terminal leaflet broadly diamond-shaped or narrowly oval, broader toward tip, 1¼–2", narrowly pointed; leaf-stalk 3½–8½"; **stem leaves in 2 whorls of 3–9 compound leaves. Flowers 2–8, ¾";** petals 4–5. Prairies, dry, open woods, pastures, roadsides, fields, banks.

5. Tall anemone *Anemone virginiana*
Plants 1–3'. Leaves trifoliate; terminal leaflet stalkless, narrowly oval or oval, broader toward tip, sharply pointed; leaf-stalk 2–14"; **stem leaves in 2 whorls of 3. Flowers 3–9, greenish white, white, or rarely red, ¾–1½".** Rocky woods; thickets; riverbanks; grasslands.

6. Rue-anemone *Thalictrum thalictroides*
Plants 4–12". **Leaves 2× trifoliate, ⅜–1", leaflets broadly oval to round, 3-lobed or toothed, ½–1", notched at base. Flowers 1–6 on umbels, ¾".** Dry or moist, open woods. See also p. 306.

a. Stem leaves absent

7. Sharp-lobed hepatica *Hepatica nobilis* var. *acuta*
Plants 4–9". **Leaves round, 3-lobed,** sometimes 5–7-lobed, ½–2½", deeply notched at base; **lobes triangular, sharp. Flowers solitary,** white, pink, or lavender, ½–1"; petals 5–12, often 6. Dry or moist, upland woods.

8. Round-lobed hepatica *Hepatica nobilis* var. *obtusa*
Plants 3–6". **Leaves round, 3-lobed,** ½–2", notched, **lobes oval, blunt, or rounded. Flowers solitary,** white, pink, or lavender, ½–1"; petals 5–12, often 6. Dry or moist, upland woods.

1. Canadian anemone

2. Wood anemone

3. Lance-leaved anemone

4. Long-fruited anemone

5. Tall anemone

6. Rue-anemone

7. Sharp-lobed hepatica

8. Round-lobed hepatica

Bugbanes, Cohosh, and Baneberries
Crowfoot family

a. Petals lobed; fruit a capsule

1. American bugbane *Cimicifuga americana*

Plants 3–7½'. Leaves 3× compound; **leaflets 32–100,** oval to oblong, coarsely toothed or cleft, 1¼–6". Flowers on lax panicles of 3–10 racemes, 4–20" tall; flowers ⅛"; **petals 5, 2-lobed.** Moist, rocky woods.

2. Black cohosh *Cimicifuga racemosa*

Plants 3–8'. Leaves 2–3× compound; **leaflets 20–70,** oval, broader toward either end, coarsely toothed or cleft, 1½–6". Flowers on slender panicles of 4–9 wand-like racemes, 8–30"; flowers ⅛"; **petals 4, 2-lobed.** Moist or dry, rich woods. Used as a source of phytoestrogens. Appalachian Azur caterpillars feed on this species.

3. Appalachian bugbane *Cimicifuga rubifolia*

Plants 1–4½'. Leaves 1–2× compound; **leaflets 3–9,** oval, broader toward tip, 3¼–12". Flowers on erect panicles of 2–6 racemes, 6–12" tall; flowers ⅛"; petals 5. Slopes, bluffs, ravines, coves.

a. Petals unlobed; fruit a berry

4. White baneberry *Actaea pachypoda*

Plants 16–32". Leaves 2–3× ternately compound; leaflets oval or oblong, ≤5", **hairless beneath.** Flowers on short cylindrical racemes, later elongate, ¾–6"; flowers ⅛"; **petals 4–10, often anther-like at apex. Fruit white with a dark spot, giving the popular name of doll's eyes.** Rich, deciduous woods; less often in coniferous woods. Differs from no. 5 in the thicker flower-stalks turning pink-red when plant is in fruit.

5. Red baneberry *Actaea rubra*

Plants 16–32". Leaves 2–3× ternately compound; leaflets oval or oblong, ≤5", **hairy beneath.** Flowers on pyramidal racemes, later elongating, ¾–6"; flowers ⅛"; **petals 4–10, pointed or blunt. Fruit red, occasionally white.** Rich, deciduous or mixed woods; banks; swamps.

1. American bugbane

2. Black cohosh

3. Appalachian bugbane

4. White baneberry

5. Red baneberry

Alum-roots, Coolwort, and Foam-flower
Saxifrage family

Alum-roots have basal, lobed leaves and panicles of small flowers in which the sepals and petals are fused together below to form a saucer-shaped or tube-shaped hypanthium.

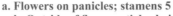

a. Flowers on panicles; stamens 5
  b. Outside of flowers sticky-hairy
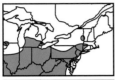
1. Common alum-root *Heuchera americana*
Plants 1–5′. **Leaves heart-shaped, ¼–½″**, 5–9-lobed, often white-mottled. Flowers on panicles; flowers drooping, sometimes purplish, ⅛–¼″, only slightly irregular; **stamens long-exserted**. Dry, upland woods.

2. Appalachian alum-root *Heuchera pubescens*
Plants 1–3′. **Leaves heart-shaped, 1½–2¾″**, 5–7-lobed. Flowers on elongate, conical panicles; flowers white, pink, or purple, ¼–⅜″, oblique; petals somewhat inflexed; **stamens barely exserted**. Upland woods.

3. Prairie alum-root *Heuchera richardsonii*
Plants 8–36″. **Leaves heart-shaped, 1¼–3″**, 7–9-lobed. Flowers on narrow panicles, greenish or greenish-white, ⅛–¼″, very oblique; **stamens included or long-exserted**. Prairies, dry woods.

4. Sullivant's coolwort *Sullivantia sullivantii*
Plants 4–14″. **Leaves kidney-shaped, 1¼–2¾″**, sparsely hairy or hairless, with numerous shallow lobes. Flowers on lax panicles, 8–16″ tall, with ascending branches, sticky; flowers ¼″. Moist, shaded cliffs. Note photo shows leaves only.

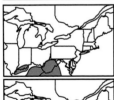
  b. Outside of flowers long-hairy, not sticky
5. Maple-leaved alum-root *Heuchera villosa*
Plants 8–32″. **Leaves broadly heart-shaped, 2¾–12″**, lobed; lobes triangular. Flowers on congested panicles, white or pink, <⅛″, not oblique; **stamens exserted**. Moist, shaded ledges; cliffs.

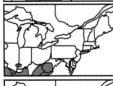
A. Small-flowered alum-root *Heuchera parviflora* (not illus.)
Similar to no. 3, but **leaves generally kidney-shaped, <5″**. Shaded cliffs and ledges.

B. Closed-flowered alum-root *Heuchera longiflora* (not illus.)
Similar to no. 2, but **generally hairless and flowers strongly distended below**. Rich woods.

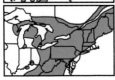
a. Flowers on racemes; stamens 10
6. Foam-flower *Tiarella cordifolia*
Plants 4–14″. Leaves **heart-shaped or round, 2–4″**, 3–5-lobed. Flowers on crowded racemes elongating to 4″; flowers ¼–⅜″. Rich woods.

1. Common alum-root

2. Appalachian alum-root

3. Prairie alum-root

4. Sullivant's coolwort

5. Maple-leaved alum-root

6. Foam-flower

Strawberries, Buckbeans, Sarsaparilla, and Harbinger-of-spring
Several families

a. Leaflets 3

1. Woodland strawberry *Fragaria vesca*
Plants 3–6″. **Leaflets 3, elliptic or narrowly oval, broader toward either end, 1–5″, silky beneath, stalkless; terminal tooth of leaflet longer than adjacent teeth.** Flowers 1–9, solitary or on racemes or panicles, taller than the leaves; flowers ¼–½″. Rocky woods, openings. Rose family.

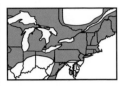

2. Common strawberry *Fragaria virginiana*
Plants 3–6″. **Leaflets 3, elliptic or narrowly oval, broader toward either end, 1–5″, silky or not beneath,** stalked; **terminal tooth of leaflet shorter than adjacent teeth.** Flowers 2–many, on umbel-like panicles, equaling the leaves; flowers ¼–1″. Fields, other open places. Rose family.

*3. Cultivated strawberry *Fragaria ×ananassa*
Plants >12″. **Leaflets 3, evergreen, elliptic or narrowly oval, broader toward either end, 1–5″, bluish and hairy beneath,** stalked; **terminal tooth of leaflet shorter than adjacent teeth.** Flowers 2-many, on umbel-like panicles, equaling the leaves; flowers ¾–1¼″. Cultivated. Rose family.

4. Buckbean *Menyanthes trifoliata*
Plants 2–12″. **Leaflets 3, elliptic or oval, broader toward either end, 1¼–2½″, smooth or wavy-margined.** Flowers on terminal racemes, white or pink, ⅜–½″; petals fringed. Quiet, cold water. Buckbean family.

a. Leaflets more than 3

5. Wild sarsaparilla *Aralia nudicaulis*
Plants 6–15″, hairless. **Leaves in 3 groups of 3–5 leaflets,** each leaflet narrowly elliptic or oval, broader toward tip, ≤6″, pointed, toothed; leaf-stalks ≤20″. Flowers usually on 3 umbels, ½–8″; flowers ¼″. Woods. See also p. 316. Ginseng family.

6. Harbinger-of-spring *Erigenia bulbosa*
Plants 2–6″. **Leaves oval, highly dissected,** 4–8″; leaflets linear or spoon-shaped. Flowers on a single, terminal umbel; flowers ¼″. Rich woods. Note the photo shows the plant in fruit. Umbel family.

1. Woodland strawberry

2. Common strawberry

*3. Cultivated strawberry

4. Buckbean

5. Wild sarsaparilla

6. Harbinger-of-spring

Page 390
LEAVES basal
LEAVES lobed or
compound
PETALS 4–8 or
irregular

Miscellaneous basal, compound-leaved species
Several families

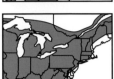

a. Petals 4, irregular; leaves dissected
1. Squirrel-corn *Dicentra canadensis*
Plants 4–12″. Leaves broadly triangular, 5½–10½″, highly dissected into linear segments. Flowers 3–12, on terminal racemes; **flowers fragrant,** nodding, ½–¾″, **narrowly oval, the spurs short and rounded**. Rich, deciduous woods. See also p. 132. Poppy family.

2. Dutchman's-breeches *Dicentra cucullaria*
Plants 4–12″. Leaves broadly triangular, 5½–6¼″, highly dissected into linear segments. Flowers on terminal racemes; **flowers not fragrant,** nodding, **pantaloon-shaped, yellow-tipped,** ½–¾″, **the spurs divergent and pointed**. Rich, deciduous woods; clearings. Poppy family.

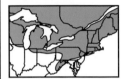

a. Petals 4–7; leaves trifoliolate or 2–3× ternately compound
3. Goldthread *Coptis trifolia*
Plants 2–6″. **Leaves trifoliolate,** oval, broader toward tip; leaflets glossy, toothed or lobed. Flowers solitary, ½″; petals 4–7. Mossy woods, swamps. The bright, yellow, sterile stamens contain nectar. Crowfoot family.

4. False rue-anemone *Enemion biternatum*
Plants 4–16″. **Leaves 2–3× ternately compound,** leaflets oval, broader toward tip, ≤¾″. Flowers 1–2, ½–¾″; petals 5. Moist woods. Crowfoot family.

a. Petals 8; leaves bifoliolate or lobed
5. Twinleaf *Jeffersonia diphylla*
Plants 4–8″, later growing to double the height. Leaves divided into 2 segments, 3¼–6″; leaf-stalks elongating to 12–20″. Flowers solitary, ⅜–1¼″. Rich woods, mostly calcareous. Barberry family.

6. Bloodroot *Sanguinaria canadensis*
Plants 2–6″ when flowering, later to 12″, sap orange-red. Leaves round, 3–9-lobed, 4–10″; lobes wavy or coarsely toothed. Flowers solitary, 1–1½″. Moist or dry, rich woods. Poppy family.

1. Squirrel-corn

2. Dutchman's-breeches

3. Goldthread

4. False rue-anemone

5. Twinleaf

6. Bloodroot

Miscellaneous aquatic species
Several families

a. Leaves floating (cont. on next page)
  b. Petals 3
1. Fanwort *Cabomba caroliniana*
Plants ≤6′. Submersed leaves round, finely dissected, 1–2″; **floating leaves linear or narrowly elliptic, ¼–1″.** Flowers solitary on stalks 1–4″; flowers white with yellow center, ½″; sepals and petals 3. Ponds, quiet waters, native only in the southern part of our range. Water-shield family.

2. American frog's-bit *Limnobium spongia*
Leaves oval, round or heart-shaped, ¾–3¼″, usually broadly pointed, spongy at base. **Flowers** 2–9 on flower-stalks 1–4″; **flowers 1″;** sepals 3, oblong or oval; petals 3, linear. Ponds, stagnant waters. Note photo shows leaves only. Frog's-bit family.

  b. Petals 4
*3. Water-chestnut *Trapa natans*
Plants ≤3′. **Leaves whorled, diamond-shaped,** 1–2″ wide, sharply toothed, slightly above water; stalks inflated; submersed leaves are feathery. Flowers axillary, held erect on short stalks, ½″; sepals 4. Quiet streams, ponds. Opens at night. Lythrum family.

  b. Petals 5
4. Little floating-heart *Nymphoides cordata*
Leaves heart-shaped, ½–2¾″. **Flowers solitary, ¼–½″,** stalked above the water ¼–2¼″; sepals 5; petals ¼″, yellow at base. Ponds, slow streams. See also p. 228. Buckbean family.

5. Big floating-heart *Nymphoides aquatica*
Leaves heart-shaped, 2–4½″. **Flowers solitary, ¾–1¼″,** stalked above the water ¼–2¼″; sepal-lobes 5; petals ¼″, yellow at base. Quiet water. Buckbean family.

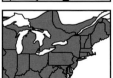

  b. Petals many
6. Common water-lily *Nymphaea odorata*
Leaves round with a narrow sinus, 2–12″. Flowers solitary, opening in the morning, 2–8″; sepals numerous, oval, 1¼–4″; petals numerous 1¼–4″, narrowly elliptic to spoon-shaped, smaller toward the center, fragrant. Lakes, ponds, ditches, slow streams and rivers. Water-lily family.

A. Pygmy water-lily *Nymphaea leibergii* (not illus.)
Leaves elliptic with a wide sinus, 2½–4½″. **Flowers** solitary, **opening in the afternoon, 1½–3″;** sepals numerous, oval, ¾–1¼″; petals numerous ¼–⅝″, narrowly elliptic to spoon-shaped, smaller toward the center. Cold ponds, lakes. Water-lily family.

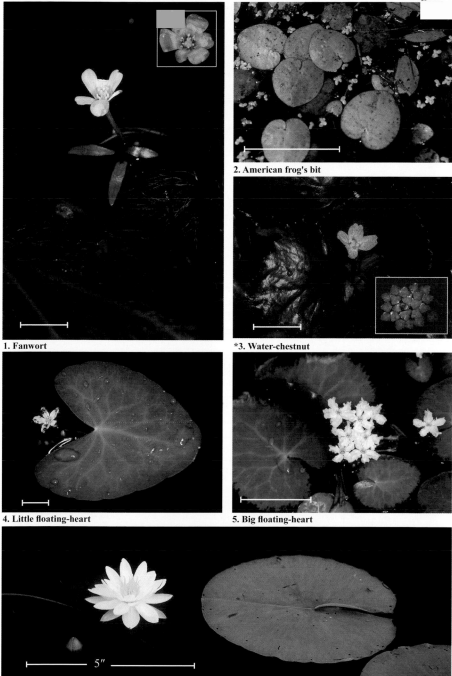

1. Fanwort

2. American frog's bit

*3. Water-chestnut

4. Little floating-heart

5. Big floating-heart

5"

6. Common water-lily

Miscellaneous aquatic species
Several families

a. Leaves submersed (cont. from previous page)
  b. Petals 3
1. Water-celery *Vallisneria americana*

Submersed leaves linear, ≤6″, blunt. Male and female flowers on separate plants; male flowers minute, developing at base of plant until they are released and become free-floating; mature female flowers, ¼–½″ with three sepals, are raised to surface on long, coiled stalks, where they are encountered by the floating males and pollinated. The coiled stalks of the fertilized flowers then contract, pulling the developing fruit underwater. Quiet waters. Tape-grass family.

2. Common water-weed *Elodea canadensis*

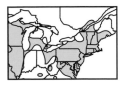

Submersed leaves in whorls of 3, linear, ¼–½″. Flowers ¼″; sepals 3, oval, broader toward tip; petals 3, oval, broader toward tip. Male and female flowers on separate plants; both develop underwater and are raised to the surface on thread-like stalks at maturity. The pollen is transported by water. Shallow, quiet, hard water. Tape-grass family.
    Western water-weed (*Elodea nuttallii*) is similar but with narrower, more crowded leaves and growing in soft-water habitats.

*3. Brazilian water-weed *Egeria densa*

Submersed leaves in whorls of 5, linear, ⅜–2½″. Flowers solitary, ¾″; sepals 3; petals 3. Lakes, streams; an escaped aquarium plant. Tape-grass family,

  b. Petals 5
4. White water-crowfoot *Ranunculus longirostris*

Submersed leaves dissected into thread-like segments, ½–1″; floating leaves sometimes produced, 3-lobed, ⅛–½″. Flowers just above the water surface, ⅜–¾″; sepals 5; petals 5. Ponds, slow streams. See also pp. 216, 222, 228. Crowfoot family.

5. Featherfoil *Hottonia inflata*

Plants ≤20″, inflated and jointed. Leaves whorled, oblong, ¾–2½″, deeply pinnately lobed, lobes linear. Flowers 3–10, axillary and on terminal umbels, ⅛″; sepals 5; petals 5. Shallow water, wet soil. Primrose family.

  b. Petals absent
6. Water-starwort *Callitriche heterophylla*

Plants 4–8″, typically submersed. Submersed leaves opposite, linear, ¼–½″; floating and emersed leaves spoon-shaped or oval, broader toward tip; 1 male and 1 female flower per axil, <⅛″; sepals and petals absent. Quiet water. Plantain family.

1. Water-celery

2. Common water-weed

*3. Brazilian water-weed

4. White water-crowfoot

5. Featherfoil

6. Water-starwort

Miscellaneous species without leaves (1)
Several families

a. Petals 4

1. Screw-stem *Bartonia paniculata*
Plants 1½–18″, erect or lax, often spiralling. Leaves scale-like, <⅛″, mostly alternate. **Flowers** on panicles 2–4″ tall, ≤⅛″; **petals 4**. Swamps, bogs. See also p. 230. Gentian family.

a. Petals 5

2. Indian-pipe *Monotropa uniflora*
Plants ¼–12″, white-waxy. Leaves scale-like. **Flowers solitary, nodding, ⅜–½″;** petals 5, broadly oblong. Flowers become dark-brown and turn upright after flowering. Rich, shady woods in humus. Indian-pipe is a saprophyte getting nutrients from soil fungi. See also p. 230. Heath family.

3. Common dodder *Cuscuta gronovii*
Plants twining and entangling other plants; stems yellow or orange, parasitic. **Flowers in irregular clusters, ≤⅛″,** 5-lobed; sepals rounded; **petals blunt**. Low ground. Common dodder is parasitic on a variety of hosts. Morning-glory family.

4. Collared dodder *Cuscuta indecora*
Plants twining and entangling other plants; stems yellow or orange, parasitic. **Flowers in irregular clusters, ⅛″,** 5-lobed; sepals blunt; **petals broadly pointed**. Parasitic on a variety of hosts. Morning-glory family.

a. Petals 6

5. Wild leek *Allium tricoccum*
Plants 4–16″. Leaves narrowly oval or narrowly elliptic, 6–12″, withering before flowering. **Flowers** 30–50 on terminal umbels, ⅛–¼″; **petals 6**, oblong or oval, blunt. Rich woods. See other blue-flowered onions, page 64; pink-flowered onions, p. 130. Onion family.

1. Screw-stem

2. Indian-pipe

3. Common dodder

4. Collared dodder

5. Wild leek

Miscellaneous species without leaves (2)
Orchid and Broom-rape families

a. Flowers solitary

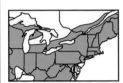

1. Cancer-root *Orobanche uniflora*
Plants 2½–8″, sticky-hairy. **Flowers solitary,** declining, ¾–1″, shaped like a curved tube, white or lilac, with yellow in throat; petals 5, triangular, sharply pointed. Moist woods, stream banks. Cancer-root is parasitic on various plants. See also p. 262. Broom-rape family.

a. Flowers on spikes

2. Squaw-root *Conopholis americana*
Plants 2–8″, stout. **Leaves scale-like, oval or narrowly oval, <¾″.** Flowers numerous on spikes; flowers ¼–½″ long. Rich woods. Parasitic on oaks. Broom-rape family.

3. Little ladies'-tresses *Spiranthes tuberosa*
Plants 2–12″. **Leaves narrowly oval, broader toward tip, ¾–2¼″, usually absent at flowering time.** Flowers on hairless spikes ¾–3″ tall; flowers spreading, ½″; sepals slightly spreading, ¼″; petals linear, ¼″, broadly pointed or blunt; lip oval to oblong, ¼″, white; apex dilated with wavy, finely torn margin. Dry, open woods; outcrops; fields; roadsides; cemeteries. See other ladies'-tresses; p. 380.

a. Flowers on racemes

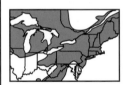

4. Spotted coral-root *Corallorrhiza maculata*
Plants 8–20″, white, pink-purple, or yellow. Flowers 6–41 on terminal racemes 1½–8″ tall; flowers ¼–½″, with sepals and petals spreading; **petals** narrowly oblong or narrowly oval, broader toward tip, ¼″, **3-nerved,** purple spotted; spur prominent; **lip ¼″, 3-lobed,** generally purple spotted; middle lobe deflected, ⅛″ wide. Dry woods. See also p. 132. Orchid family.

5. Pale coral-root *Corallorrhiza trifida*
Plants 4–12″. Flowers 2–15 on racemes ¾–3″ tall, yellow-green to purple, ¼–⅜″; sepals and lateral **petals** narrowly oval, ¼″, **1-nerved;** lateral sepals extending forward; **lip ⅛–¼″, 3-lobed,** white. Wet woods, thickets, bogs. Orchid family.

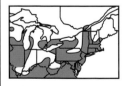

6. Fall coral-root *Corallorrhiza odontorhiza*
Plants ½–7″, purple to brown. Flowers 6–26 on terminal racemes ½–2¾″ tall; flowers ¼″, with sepals and petals extending forward over the lip, oblong, ⅛″, various colors; **lip <⅛″, not lobed,** declined, narrowed to base, expanded to a round blade, ⅛″, usually white with purple margins and 2 purple spots. Woods. Orchid family.

1. Cancer-root

2. Squaw-root

3. Little ladies'-tresses

4. Spotted coral-root

5. Pale coral-root

6. Fall coral-root

Glossary

Alternate: Arranged singly at a node or point of attachment. Used here to refer to leaves.

Areoles: The round areas with thorns on cacti.

Ascending: Growing upward but at an angle, as opposed to erect, which is growing straight upward.

Axil, axillary: The position between the leaf and the stem. Usually used here to refer to the position of a flower or inflorescence.

Banner: The upper, usually enlarged, petal of a bean family flower; sometimes referred to as the standard.

Basal: The position at the base of the plant (at soil level). Used here to refer to leaves.

Bisexual: A flower with both male and female parts (fertile stamens and pistils).

Bract, involucral bract: A leaf-like structure, usually smaller than the regular leaves and associated with the flower or inflorescence. Involucral bracts are the bracts beneath the heads of composite flowers.

Bulblet: A small, bulb-like structure in the axils of leaves or in an inflorescence.

Calyx: A collective term for the sepals as a whole.

Clasping: Surrounding or partially surrounding something. Used here generally to refer to leaves surrounding the stem.

Column: The united filaments and style of an orchid flower.

Compound: Used here to refer to a leaf comprised of more than two leaflets.

Corolla: A collective term for the petals as a whole.

Elliptic: A two-dimensional shape equivalent to the geometrical ellipse.

Entire: Used here to refer to a leaf without lobes or teeth.

Fringe: Refers to hairs on the margin, particularly of a leaf or petal. Equivalent to the botanical term ciliate.

Globose: Globe-shaped.

Grain: A small, seed-like attachment on the valves of some Rumex fruit.

Habit: The general appearance of the plant; e.g., shrub, herb, etc.

Hemisperical: Dome-shaped or half-spherical.

Inflorescence: The grouping of flowers, often set off from the leaves by a section of stem or by reduced leaves.

Inrolled: The margin of a leaf that is curled under.

Keel: The lower two, united petals of a bean flower.

Kidney-shaped: A broad, rounded shape with rounded apex and notched base, like the outline of a human kidney.

Linear: A two-dimensional shape that is long and narrow, grass-like.

Lip: The third petal of an orchid flower, which is modified and differs from the two lateral ones. The pouch of a lady-slipper orchid is the lip.

Lobe: A projecting part of a leaf that is too large to be called a tooth.

Oblong: A two-dimensional shape essentially retangular with rounded corners. Similar to elliptic, but with straight sides.

Opposite: Arranged in pairs at a node or point of attachment. Used here to refer to leaves.

Oval, narrowly oval: A two-dimensional shape like the outline of a chicken egg. Unless otherwise indicated, the broader end is toward the base (point of attachment). Equivalent to the botanical term *ovate*; narrowly oval is equivalent to *lanceolate*, and "oval, broader toward tip" is equivalent to *obovate*.

Ovary: The lower part of the pistil, where the ovules and eggs are found.

Palmate: With four or more leaflets or lobes from a common point. Compare with pinnate.

Panicle: A branched inflorescence. For simplicity, in this book this term includes the botanical terms *panicle, cyme* (in part), *thyrse,* etc.

Pappus: The modified calyx of a composite flower. Often represented as hairs.

Pentagonal: Five-sided, like a pentagon.

Petal: The inner, showy part of a flower. As used here, this term also refers to the petal-lobe of a sympetalous (tubular or bell-shaped) flower.

Pinnate: With four or more leaves or leaflets attached to the opposite sides of an axis. Like the barbs of a feather. Compare with palmate.

Pistil: The central female part of a flower; it has three parts: ovary, style, and stigma.

Pleat: Something that is folded like a fan.

Pollen: The very small, dust-like bodies shed from the anthers. Pollen contains the male reproductive cells.

Prostrate: Lying on the ground.

Raceme: An unbranched inflorescence in which the flowers are stalked. As used here, this term is equivalent to the botanical terms *raceme* and *cyme* (in part).

Rays: The outer, petal-like flowers of a composite head.

Recurved: Curved backwards.

Reflexed: Bent backwards more abruptly than *recurved*.

Scrambling: Running along the ground or over small bushes as opposed to going up a tree. Used here to refer to vines.

Sepal: The outer, green and leaf-like, broad part of a flower. As used here this term also refers to the sepal-lobe of a symsepalous flower.

Simple: Used here to refer to a leaf that is not compound. We have occasionally indicated a distinction between "simple" and "lobed" in the quick guides, but in fact a simple leaf can be deeply lobed.

Spike: An unbranched inflorescence in which the flowers are stalkless.

Spoon-shaped: Shaped like a spoon or spatula, rounded toward the tip and narrowed at the base.

Spur: A hollow appendage of a sepal or petal, usually holding nectar for insect visitors.

Stamen: The male part of the flower, located inside the petals and outside the pistil. Made up of two parts: the filament, to which the anthers are attached at the tip.

Stigma: The terminal part of the pistil, where the pollen lands.

Stipule: A leaf-like structure at the base of the leaf-stalk where it attaches to the stem.

Style: The stalk-like part of the pistil, located between the ovary and the stigma.

Throat: The opening of a sympetalous flower.

Trifoliate: With three leaves. Compare with palmate and pinnate.

Truncate: Straight across. Used to refer to leaf apices and bases in which it appears that the leaf is cut straight across.

Umbel: An inflorescence in which the stalked flowers arise from a single point.

Unisexual: A flower with either fertile male or female parts, although there may be sterile parts of the opposite sex.

Wing: A thin, flat projection from the stem. Also refers to the lateral petals of a bean flower.

Bibliography

Andreas, B. K. 1989. The vascular flora of the glaciated Allegheny Plateau region of Ohio. *Ohio Biol. Surv. Bull. New Series* 8(1):1–191.

Baldwin, W. K. W. 1958. *Plants of the clay belt of northern Ontario and Quebec.* Bulletin (National Museum of Canada) no. 156. [Ottawa]: Canada Dept. of Northern Affairs and National Resources.

Botham, W. 1981. *Plants of Essex County: a preliminary list / compiled by Wilfred Botham.* Essex, Ont.: Essex Region Conservation Authority.

Braun, L. 1967. *The Monocotyledoneae: cat-tails to orchids.* Columbus: The Ohio State University Press.

Bruce-Grey Plant Committee. 1997. *A checklist of vascular plants for Bruce and Grey counties, Ontario,* 2nd edition [Ontario]: The Committee.

Cooperrider, T. A. 1995. *The Dicotyledoneae of Ohio: Part 2. Linaceae through Campanulaceae.* Columbus: Ohio State University Press.

Cullina, W. 2000. *The New England Wild Flower Society guide to growing and propagating wildflowers of the United States and Canada.* Boston: Houghton Mifflin.

Cusick, A. W. and G. Silberhorn. 1977. The Vascular Plants of Unglaciated Ohio. *Ohio Biological Survey Bulletin. New Series* 5(4):1–153.

Deam, C. C. 1940. *Flora of Indiana.* Repr., Caldwell, NJ: The Blackburn Press.

Dole, C. H. 2003. *The butterfly gardener's guide.* Brooklyn: Brooklyn Botanic Garden.

Easterly, N. W. 1964. Distribution patterns of Ohio Cruciferae. *Castanea* 29:164–173.

Environment Canada. *Biodiversity Portrait of the St. Lawrence.* www.qc.ec.gc.ca/faune/biodiv/index.html. (accessed between 2001 and 2005.)

Fisher, T. R. 1988. *The Dicotyledoneae of Ohio: Part 3. Asteraceae.* Columbus: Ohio State University Press.

Flora of North America Editorial Committee, eds. 1997. *Flora of North America North of Mexico. Vol. 3. Magnoliophyta: Magnoliidae and Hamamelidae.* New York: Oxford University Press.

Flora of North America Editorial Committee, eds. 2000. *Flora of North America North of Mexico. Vol. 22. Magnoliophyta: Alismatidae, Arecidae, Commelinidae (in part), and Zingiberidae.* New York: Oxford University Press.

Flora of North America Editorial Committee, eds. 2002. *Flora of North America North of Mexico. Vol. 26. Magnoliophyta: Liliidae: Liliales and Orchidales.* New York: Oxford University Press.

Flora of North America Editorial Committee, eds. 2003. *Flora of North America North of Mexico. Vol. 4. Magnoliophyta: Caryophyllidae,* part 1. New York: Oxford University Press.

Gaiser, L. O. 1966. *A survey of the vascular plants of Lambton County, Ontario.* Ottawa: Plant Research Institute, Research Branch, Canada Dept. of Agriculture.

Gleason, H. A. and A. Cronquist. 2004. Manual of the vascular plants of northeastern United States and adjacent Canada. 2nd edition, revised. Bronx: New York Botanical Garden Press.

Hinds, H. R. 2000. *Flora of New Brunswick: a manual for the identification of the vascular plants of New Brunswick,* 2nd edition. Fredericton, NB: Dept. of Biology, University of New Brunswick.

Holmgren, N. H. and P. Holmgren. 1998. *Illustrated companion to Gleason and Cronquist's Manual: illustrations of the vascular plants of Northeastern United States and adjacent Canada.* Bronx: New York Botanical Garden Press.

Hough, M. Y. 1983. *New Jersey wild plants.* Harmony, NJ: Harmony Press.

Kress, S. W. 2000. *Hummingbird gardens: turning your yard into hummingbird heaven.* Brooklyn: Brooklyn Botanic Garden.

Magee, D. W. and H. E. Ahles. 1999. *Flora of the Northeast: a manual of the vascular flora of New England and adjacent New York.* Amherst: University of Massachusetts Press.

Mohlenbrock, R. H. and D. M. Ladd. 1978. *Distribution of Illinois vascular plants.* Carbondale: Southern Illinois University Press.

Morton, J. K. and J, M, Venn. 2000. *The flora of Manitoulin Island and the adjacent islands of Lake Huron, Georgian Bay and the Northern Channel.* University of Waterloo biology series no. 40. Waterloo, Ont.: Department of Biology, University of Waterloo.

NatureServe. 2004. *NatureServe Explorer: an online encyclopedia of life.* version 4.1. Arlington, VA: NatureServe. www.natureserve.org/explorer. (accessed between 1999 and 2005).

New York Flora Association. 1990. *Preliminary vouchered Atlas of New York State flora.* Albany: New York State Museum Institute.

Oldham, M. J. 1999. *Natural Heritage Resources of Ontario: rare vascular plants.* 3rd edition. Peterborough, Ont.: Natural Heritage Information Centre, Ontario Ministry of Natural Resources. Also available online at www.mnr.gov.on.ca/ MNR/nhic/species/lists/rarevascular.pdf (accessed April 2005)

Radford, A. E., H. A. Ahles, and C. R. Bell. 1968. *Manual of the vascular flora of the Carolinas.* Chapel Hill: The University of North Carolina Press.

Rhoads, A. F. and W. M. Klein, Jr. 1993. *The vascular flora of Pennsylvania: annotated checklist and atlas.* Philadelphia: American Philosophical Society.

Rhoads, A. F. and T. A. Block. 2000. *The plants of Pennsylvania: an illustrated manual.* Philadelphia: University of Pennsylvania Press.

Rousseau, C. 1974. *Géographie floristique du Québec-Labrador: distribution des principales espèces vasculaires.* Travaux et documents du Centre d'études nordiques, no. 7. Québec: Presses de l'Université Laval.

Soper, J. H., C. E. Garton, and D. R. Given. 1989. *Flora of the north shore of Lake Superior: (vascular plants of the Ontario portion of the Lake Superior drainage basin).* Syllogeus no. 63. Ottawa: National Museums of Natural Sciences, National Museums of Canada.

Stein, S. 1997. *Planting Noah's garden: further adventures in backyard ecology.* Boston: Houghton Mifflin.

Stewart, W. G. and L. E. James. 1969. *A guide to the flora of Elgin County, Ontario: an annotated list of ferns, fern allies, conifers, and flowering plants growing without cultivation in Elgin County, Ontario.* [St. Thomas, Ont.]: Catfish Creek Conservation Authority.

Steyermark, J. A. 1963. *Flora of Missouri.* Ames: Iowa State University Press.

Tatnall, R. 1946. *Flora of Delaware and the Eastern Shore; an annotated list of the ferns and flowering plants of the peninsula of Delaware, Maryland and Virginia.* [Wilmington]: Society of Natural History of Delaware.

USDA, NRCS. 2004. *The PLANTS Database,* version 3.5. Baton Rouge, LA: National Plant Data Center. www.plants.usda.gov. (accessed between 1999 and 2005.)

Voss, E. G. 1972. *Michigan Flora: Part I Gymnosperms and Monocots.* Cranbrook Institute of Science Bulletin 55. Ann Arbor: Cranbrook Institute of Science.

Voss, E. G. 1985. *Michigan Flora: Part II Dicots (Saururaceae-Cornaceae).* Cranbrook Institute of Science Bulletin 59. Ann Arbor: Cranbrook Institute of Science.

Voss, E. G. 1996. *Michigan Flora: Part III Dicots (Pyrolaceae-Compositae).* Cranbrook Institute of Science Bulletin 61. Ann Arbor: Cranbrook Institute of Science.

Weldy, T., R. Mitchell, and R. Ingalls. 2002. *New York flora atlas.* Albany: New York Flora Association, New York State Museum. www.nyflora.org/atlas/atlas.htm. (accessed between 2002 and 2005).

Wisconsin State Herbarium. *Wisflora: Wisconsin Vascular Plant Species.* www.botany.wisc.edu/wisflora/. (accessed between 1999 and 2005.)

Yatskievych, G. 1999. *Steyermark's flora of Missouri–Volume 1.* Jefferson City, MO: Missouri Department of Conservation.

Yatskievych, K. 2000. *Field Guide to Indiana Wildflowers.* Bloomington: Indiana University Press.

Photographic Credits and Map Sources

Except for the 24 images noted below, all photographs are by Carol Gracie. Photographs of pressed leaves are by Steve Clemants.

Credits

p. 5.7 *Gentiana linearis* Ed Kanze; p. 33.3 *Acontium noveboracense*, Steven Clemants; p. 47.2 *Collinsia verna*, Virginia Weinland; p. 111.6 *Asclepias sullivantii*, Perry Peskin; p. 130.8 *Calypso bulbosa*, Kenneth M. Cameron; p. 138.8 *Platanthera integra*, Rob Naczi; p. 159.6 *Solidago ohioensis*, Virginia Weinland; p. 166.1b *Cypripedium parviflorum* var. *makasin*, Charles Sheviak; p. 205.5 *Arnica cordifolia*, Virginia Weinland; p. 223.8 *Potentilla robbinsiana*, Walter Brust; p. 259.1 *Cleistes divaricata*, Kenneth M. Cameron; p. 263.1 *Pterospora andromedea*, Virginia Weinland; p. 299.1 *Cypripedium candidum*, Perry Peskin; p. 319.5 *Cirsium pitcheri*, Jacqueline Kallunki; p. 361.2 *Plantago cordata* plant, Steven Clemants; p. 361.2 *Plantago cordata* inset, Steven Clemants; p. 365.2 *Pyrola secunda* , A. Scott Earle/Larkspur Books; p. 365.5 *Pyrola chlorantha*, A. Scott Earle/Larkspur Books; p. 265.5 *Pyrola chlorantha* inset, A. Scott Earle/Larkspur Books; p. 369.6 *Zigadenus elegans* plant, Jacqueline Kallunki; p. 369.6 *Zigadenus elegans* inset, Virginia Weinland; p. 371.5 *Amianthium muscitoxicum*, Rich Kelly; p. 379.2 *Goodyera tesselata*, Virginia Weinland; p. 381.6 *Spiranthes romanzoffiana*, Rich Kelly.

Species Distribution Map Sources

Distribution maps are provided for each species treated in this book. The maps were compiled from a variety of sources. The primary sources for the region were Flora of North America Editorial Committee 1997, 2000, 2002, 2003; Natureserve 2004; USDA 2004. For the Canadian Provinces, additional sources were New Brunswick (Hinds 2000), Ontario (Baldwin 1958; Botham 1981; Bruce-Grey Plant Committee 1997; Gaiser 1966; Morton and Venn 2000; Oldham 1999; Soper et al. 1989; Stewart and James 1969), and Quebec (Environment Canada n.d.; Rousseau 1974). For the United States, additional sources were Connecticut (Magee & Ahles 1999; USDA 2004), Delaware (Tatnall 1946), Illinois (Mohlenbrock and Ladd 1978; USDA 2004), Indiana (Deam 1940; Yatskievych 2000), Kentucky (USDA 2004), Maine (Magee and Ahles 1999; USDA 2004), Maryland (Tatnall 1946), Massachusetts (Magee and Ahles 1999; USDA 2004); Michigan (Voss 1972, 1985, 1996; USDA 2004), Missouri (Steyermark 1963; Yatskievych 1999), New Hampshire (Magee and Ahles 1999; USDA 2004), New Jersey (Hough 1983), New York (New York Flora Association 1990; Weldy et al. 2002), Ohio (Andreas 1989; Braun 1967; Cooperrider 1995; Cusick and Silberhorn 1977; Easterly 1964; Fisher 1988), Pennsylvania (Rhoads and Klein 1993), Rhode Island (Magee and Ahles 1999; USDA 2004), Vermont (Magee and Ahles 1999; USDA 2004), Virginia (Tatnall 1946; USDA 2004), West Virginia (USDA 2004), and Wisconsin (Wisconsin State Herbarium n.d.).

Family Index

The following list contains the scientific and English names for the families mentioned in the text. The families follow the circumscription found on the Angiosperm Phylogeny Group's website: *www.systbot.uu.se/classification/APG.html* (accessed April 2005).

Acanthaceae—Acanthus family
Acanthus family—Acanthaceae
Acoraceae—Sweet-flag family
Agavaceae—Agave family
Agave family—Agavaceae
Alismataceae—Water-plantain family
Alliaceae—Onion family
Amaranth family—Amaranthaceae
Amaranthaceae—Amaranth family
Amaryllidaceae—Amaryllis family
Amaryllis family—Amaryllidaceae
Apiaceae—Umbel family
Apocynaceae—Dogbane family
Araceae—Arum family
Araliaceae—Ginseng family
Aristolochiaceae—Dutchman's-pipe family
Arrow-grass family—Juncaginaceae
Arum family—Araceae
Asparagaceae—Asparagus family
Asparagus family—Asparagaceae
Asteraceae—Composite family
Balsaminaceae—Touch-me-not family
Barberry family—Berberidaceae
Bean family—Fabaceae
Bellflower family—Campanulaceae

Berberidaceae—Barberry family
Bladderwort family—Lentibulariaceae
Bloodwort family—Haemodoraceae
Bog-asphodel family—Nartheciaceae
Borage family—Boraginaceae
Boraginaceae—Borage family
Boxwood family—Buxaceae
Brassicaceae—Mustard family
Broom-rape family—Orobanchaceae
Buckbean family—Menyanthaceae
Bunch-flower family—Melanthiaceae
Bur-reed family—Sparganiaceae
Bush-honeysuckle family—Diervillaceae
Butomaceae—Flowering-rush family
Buxaceae—Boxwood family
Cabombaceae—Water-shield family
Cactaceae—Cactus family
Cactus family—Cactaceae
Campanulaceae—Bellflower family
Cannabaceae—Hemp family
Caprifoliaceae—Honeysuckle family
Carpet-weed family—Molluginaceae
Caryophyllaceae—Pink family
Catbrier family—Smilacaceae
Cat-tail family—Typhaceae
Cistaceae—Rock-rose family
Cleomaceae—Cleome family
Cleome family—Cleomaceae
Colchicaceae—Colchicum family
Colchicum family—Colchicaceae
Commelinaceae—Spiderwort family
Composite family—Asteraceae
Convolvulaceae—Morning-glory family
Cornaceae—Dogwood family
Crassulaceae—Stonecrop family
Crowfoot family—Ranunculaceae
Cucurbitaceae—Gourd family
Daylily family—Hemerocallidaceae

Index